Isoelectric focusing: theory, methodology and applications

LABORATORY TECHNIQUES IN BIOCHEMISTRY AND MOLECULAR BIOLOGY

Edited by

T.S. WORK – *Cowes, Isle of Wight (formerly N.I.M.R., Mill Hill, London)*
R.H. BURDON – *Department of Biochemistry, University of Glasgow*

Advisory board

P. BORST – *University of Amsterdam*
D.C. BURKE – *University of Warwick*
P.B. GARLAND – *University of Dundee*
M. KATES – *University of Ottawa*
W. SZYBALSKI – *University of Wisconsin*
H.G. WITTMAN – *Max-Planck Institut für Molekulaire Genetik, Berlin*

ELSEVIER BIOMEDICAL PRESS
AMSTERDAM · NEW YORK · OXFORD

ISOELECTRIC FOCUSING: THEORY, METHODOLOGY AND APPLICATIONS

Pier Giorgio Righetti

Professor of Biochemistry, Faculty of Pharmaceutical Sciences, University of Milano, Via Celoria, 2 Milano 20133, Italy

ELSEVIER BIOMEDICAL PRESS
AMSTERDAM · NEW YORK · OXFORD

ISBN — series: 0 7204 4200 1
   — paperback: 0 444 80467 6
   — hardbound: 0 444 80498 6

First edition: 1983
Second printing: 1984
Third printing: 1985
Fourth printing: 1986

Published by:

ELSEVIER BIOMEDICAL PRESS
1 MOLENWERF, P.O. BOX 211
1014 AG AMSTERDAM, THE NETHERLANDS

Sole distributors for the U.S.A. and Canada:

ELSEVIER-NORTH HOLLAND INC.
52 VANDERBILT AVENUE
NEW YORK, N.Y. 10017

Library of Congress Cataloging in Publication Data

Righetti, P. G.
   Isoelectric focusing.

   (Laboratory techniques in biochemistry and molecular biology; Volume 11)
   Bibliography: p.
   Includes index.
   1. Isoelectric focusing. I. Title. II. Series.
QP519.9.18R53 1983   574.19'285   82-24196
ISBN 0-444-80467-6 (U.S. : pbk.)

Printed in The Netherlands

Contents

List of abbreviations

α_1 AT = α_1-antitrypsin
AC = acrylic acid
ACES = (N-2-acetamido-2-amino ethane) sulfonic acid
ADC = ampholyte displacement chromatography
ADP = adenosine diphosphate
ATP = adenosine triphosphate
BAC = N,N'-bisacrylyl-cystamine
BEF = buffer isoelectric focusing
Bis = N,N'-methylene bis-acrylamide
BSA = bovine serum albumin
% C = grams of bisacrylamide/ % T
CA = carrier ampholyte
CBB = Coomassie Brilliant Blue
CM = carboxymethyl
CMC = critical micellar concentration
Cyt b_s = cytochrome b_s
D = translational diffusion coefficient
2-D = two-dimensional
DATD = N,N'-diallyltartar diamide
DEAE = diethylaminoethyl
DHEBA = N,N'-(1,2-dihydroxyethylene) bisacrylamide
DMAPMA = 3-dimethylaminopropyl methacrylamide
DMF = dimethylformamide

DMSO = dimethylsulphoxide
DNA = deoxyribonucleic acid
DNase = deoxyribonuclease
d.p.m. = disintegrations per minute
EDA = ethylene diacrylate
EDTA = ethylendiamino tetra-acetic acid
EtOH = ethanol
FA = formamide
FMN = flavin mono-nucleotide
A_γ = fetal chains of hemoglobin with alanine in residue No. 136
G_γ = fetal chains of hemoglobin with glycine in residue No. 136
HAc = acetic acid
Hb = hemoglobin (A = adult; S = sickle cell; A_{1c} = glycosilated Hb)
HMTA = hexamethylene-tetramine
Hp = haptoglobin
I = ionic strength
i.d. = inner diameter
IEF = isoelectric focusing
IgG = immunoglobulin G
IHP = inositol hexa phosphate
IHS = inositol hexa sulphate
ISO-DALT = isoelectric focusing followed by SDS-electrophoresis (charge-size fractionation)
ITP = isotachophoresis
kV = kilovolts

LDH	= lactic dehydrogenase	pK_{int}	= intrinsic dissociation constant
mA	= milliamperes		
MAPS	= (3-sulfopropyl)dimethyl (3-methacrylamido-propyl) ammonium inner salt	PPO	= 2,5-diphenyloxazole
		QAE	= quaternary amino ethyl
		RBC	= red blood cells
		RIP	= rat incisor phospho-protein
MAPTAC	= methacrylamidopropyl trimethylammonium chloride	RNA	= ribonucleic acid
		RNase	= ribonuclease
Mb	= myoglobin	rev./min	= revolutions per minute
μCi	= micro Curie	SB_{14}	= N-tetradecyl-N,N-dimethyl-3-amino propane sulfonic acid
MDPF	= 2-methoxy-2,4-diphenyl-3(2H)-furanone		
met Mb	= Fe^{3+} myoglobin	SDS	= sodium dodecyl sulphate
mequiv.	= milli equivalents	$\% T$	= (grams acrylamide + grams cross-linker)/ 100 ml solution
MM	= macromolecular mapping		
$\overline{M}_r$	= molecular mass (in daltons)	TACT	= N,N',N''-triallylcitric triamide
NADH	= reduced nicotinamide adenine dinucleotide	TCA	= trichloroacetic acid
		TEMED	= N,N,N',N'-tetramethyl ethylene diamine
NP-40	= Nonidet P-40		
o.d.	= outer diameter	TEPA	= Tetraethylene penta-amine
ORD	= optical rotatory dispersion	TETA	= Tetraethylene tetra-amine
PAGIEF	= polyacrylamide gel isoelectric focusing	TLIEF	= thin-layer isoelectric focusing
PAS	= periodic acid schiff stain	TMU	= tetramethyl urea
PBE	= polybuffer exchanger	TRANSIEF	= transient state isoelectric focusing
PCA	= perchloric acid		
PEHA	= pentaethylene hexa-amine	Tris	= tris(hydroxymethyl) amino methane
PEI	= polyethylenimine	tRNA	= transfer ribonucleic acid
pI	= isoelectric point		
pI_{app}	= apparent isoelectric point	V	= volts
		W	= watts

Preface

Prefaces are assumed to be pedantic and not very informative, so the smart reader tends to skip a few pages and to plunge straight into the text. This preface does not conform to pattern and will not, I hope, kill your enthusiasm for the rest of my book.

Why should it be me writing a manual and not Professor H. Svensson-Rilbe or Dr. O. Vesterberg who were largely responsible for development of the technique? It can only be because of 'being there'. I was working at M.I.T., Department of Nutrition and Food Science, when, in 1968, a Japanese scientist gave a lecture on the purification of tissue ferritins by isoelectric focusing. The seminar was the smoothest ever; no questions, no cross examination since nobody had ever heard of the technique and nobody understood how it worked! The year 1968 was revolutionary in more senses than one; the university students of Paris, indeed of all France and then of all Western Europe were vigorously stirring the academic pot so that, at the time, the revolution in separation techniques represented by that new star – isoelectric focusing – was hardly noticed.

I shall give an account of the gestation of this new biochemical tool based on a very private meeting (October 1980) held in the Tre Travare (three running horses) restaurant; the one behind the Solvalla racing tracks in Stockholm. The authority for my account is that H. Svensson-Rilbe, O. Vesterberg, H. Haglund, H. Davies, B. Bjellqvist and K. Ek were present and I was the recorder.

In 1959 Svensson (who in 1968 changed his name to Rilbe) was on a visit to Pauling in California Institute of Technology. He was already outlining the basic theory of isoelectric focusing but was frustrated to find that even in the USA the chemical catalogues listed no suitable ampholytes. After his return to Sweden as a freelance at the Karolinska

Institute (note that professorships were just as hard to come by then as now) Svensson teamed with Vesterberg (a medical student) to solve the practical aspects of the problem. After three years of slow progress, Svensson became Professor at Gothenburg and the I.E.F. team broke up. Vesterberg continued with the problem in Stockholm and in the spring of 1964 Svensson received a phone call from an excited Vesterberg who appeared to have devised a satisfactory synthesis of carrier ampholytes. After careful checking of the data, Svensson and Vesterberg were convinced that a breakthrough had been achieved and they approached LKB Produkter with a proposal for commercial production. At that time Uppsala was the Mecca of separation techniques so two pilgrims from LKB were sent to consult the high priest of electrophoresis Professor Tiselius. They were disconcerted to be told that an electrophoretic separation technique in which macromolecules would be driven to their isoelectric points could never work as the macroions would aggregate and precipitate.

After such a discouraging verdict the credit for faith and persistence goes to H. Haglund (like Svensson a former pupil of Tiselius) who was head of a small team at LKB involved in separation techniques. Knowing of the Svensson-Vesterberg experiments and of the discouraging assessment of the 'master', Haglund with his colleagues Holmström and Davies decided to try to market the 'gimmick'. They successfully squeezed from a reluctant LKB a bare 60,000 SKr and went into production making Svensson's vertical density gradient columns and the first batches of Vesterberg's carrier ampholytes. These were offered on a free trial basis to the scientific community. It seems only yesterday that the trials were launched; now only 15 years later I am forced to scribble my life away digesting over 4,500 scientific papers.

You will find, and I trust enjoy the results in this book.

Sic transit gloria mundi
Milano, April 1st 1982
PIER GIORGIO RIGHETTI

Nihil obstat quominus imprimatur
(i.e. read and approved by H. Svensson-Rilbe
and H. Haglund at the Viking's ship, Erik
restaurant in the Stockholm harbor).

2nd *imprimatur*
by Dr. T.S. Work and Professor R.H. Burden

Acknowledgements

I owe many thanks to Dr. E. Gianazza, whose enthusiasm and dedication have been very valuable to the research in our laboratory. Mrs. G. Tudor has survived the ordeal of going through my handwriting and typing and correcting all the book. LKB Produkter AB kindly supplied me with a set of Acta Ampholinae, a valuable reference source indeed. Most of the theory and equations reported in the text belong to principles developed by Professor H. Svensson-Rilbe, whose help in these matters is gratefully acknowledged. I am greatly indebted to Professors H.R. Burdon and T.S. Work, who have patiently and competently checked the entire manuscript and burnished it to the present form. Finally I should like to thank all the colleagues who have supplied me with figures and experimental material collected in the text.

CHAPTER 1

Theory and fundamental aspects of isoelectric focusing

This book deals with theoretical and practical developments of IEF and offers detailed methodology for many of the commonly used procedures, such as IEF in gels. It is intended both as a reference guide and a practical manual. The first section of the book deals with the theory and development of the technique of isoelectric focusing. The second section discusses tactical and practical considerations of analytical and preparative aspects and also details current experimental methodology. The third section reviews some applications of IEF in biological and biomedical research which indicate the wide applicability of this exciting technique.

1.1. Historical

Some authors trace back isoelectric fractionations to early, and rather crude, beginnings when in 1912 Ikeda and Suzuki described a method for production of sodium glutamate in an electrolysis apparatus consisting of three compartments separated by membranes permeable to ions. Modifications of their chamber were utilized subsequently by many investigators for separations of amino acids and other ampholytes (Foster and Schmidt, 1926; Cox et al., 1929; Albanese, 1940; Theorell and Akesson, 1941; Sperber, 1946). An important step forward was taken by Williams and Waterman (1929) who increased the number of intermediate compartments to 12 and evolved a clear concept of establishing a stepwise variation of pH from one compartment to the next, with a steady-state pH distribution and accumulation of ampholytes at their isoelectric point (pI). Their insight on the separation process, as stated in their article, was quite remarkable: 'if the substance sought is an ampholyte, it may be concluded that the pH of that portion of the solution which contains the substance in maximum concentration approximates the pI of the substance, though at any pH greater or less than that of the pI, the substance will be ionized more strongly as a base than as an acid, or vice versa, and will therefore migrate toward the region at which ionization as acid and as a base are equal, namely the region of its pI'. This principle was used by Du Vigneaud et al. (1938) to isolate vasopressin and oxytocin hormones from pituitary gland extracts. As a variant to this multimembrane apparatus, Williams and Truesdail (1931) used siphons to connect adjacent compartments, while Williams (1935) built a tubular chamber subdivided into 61 individual segments or compartments by means of baffles. Spies et al. (1941) devised yet another version consisting of six Erlenmeyer flasks provided with lateral horizontal tube nipples approximately at mid-height. These nipples made it possible to connect the flasks in series so that they acted in much the same fashion as a multicompartment electrolyzer. A thorough account of many of these systems and procedures can be found in a review article by Svensson (1948) on preparative electrophoresis.

Another important contribution was made by Tiselius (1941) with his work on stationary electrolysis of ampholyte solutions. He recognized the necessity of achieving a stationary concentration distribution of electrolytes and the achievement of an equilibrium between electromigration and diffusion by prolonged electrolysis. He described his experiments as follows: 'If an ampholyte is added (to a multimembrane electrolyzer) it must be repelled both from the anode and the cathode as acid solutions will give it a positive charge, alkaline solutions a negative charge. Consequently, the ampholyte should move until it finds one or more compartments of a pH equal to its pI and remain there as long as electrolysis is continued ... In the stationary state therefore all anions (in the form of their acids) are found in the compartments adjoining the anode, all cations (as hydroxides) in those adjoining the cathode and the ampholytes in the intermediate compartments, according to the pH value of their pI value's ... The realization of this principle for separation of ampholytes should have several advantages. In separation by migration, it is usually difficult to avoid a considerable dilution of the components since as a rule the front migrates faster than the rear, due to the so-called boundary anomalies. In the above procedure, however, a concentration of each component must be expected.'

Given the fact that, already in 1941, the knowledge of isoelectric fractionation was apparently so advanced, why had we to wait some 25 years before it became a reality? The reason could lie in a phenomenon which Kolin (1977) has designated as 'defocusing of the origins of ideas. It consists of blurring of historical perspectives and usually takes the form of shifting the responsibility for discoveries and ideas to (preferably multiple) sources in the remote past'. Indeed, this early knowledge on electrolysis of ampholytes could not have possibly led to present-day isoelectric focusing, since these experiments had barely scratched the surface of such a vast and deep field. A new contribution to the IEF methodology was made in 1954–1955 by Kolin himself (Kolin 1954, 1955a,b, 1958, 1970). He conceived the idea of 'focusing ions in a continuous pH gradient', stabilized by a sucrose density gradient. He used the term 'isoelectric spectrum' to denote a distribution pattern established by a sorting process. The pH gradient was generated by placing the substance to be separated at the interface between an acidic and a basic buffer, in a Tiselius like apparatus, and allowing diffusion to proceed under an electric field. In these 'artificial' pH gradients, Kolin was able to obtain 'isoelectric line spectra' of dyes, proteins, cells, microorganisms and viruses on a time scale ranging from 40 s up to a few minutes, a rapidity still unmatched in the field of electrophoresis. Concomitant with the pH gradient, a density gradient, an electrical conductivity gradient and a vertical temperature gradient were acting upon the separation cell.

The design of the cell, shown in Fig. 1.1, indicates that Kolin's method was quite an advanced technique. A U-cell was machined out of a Lucite block, with the electrode compartments quite removed from the separation cell (one electrode can be seen in the left side). The front and back Lucite plates could be lifted from the cell body and were provided with a glass or quartz window covering the left leg of the cell. Separation would usually take place on this left limb of the U-cell, which was also provided with tubes D1, D' and D'' for sample loading and harvesting at the end of the separation. Since these three capillaries were in a fixed position along the column length, thus complicating the retrieval of samples which had migrated away from the application point, an improved version was built by Kolin (Fig. 1.2), incorporating a harvesting plunger. By vertical excursion, the plunger could be moved down the

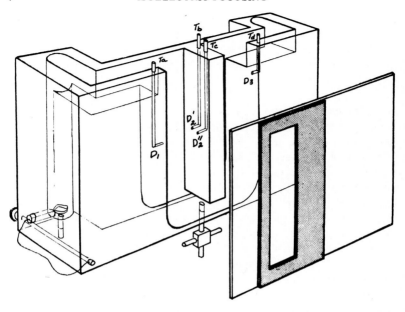

Fig. 1.1. Perspective view of the electrophoretic fractionation cell. T_a, T_b, T_c and T_d: hypodermic syringe tubings. D_1, D_2', D_2'' and D_3: corresponding horizontal ducts (from Kolin, 1958).

left limb of the cell thus fractionating its entire content. As an additional advantage, when the plunger is lowered in the column, its cylindrical liquid volume is raised and as a result the separation between two sample ions will be increased in the ratio of the initial to the final cross-section areas.

Even the optical detection method was quite advanced. For colored samples, the image of the left limb was projected on a screen or on a small ground glass plate, and directly photographed. For colorless fractions, advantage was taken of the strong refractive index gradient formed by 'isoelectric lines'. A grid (Ronchi grating) of fine black lines, inclined at 45 degree angle toward the axis of the electrophoretic column, was placed against the back wall of the left leg of the cell, and the objective focused to project a sharp image of the rulings. The lines were conspic-

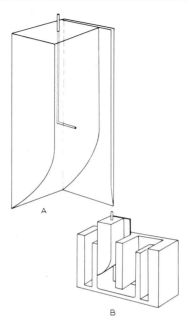

Fig. 1.2. Sample harvesting in the Kolin U-cell for 'isoelectric line spectra'. (A) Perspective view of plunger used as a pipet for withdrawal of desired fractions. (B) Mode of insertion of plunger (from Kolin, 1958).

uously distorted by sufficiently steep refractive index gradients at the boundaries of each separated sample zone. This smart and effective detection principle has been recently re-discovered by Edwards and Anderson (1981).

The question then rises again: if, by 1954, the knowledge of IEF was so advanced, both theoretically and experimentally, why did we have to wait another 12 years before it could be of general use in the field of separation techniques? Probably, because there was still a 'missing link'. I believe that this link could be the concept of 'carrier ampholyte', introduced by Svensson in 1961. With the knowledge of hindsight, this idea might appear rather simple today, but it took decades of failures to develop it. Kolin's pH gradients, obtained by diffusion of non-amphoteric buffers of different pH under an electric field, were indeed 'artificial' pH gradients (and were so termed by Svensson, 1961), since their slope was continuously changed by electric migration and diffusion of the buffer ions. The most favourable conditions were obtained with pairs of buffers of the same ionic species, e.g., two phosphate buffers whose initial pH values encompass the pI values of the substances of interest. Such gradients could never be expected to give more than quasi-equilibrium position of amphoteric molecules. Svensson's (1961, 1962a,b) brilliant theoretical work synthesized the minimum basic requirement for stable pH gradients in an electric field in the term 'carrier ampholyte': the buffers used in this system had to have two fundamental properties: (a) to be amphoteric (so that they would also reach an equilibrium position along the separation column) and (b) to be 'carrier'. This last concept is more subtle but just as fundamental. Any ampholyte could not be used for IEF, only a carrier ampholyte, i.e. a compound capable of 'carrying' the current (a good conducting species) and capable of 'carrying' the pH (a good buffering species). With these concepts, and with Vesterberg's (1969a) elegant synthesis of such ampholytes, present-day isoelectric focusing was born.

1.2. The principle of IEF

The principle of IEF, as compared with conventional electrophoresis, is outlined in Fig. 1.3. In the latter case, the mixture of proteins to be separated is applied as a very thin starting zone in proximity of the cathodic compartment, and forced to migrate into an inert zonal support (cellulose acetate, starch, agarose, polyacrylamide gel, Sephadex slurry, etc.) buffered throughout at a constant and single pH value. In the absence of appreciable sieving, the protein ions will migrate at constant velocity, since they acquire, in the given buffer, a constant surface charge. In the case depicted in Fig. 1.3 for the two species A and B, this is equivalent to saying that they migrate at the constant charge which can be found in a single, specified point (pH = 9) of their titration curve, that would impart a net surface charge of (-3) to protein A and (-1) to protein B. As the two species migrate away from the application point toward the anode, A will always run ahead of B, and both macro-ion zones will widen as electrophoresis progresses, since no force is present

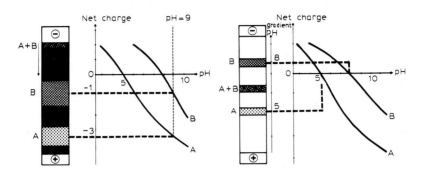

Fig. 1.3. Comparison between zone electrophoresis (left) and IEF (right). A mixture of two proteins A (pI = 5) and B (pI = 8), whose titration curves are shown on the right of each drawing, are separated by each method. In the former case, since the anticonvective medium is buffered at constant pH (=9) the two macroions move with a constant net negative charge and at constant velocity (in absence of differential sieving). As time progresses, the zones diffuse. In the latter technique, the proteins are continuously decelerated to an equilibrium zone of zero net charge (pH = pI) (courtesy of LKB Produkter AB).

in this system to counteract diffusion. In conventional electrophoresis, there is no end point, and the potential differential has to be discontinued before the migrating species are eluted in the anodic compartment.

Conversely, in IEF, a stable pH gradient increasing progressively from anode to cathode is established by electrophoretic 'sorting out' of carrier ampholytes in a suitable anticonvective liquid medium. When introduced into this system, a protein or other amphoteric molecule will migrate according to its surface charge in the electric field. Should its initial charges and gain negative charges, e.g., through deprotonation of carboxyl higher pH. As it does so, the molecule will gradually loose positive charge and gain negative charges, e.g., through deprotonation of carboxyl or amino functions. Eventually, it will reach a zone where its net electrical charge is zero, i.e., its pI. Thus, during an IEF separation, as the protein moves along the prefocused pH gradient, its velocity keeps changing, as its net surface charge keeps diminishing toward a minimum value, according to a proton equilibrium pathway determined by its titra-

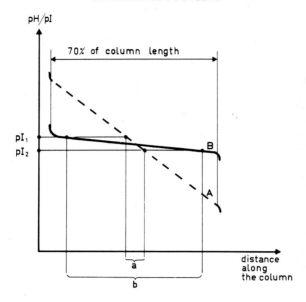

Fig. 1.4. Separation of ampholytes as a function of the slope of the pH gradient. Two amphoteric molecules, represented by pI_1 and pI_2, are separated only by the distance (a) in a steep pH gradient (broken line A) but by a greater distance (b) in a shallower pH gradient (solid line B) (courtesy of LKB Produkter AB).

tion curve, until it comes to a full stop. Should the molecule migrate or diffuse away from its pI, it will develop a net charge and be repelled back to its pI. Thus, by countering back diffusion with an appropriate electrical field, a protein or other amphoteric macromolecule will reach an equilibrium position where it may be concentrated into an extremely sharp band. As might be expected, the degree of separation of two ampholytes is a function, among other parameters, of the slope of the pH gradient in which they are focused. The shallower the pH gradient, the better is the separation for a fixed column geometry (migration path), as illustrated in Fig. 1.4. This principle is discussed more fully later (§ 1.5.7).

The effect of 'focusing' or 'condensation' at the isoelectric point is illustrated in Fig. 1.5. If we apply the same protein ion (pI = 8), to a

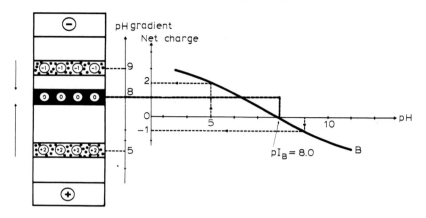

Fig. 1.5. Focusing or concentrating effect in IEF. The same protein ion (B, pI = 8.0) is applied to a prefocused gel slab near the cathode (negatively charged zone) and near the anode (positively charged species). The two protein fronts will migrate towards each other and condense, or focus, in a single zone of zero net charge and zero mobility. On the right, the pH/mobility curve of the protein is shown (courtesy of LKB Produkter AB).

preformed pH gradient, in two different zones, one close to the cathode (pH 9), the other at the anodic side (pH 5), they will simultaneously acquire opposite charges, negative for the pH 9 zone and positive for the pH 5 zone. This will result in a 'head-on' migration, until the two species converge and merge in the same zone, having zero mobility and zero net charge (pH = pI = 8). This stationary zone will represent a dynamic equilibrium between pI-directed electrophoretic migration and back diffusion.

IEF was originally developed as a preparative technique. Fractionations were conducted in sucrose density gradients which served as an anti-convective medium to stabilize the pH gradient and focused protein zones. Experiments were usually performed in column volumes of 110 or 440 ml and usually required 2–4 days to reach equilibrium. These systems were, however, rather expensive in time and required careful standardization. They also suffered from some practical problems arising from isoelectric precipitation and excessive diffusion during elution of the column in the absence of the electrical field. Finally, sample detection was laborious and often difficult. Consequently, these early systems were not convenient for routine analytical procedures. Fortunately, many of the problems inherent in IEF in sucrose density gradients have now been overcome with the use of more suitable anti-convective media such as polyacrylamide gels, Sephadex beds and agarose matrices. With these media many of the potential advantages of IEF for high resolution separations of amphoteric substances can be realised. The methodology for IEF in gels has now been well standardized for both analytical and preparative procedures. Lastly, IEF in gels is a remarkably forgiving technique in which a minimum of technical skill and effort is amply rewarded in the quality of the fractionation achieved.

1.3. Extension of the IEF principle

In a recent historical article Kolin (1977) has pointed out that, in a sense, IEF can be considered as a null method in which particles reach equi-

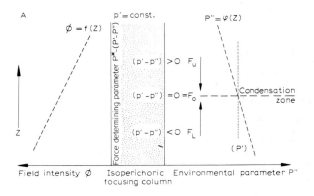

Fig. 1.6. Isoperichoric focusing. A chemical environment is generated in which an environmental property described by parameter p'' is changing from point to point along the z-axis ($p'' = \phi(z)$). A species of particles characterized by the parameter value p', corresponding to the parameter p'' of the suspension fluid, is dispersed with an arbitrary distribution in the fluid column. It is assumed that there is a point within the gradient of the p'' distribution at which $p' = p''$, where the particle parameter equals the environmental parameter (focusing or condensation zone (from Kolin, 1977)).

librium in a focal zone of vanishing force. He has generalized the idea of IEF and shown that it may be considered as a member of a family of phenomena – isoperichoric (iso = equal and perichoron = environment) focusing effects – 'in which particles migrating in a force field along a chemical concentration gradient are swept into sharp, stable stationary zones of vanishing force in which a physical parameter of the suspension medium matches that of the particle. Examples of such condensation phenomena are isopycnic-, isoconductivity-, isodielectric-, isomagnetic-, isoparamagnetic- and isodiamagnetic-focusing'. Figure 1.6 gives a general example of this concept: a force field having an intensity $\Phi = f(z)$ is applied externally to the separation cell (isoperichoric focusing column). Inside the column a chemical environment $p'' = \phi(z)$, either generated by the external force field or pre-existing to it, acts in practice as a counter-force which, by matching some physico-chemical parameter p' of the particle to be separated, nullifies the external force Φ acting upon it, so that the particle comes to rest into a zone of dynamic equilibrium between the opposite forces F_U and F_L. Table 1.1 summarizes the equilibrium separation methods today known in the field of biochemistry: isopycnic centrifugation, isoelectric focusing and pore gradient electro-

TABLE 1.1
Comparison among equilibrium techniques

Technique	Support	Parameter measured	Resolution limits
Isoelectric focusing	Sucrose, polyacrylamide, agarose gels	pI [a]	0.005 pI [d]
Isopycnic centrifugation	Sucrose, CsCl	ρ [b]	0.003 ρ [e]
Polyacrylamide gel gradient	Polyacrylamide gel	R_s [c]	3,000 daltons [f]

[a] Isoelectric point.
[b] Density (g/cm³).
[c] Stoke's radius.
[d] Allen, R.C., Hasley, R.A. and Talamo, R.C. (1974) Am. J. Clin. Pathol. *62*, 732–739.
[e] Dawin, I.B. and Wolstenholme, D.R. (1968) Biophys. J. *8*, 65–70.
[f] For globular proteins. Margolis, J. and Kenrick, K.G. (1968) Anal. Biochem. *25*, 347–362.

phoresis. The first two methods act on such a similar principle that Giddings and Dahlgren (1971) have derived similar equations for the resolving power and peak capacity of both techniques. The times must have been ripe for both these ideas, since they were described independently and within a short time lapse by Meselson et al. (1957) (isopycnic centrifugation) and by Svensson (1961) (IEF). In both methods an externally applied force field is used to create a variable chemical environment (a density gradient and pH gradient) within the separation column. It should be noticed that, in both cases, the slope of the environmental parameter is opposite to the slope of the external force field: if both were parallel, the particle would be swept away from the focusing zone (Kolin, 1977).

It might be argued that the third method, pore gradient electrophoresis can hardly be classified among equilibrium techniques (Chrambach and Robard, 1971) since no end point is ever reached, but at best an asymptotic mobility decrement. Yet, as long as the migration is performed in a gel slab along with a proper set of molecular mass standards (especially in pore gradient electrophoresis in presence of sodium dodecyl sulphate) a true determination of molecular mass and hydrodynamic radius can be obtained (Gianazza and Righetti, 1979a).

All three techniques share a very high resolution capability, much superior than any other non-equilibrium separation method, and, in addition to purifying the macromolecule under study, are also able to measure, within a single experiment, some physico chemical parameters intrinsic to the species under fractionation (such as its density, its isoelectric point, its molecular mass and its Stoke's radius). These physical parameters depend on the chemical composition of the macromolecule (primary structure) and on its three-dimensional configuration, thus can be used to characterize unquivocally a given macromolecular species.

1.4. Books and reviews on IEF

Needless to say, this monograph cannot possibly cover all aspects of IEF in detail. Since the advent of present day techniques and up to the end of 1981, more than 4,500 papers have appeared on the use of IEF and eight International Symposia on this and other electrophoretic techniques have been held. For additional information the reader is referred to the publications of these meetings (Catsimpoolas, 1973a; Allen and Maurer, 1974; Arbuthnott and Beeley, 1975; Righetti, 1975a; Radola and Graesslin, 1977; Catsimpoolas, 1978; Radola, 1980c; Allen and Arnaud, 1981). In April 1982, a ninth international symposium on electrophoretic methods, organized by D. Stathakos, was held at the University of Athens in Greece. A tenth symposium is scheduled for May 1983 in Tokyo (Professor H. Hirai). The proceedings of both meetings will be published by De Gruyter. In addition to these proceedings of meetings, other books, covering general theoretical and methodological aspects of the technique (Catsimpoolas, 1976) as well as its biomedical and biological applications (Catsimpoolas and Drysdale, 1977) have been published. An interesting treatise, devoted partly to the electrophoretic, partly to the isoelectric analysis of mammalian cells, subcellular organelles, bacteria and viruses has also recently appeared (Sherbet, 1978). As a general reading, a new book, *Electrokinetic Separation Methods* (Righetti et al., 1979a) has also been published, which covers practically all aspects of electrophoresis in 21 chapters. Several chapters deal with different aspects of IEF: continuous-flow methods (Just and Werner, 1979); IEF in free solution (Quast, 1979); preparative aspects (Chrambach and Nguyen, 1979); protein-carrier ampholyte interactions (Cann, 1979) and a general review of the field (Righetti, 1979). The only manual fully devoted to all facets of IEF, published before the present treatise, is the one written in 1976 by Righetti and Drysdale. A host of reviews covering practically all aspects of IEF have also been published over the years. Perhaps the one which served as a basic training to most biochemists is the review published by Haglund in 1971. In breadth and depth, it was unsurpassed for many years and, at that time, it was a *Summa Theologica* of all the knowledge available in the field. The following fields have been reviewed: general methodology (Vesterberg, 1968, 1970a, 1971a, 1973a,b, 1978; Righetti and Drysdale, 1974); theoretical aspects (Catsimpoolas, 1970, 1973a, 1975a; Rilbe, 1973a); IEF of immunoglobulins (Williamson, 1973); IEF of cereal proteins and food stuff (Du Gros and Wrigley, 1979; Wrigley, 1973, 1977; Bishop, 1979; Righetti and Bianchi Bosisio, 1981); preparative aspects of IEF (Rilbe, 1970; Righetti, 1975a); biochemical and clinical applications (Latner, 1975; Righetti et al., 1976; Allen, 1978; Righetti et al., 1979); methodological aspects in comparison with electrophoresis and isotachophoresis (Chrambach, 1980); recent developments (Righetti et al., 1980a,b); zymogram techniques in IEF (Wadström and Smyth, 1973); use of pH markers in IEF (Bours, 1973d). General reviews of the IEF methodology can also be found in manuals covering general biochemical techniques, such as in Morris and Morris (1976) and the chapter of Leaback and Wrigley (1976) in I. Smith's book. Other interesting compilations are two tables published in 1976 (Righetti and Caravaggio) and in 1981 (Righetti et al. 1981a). These tables list the isoelectric points, as determined by IEF, together with native molecular mass, subunit molecular mass and number, of more than 1,200 enzymes and structural proteins isolated from prokaryotes and eukaryotes. For all the entries, the source and, for eukaryotic species, the organ and subcellular location of the purified enzyme are given. These two compilations cover the years from the origin of IEF (ca. 1966) until the end of 1979. They should be a useful vademecum for protein chemists.

The year 1980 has also witnessed the birth of the Electrophoresis Society with the companion, official journal *Electrophoresis*, published by Verlag-Chemie. Thus, it is to be expected that many articles dealing with new methodological aspects and applications of IEF will appear in this journal, as well as in other methodological journals such as *Journal of Biochemical and Biophysical Methods* (Elsevier); *Analytical Biochemistry* (Academic Press) and, to a lesser extent, *J. Chromatography* (Elsevier). In addition to this extensive literature, a very useful reference list, called *Acta Ampholinae*, is published by LKB Produkter AB. A volume covering the years 1960–1974 is now available. Supplements are published yearly covering the field of IEF and isotachophoresis. At the time of preparation of this monograph (end of 1981), more than 4,500 articles on IEF have been listed.

1.5. Theory and fundamental aspects of IEF

This section can only be a brief survey of the fundamental equations pertaining to IEF. For a detailed treatise, the reader is referred to the early articles by Svensson (1961, 1962a,b), as well as to more recent surveys and extensions of his theory (Rilbe, 1971, 1973a, 1976; Almgren, 1971; Giddings and Dahlgren, 1971). It must be emphasized that, even though Rilbe's theories have been challenged (especially on account of steady-state decay in gel IEF, as well as lack of uniform conductivity, buffering capacity and pH gradients in present-day IEF) no alternative theory has been presented to date, and Rilbe's basic concepts remain valid.

1.5.1. Artificial pH gradients

The pH gradients obtained by diffusion of non-amphoteric buffers of different pH under an electric field (Kolin, 1954, 1955, 1958) were termed artificial pH gradients by Svensson (1961), since the gradient is affected by changes in electric migration and diffusion of the buffer ions. The most favorable conditions are obtained with pairs of buffers of the same ionic species, e.g. two acetate buffers whose initial pH values encompass the pI values of substances of interest. Such gradients can never be expected to give more than quasi-equilibrium positions of amphoteric molecules.

1.5.2. Natural pH gradients

In the early 1960s Rilbe (Svensson, 1961) introduced the concept of 'natural' pH gradients and established a theoretical basis for IEF in the present form. If the natural pH gradient were created by the current itself it should be stable over long periods if protected from convective disturbances. IEF should be, therefore, a true equilibrium method. This concept, as just mentioned, has similarities with equilibrium (isopycnic) density gradient centrifugation and, indeed, Giddings and Dahlgren (1971) have derived similar equations for the resolving power and peak capacity of both methods. Svensson predicted that stable pH gradients would be obtained by isoelectric stacking of a large series of carrier ampholytes, arranged under the electric current in order of increasing pI from anode to cathode. The pH in every part of the gradient would then be defined by the buffering capacity and conductivity of the specific isoelectric carrier ampholyte located in that particular region. The stability of such natural pH gradients is, however, contingent on other factors such as convection and diffusion. The electrical load should not exceed the cooling capacity of the column to prevent convective mixing. Ideally, the conductivity should be uniform throughout the gradient to prevent local overheating in regions containing low levels of ampholyte or ampholytes exhibiting poor conductivity and buffering capacity. Instability of the pH gradient could also arise by anodic oxidation or cathodic reduction of carrier ampholytes. Consequently, strong acids and bases were recommended for use at the electrodes to repel the ampholytes from the electrode compartments.

1.5.3. The differential equation of IEF

Under steady-state conditions, as in ideal IEF, the following differential equation (which describes the equilibrium conditions between simultaneous electrophoretic and diffusional mass transport) can be derived (Svensson, 1961):

$$\frac{d(CuE)}{dx} = \frac{d}{dx} D \frac{dC}{dx} \tag{1}$$

where C is the protein concentration at the level x in the separation column, u is its mobility at that point, E is the field strength and D is its diffusion coefficient. Equation (1), upon integration gives:

$$CuE = D \frac{dC}{dx} + \text{integration constant} \tag{2}$$

This equation must be valid everywhere in the column, even outside the focused protein zone, where the protein under consideration is absent. Thus the integration constant is zero, and we obtain:

$$CuE = (dC/dx) D \tag{3}$$

which is the general differential equation for the final steady-state in IEF. Equation (3) is not applicable to components which cannot reach a steady-state (such as salts, which are split up in their ionic constituents) nor to non-electrolytes (such as sucrose, sorbitol, glycerol, often used in IEF as anti-convective media). However, a constant concentration gradient of a non-electrolyte in IEF does not violate Eq. (3) if it can be kept reasonably constant with time. In theory, Eq. (3) can also be applied to acids and bases which migrate to the anode and cathode, respectively, and are concentrated there. Therefore, one free acid at the anode and one free base at the cathode are compatible with steady-state conditions. However, if mixtures of acids and/or bases are used, they will not necessarily come to a steady-state. For instance, when one strong and one weak acid simultaneously migrate to the anode, the strong one collects there more rapidly than the weak one, since the former suppresses the dissociation of the latter. Thus, the electrophoretic migration of the weak acid is stopped, but its diffusion proceeds through the strong acid layer toward the anode. Since no electric mass transport can balance this diffusional mass transport, a steady-state cannot be reached (see also §1.11.5). Components unable to reach the steady-state may slowly alter the conditions underlaying the steady-state of other components. The latter states then become quasi-stationary, subjected to slow drift.

1.5.4. Buffering capacity in the isoprotic state

The molar buffer capacity (B_i) of an ampholyte in the isoprotic state (isoprotic and isoelectric states coincide when the absolute mobilities of the cationic and anionic species of an amphoteric substance are equal) is given by (Rilbe, 1976):

$$B_i = (\ln 10)/(1 + s) \tag{4}$$

where the parameter s is defined as:

$$s = f(K_1/4K_2)^{1/2}, \log s = \tfrac{1}{2} \Delta pK + \log (f/2) \tag{5}$$

i.e., it is a function of the activity factor f and of the two dissociation constants, K_1 and K_2, of a bivalent ampholyte. The maximum buffer capacity of a monovalent weak acid or base, i.e., its buffer capacity at pH = pK, is:

$$B_i = (\ln 10)/4 \tag{6}$$

The relative buffer capacity (b_i), defined as the ratio between (4) and (6), is thus:

$$b_i = 4/(1 + s) \tag{7}$$

Since this cannot possibly be more than 2, it follows that s cannot possibly be smaller than unity ($s \geqslant 1$). By virtue of Eq. (5) this is equivalent to the conditions for bivalent protolytes:

$$\Delta pK' = pK_2' - pK_1' \geqslant \log 4 \qquad (8)$$

where $K_1' = fK_1$ and $K_2' = K_2/f$ are the stoichiometric dissociation constants. Equations (7) and (8) reveal that the buffer capacity in the isoprotic state decreases with increasing s, i.e., with increasing ΔpK across the isoprotic point, linearly at first, then exponentially. For example, the illustration in Fig. 1.7 shows that when pK_2 and pK_1 lie 1.5 pH units apart, 50% of the limiting buffer capacity is still retained. However, when ($pK_2 - pK_1$) = 3.2 pH units, only 1/10 of the maximum buffering capacity remains and the compound can be considered a poor carrier ampholyte. Thus, good carrier ampholytes should have a pair of pK

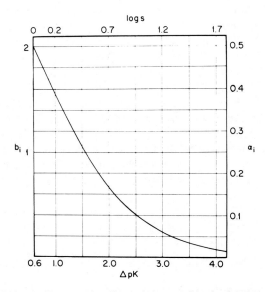

Fig. 1.7. Relative buffer capacity (b_i) and degree of ionization (α_i) in the isoprotic state as a function of ΔpK across the pI of a divalent ampholyte and of the parameter s (as defined in Eq. 5) (from Rilbe, 1976).

values (for diprotic species, only one pK value for polyprotic compounds, such as the Vesterberg series, 1969) which lie as close as possible to the pI of the ampholyte itself. Notice, from Fig. 1.7, that, as s tends to its limit value of 1 (log $s = 0$) the three variables b_i, α_i and ΔpK come to a coincidence in point which also represents their limit values: a maximum degree of ionization of 0.5, a maximum buffer capacity of 2 and a minimum ΔpK value, across the pI of a divalent ampholyte, of 0.6. Below this limit value of Δp$K = 0.6$, the dissociation constants pK'_1 and pK'_2 no longer represent true, independent protolytic equilibria, but become 'hybrid' dissociation constants (MacInnes, 1939).

1.5.5. Conductivity in the isoprotic state

An important prerequisite for a good carrier ampholyte is that it has a good conductivity at its pI. Regions of low conductivity will not only cause local overheating, but will also absorb much of the applied voltage. This reduces the field strength and hence the potential resolution in other parts of the gradient.

The degree of ionization, α, of a diprotic ampholyte can be written:

$$\alpha = \frac{C_{amph+} + C_{amph-}}{C_{amph}} \qquad (9)$$

where C_{amph+} is the molar concentration of positive ions, C_{amph-} is the molar concentration of negative species and C_{amph} is the total molar concentration of ionic and undissociated ampholyte. Neglecting activity coefficients, the following equations follow from the law of mass action:

$$C_{amph+} = C_{amph_0} (10^{pK_1 - pH}) \qquad (10)$$

$$C_{amph-} = C_{amph_0} (10^{pH - pK_2}) \qquad (11)$$

where C_{amph_0} is the molarity of zwitterionic and undissociated forms.

The total molar concentration, C_{amph}, can be written:

$$C_{amph} = C_{amph+} + C_{amph-} + C_{amph_0} \qquad (12)$$

Elimination of C_{amph_0} gives, after rearrangement:

$$C_{amph}/C_{amph+} = 1 + 10^{pH - pK_1} + 10^{2(pH - pI)} \tag{13}$$

$$C_{amph}/C_{amph-} = 1 + 10^{pK_2 - pH} + 10^{2(pI - pH)} \tag{14}$$

The relationship between the pK values of a diprotic ampholyte and the isoionic point (pI) is:

$$pI = \frac{pK_1 + pK_2}{2} \tag{15}$$

where pK_1 is defined to be smaller than pK_2.

By substituting the proper values of Eqs. (13), (14) and (15) into (9), the degree of protolysis at the isoelectric state (pH = pI) will be given by:

$$\alpha = \frac{2}{2 + 10^{(pI - pK_1)}} \tag{16}$$

The highest possible value for α at maximum conductivity is 1/2, since pI $-$ pK_1 has a lower limit of log2 (Svensson, 1962a; Rilbe, 1971). Equation (16) shows that good conductivity is associated with small values of pI $-$ pK_1. This is also true for the buffering capacity of an ampholyte. Thus the parameter pI $-$ pK_1 becomes the most important factor in selecting carrier ampholytes exhibiting both good conductivity and buffering capacity. As shown in Fig. 1.8 glycine, which is isoelectric between pH 4 and 8 (ΔpK = 7.4, corresponding to a degree of ionization of 0.00038 in the isoprotic state) is a very poor carrier ampholyte, while histidine, lysine and glutamic acid, with sharp isoelectric points (low pI $-$ pK_1 values) are quite useful ampholytes. Thus, buffering capacity and conductivity in the isoprotic state are always complementary and are indeed the two fundamental properties of the compounds which Svensson (1962a) has defined as carrier ampholytes (see also §1.1).

1.5.6. Law of pH monotony

Formulated by Svensson (1967), this law states that a natural pH gradient, developed by the current itself, is positive throughout the gradient as the pH increases steadily and monotonically from the anode to cathode.

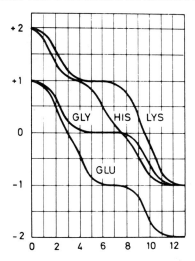

Fig. 1.8. Titration curves of glutamic acid, glycine, histidine and lysine. The three amino acids with sharp isoionic points (low pI − pK_1 values) are useful as carrier ampholytes, while glycine, which is isoelectric over a wide pH range, is a useless ampholyte (from Svensson, 1962a).

Reversal of the pH gradient at any position between the electrodes is thus incompatible with the steady state. Thus in stationary electrolysis, two ampholytes can never be completely separated from one another, unless a third ampholyte, with intermediate pI, is present in the system. For instance, in the case of Asp (pI 2.8) and Glu (pI 3.2) a complete separation could only be achieved if a zone of pure water (pI 7) were to develop between them. This situation is excluded by the law of pH monotony, therefore, the Gaussian distributions of two amino acids must overlap in the steady state if no intermediate ampholyte is present.

Ideally the ampholytes should be evenly distributed over the entire pH range. It is particularly important to have a suitable ampholyte in the neutral pH range to prevent formation of a zone of pure water. Development of a region of pure water within the electrofocusing column will have undesirable effects, notably in creating local hot spots. The theoretical conductance of pure water (at 0 °C) is:

$$\chi_W = 1.8 \times 10^{-8} \text{ ohm}^{-1} \text{ cm}^{-1}$$

The field strength is given by the equation:

$$E = I/\chi$$

and the Joule heat per unit volume by the equation:

$$q = I^2/\chi$$

where I is the current density and χ the conductance along the separation column. Even with I as low as 10^{-4} A/cm^2, the field strength in pure water would be 10,000 V/cm, and the Joule heat 1 W/cm^3. Any region of pure water will absorb almost all of the available voltage, thereby decreasing the focusing effect and allowing neighboring electrolytes to diffuse with the developing region of pure water. Theoretically, and compatible with the law of pH monotony, a zone of pure water could develop at a neutral pH and, in fact, water is an ampholyte, isoelectric at pH 7. Cationic water (H_3O^+) is repelled from the anode and anionic water (OH^-) from the cathode, so that water as a whole tends to concentrate at pH 7. However, it is an extremely poor carrier ampholyte, with little, if any, buffering capacity and conductivity at its pI, and it is, therefore, important to have other ampholytes in the neutral pH region.

1.5.7. Resolving power

In general, focused proteins zones are usually very narrow, i.e., are confined within a narrow pH interval, where the pH gradient, $d(\text{pH})/dx$, and the mobility slope, $du/d(\text{pH})$, can be regarded as constant. With the x axis pointing in the direction of the current, the pH gradient becomes a positive quantity, but the mobility slope is always negative. Thus p, defined as:

$$p = -\frac{du}{dx} = -\frac{du}{d(\text{pH})}\frac{d(\text{pH})}{dx} \tag{17}$$

is a positive quantity and, since $u = 0$ for $x = 0$, the relation between u and x becomes: $u = -px$. Inserted into Eq. (3), this gives:

$$dC/C = -(pE/D)\, x\, dx \tag{18}$$

If E and D are also considered constant within the focused zone, this equation has the solution:

$$C = C_{(0)} \exp\left(-pEx^2/2D\right) \tag{19}$$

where $C_{(0)}$, the integration constant, physically means the local concentration maximum in the region where the protein is isoelectric. Equation (19) expresses a gaussian concentration distribution with inflection points lying at:

$$\sigma = \pm \sqrt{D/pE} \tag{20}$$

By substituting, into Eq. (20) the value of the proportionality factor p as defined by Eq. (17), then Eq. (20) can be given in the form:

$$\sigma = \pm \sqrt{DE^{-1}\left[-du/d(pH)\right]^{-1}\left[d(pH)/dx\right]^{-1}} \tag{21}$$

If two adjacent zones of equal mass have a peak to peak distance three times larger than the distance from peak to inflection point, there will be a concentration minimum approximating the two outer inflection points. Taking this criterion for resolved adjacent proteins, Rilbe (1973a) has derived the following equation for minimally but definitely resolved zones:

$$\Delta pH = \Delta x\left[d(pH)/dx\right] = 3\sigma\left[d(pH)/dx\right] \tag{22}$$

By inserting Eq. (21) into (22) one obtains, at the pI:

$$\Delta(pI) = 3\sqrt{\frac{D\left[d(pH)/dx\right]}{E\left[-du/d(pH)\right]}} \tag{23}$$

Equation (23) shows that good resolution should be obtained with substances with a low diffusion coefficient and a high mobility slope $[du/d(pH)]$ at the isoelectric point — conditions that are satisfied by all proteins. Good resolution is also favored by a high field strength and a shallow pH gradient. If the pH gradient $[d(pH)/dx]$ is known and the value of σ is obtained from the zone breadth at theoretical ordinates $e^{-\frac{1}{2}} = 0.61$ of the peak height, the numerical value of ΔpH can be estimated. Such calculations have shown that the resolving power of IEF allows separation of molecules which differ in pI by as little as 0.02 pH unit (Vesterberg and Svensson, 1966).

A similar equation has been derived by Giddings and Dahlgren (1971) for the resolving power (R_s) in IEF:

$$R_s = \frac{\Delta pH}{4} \sqrt{\frac{-FE\,[dq/d(pH)]}{RT[d(pH)/dx]}} \qquad (24)$$

where F is the faraday (96,500 coulombs) E the electric field strength, R the gas constant, T the absolute temperature, q the effective charge of the particle and x the distance moved.

1.5.8. Peak capacity

Giddings and Dahlgren (1971) have defined the peak capacity n as the maximum number of components resolvable by a given technique under specified conditions. In IEF the peak capacity is given by:

$$n = \sqrt{\frac{-FE[dq/d(pH)]\,[d(pH)/dx]\,L^2}{L6RT}} \qquad (25)$$

where L is the total path length. This equation shows that peak capacity increases in proportion to total path length and is a function of the square root of both the slope of the pH gradient and the protein mobility slope. The alternate form is obtained by replacing $[d(pH)/dx]$ with $(pH_L - pH_0)$, the total pH increment. In this last case, assuming a uniform gradient model, i.e., a constant value of $[d(pH)/dx]$, we have:

$$n = \sqrt{\frac{-FE[dq/d(pH)]\,[d(pH)/dx]\,L^2}{16RT}} \qquad (26)$$

Thus the peak capacity is directly proportional to the square root of the electric field, path length and total pH increment. The three parameters may be varied simultaneously or independently according to experimental needs. Just as an example, Rapaport et al. (1980) have recently described IEF separations in 60 cm long polyacrylamide gel rods (a standard, conventional length is around 10–12 cm). Peak capacity and resolving power were definitely improved, but at the expense of having to handle (casting, staining, destaining etc.) such long and slim gel cylinders.

1.5.9. Mass content of a protein zone

Rilbe and Pettersson (1975) have derived a relationship for the mass load (m) of a protein zone in IEF:

$$m < 5q\sigma^2 \ (dc/dx) \tag{27}$$

where q is the cross-sectional area of the separation column, σ the inflection points of the Gaussian protein zone, and dc/dx the sucrose density gradient along the column. We now introduce the volume V, the height H and the zone width r, taken as 2σ in units of column length:

$$i.e. \ r = 2\sigma/H \quad \sigma = rH/2 \tag{28}$$

Inequality (27) can be written in the form:

$$m < 1.25 \ VHr^2 \ (dc/dx) \tag{29}$$

If we now assume a linear sucrose concentration gradient from 0 to 0.5 g/cm^3, dc/dx will have the numerical value:

$$dc/dx = (1/2H) \ g/cm^3 \tag{30}$$

by substituting this value in Eq. (29), we obtain:

$$m < 0.625 \ Vr^2 \ g/cm^3 \tag{31}$$

for a 100 cm^3 column, one thus has:

$$m < 62.5 \ r^2 \ grams \tag{32}$$

The carrying capacity thus rises with the square of the zone width. For a 1 mm thick zone, in a 25 cm high column, Eq. (32) shows that the mass content cannot be more than 1 mg. But if in the same column the protein zone is allowed to extend over 1/10 of the column height, its maximum protein content will be 625 mg, whereas a zone extending over one fifth of the column can support up to 2.5 g protein.

By combining Eqs. (20), (17), (23) and (27) the mass load of the protein zone can be related to the resolving power, thus obtaining the inequality:

$$m < \frac{45qD^2\,(dc/dx)}{E^2\,(pI)^2\,[du/d(pH)]^2} \tag{33}$$

Thus, the capacity of a density gradient column rises with the square of the resolving power with which it operates for a given protein system. This means that an experiment with a pH gradient favoring the purity of protein fractions obtained also favors the quantity of pure protein that can be separated. From this standpoint, IEF has considerable advantages over other separation techniques in which a good yield axiomatically excludes high purity and vice versa.

No equations have yet been derived for the load capacity of IEF in gel media. Radola (1975) studied this aspect experimentally in Sephadex gel layers, and found an upper load limit of approximately 10−12 mg protein/ml gel suspension. In these experiments, he defined load capacity as the weight (mg) of protein per ml gel suspension used for preparing the gel layers. He was able to fractionate 10 g pronase E using a pH range 3−10 in a 400 × 200 mm trough with a 10 mm thick gel layer (total gel volume of 800 ml). The resolution was comparable to that of analytical systems. At these remarkably high protein loads, the focused bands could be seen directly in the gel layer as translucent zones. This load limit may only apply to wide pH range gradients (3−10). From theoretical considerations (Eq. 33) Radola had predicted much higher protein loads in narrower pH ranges. These figures dramatize one of the major advantages of IEF in gel media in stabilizing protein zones. The load capacity will of course also depend on the actual number of components and their solubility at their pI. Experimentally, Chrambach (1980) has derived upper limits for the load capacity of different electrophoretic techniques. He has reported that the load capacity of poly-acrylamide gel electrophoresis (PAGE) seems to be limited to 0.1 mg per component/cm of gel, while that of IEF is 1–2 orders of magnitude higher, and that of isotachophoresis (ITP) 2–3 orders greater than that of PAGE. Thus maximum protein loads in IEF should be of the order of 1–10 mg per component/cm of gel, while that of ITP should be in the range of 10–100 mg. It seems to be difficult to be able to focus or stack more than 100 mg protein/ml because at such high concentrations proteins withdraw water and salt from the medium to such a degree that voltage gradients across the protein zones increase enormously, with concomitant heat denaturation. Presumably the conductivity within the protein zone becomes so low because the buffer constituents and solvent are largely protein-bound.

1.5.10. Conditions for a stepless pH course

If too few carrier ampholytes are present in a pH gradient in IEF (5 to 10 species per pH unit), with a high ΔpK value, the pH course, rather than smooth, will be more like a staircase. The question of how many individual carrier ampholytes are really necessary in order to obtain a stepless pH course can be answered by formulating a condition for completely unresolved double zones of adjacent carrier ampholyte peaks. As demonstrated by Svensson (1966), the sum of two similar gaussian curves has one flat maximum and no minimum at all for a peak to peak distance equal or less than two standard deviations. If all individual carrier ampholytes lie as closely spaced as that, all of them are completely unresolved from their neighbors. No plateau in the pH course can then be expected. Therefore, the greatest allowable pI difference between adjacent carrier ampholytes compatible with a smooth pH course is given by the inequality [partly derived from Eq. (24)] :

$$\Delta(\text{pI}) < 2\sqrt{\left(\frac{D(\text{d}(\text{pH})/\text{d}x)\,(1+s)}{Eu\,\ln 10}\right)} \tag{34}$$

This inequality is in full agreement with Almgren's (1971) conclusions. The parameter s is the quality mark of ampholytes: the lower the s value, the better are the conductivity and buffering capacity. Condition (34) shows that ampholytes of high quality must lie more closely spaced on the pH axis than ampholytes of lower quality. Ampholytes with low s values are especially needed in the pH region around the neutral point in order to compensate for the very low conductivity of water ions in that region. According to Rilbe (1976), high quality as well as low quality carrier ampholytes, with a preponderance for the high quality in the neutral range, are essential for a successful isoelectric focusing. According to Almgren (1971), the use of ampholytes with widely separated pK values reduces the number of ampholytes required to give a good pH gradient. However, if ΔpK is too large, more ampholyte must be added to give a certain conductivity and buffering capacity and the system will approach the steady-state more slowly.

Can we actually calculate how many different species of carrier ampholytes should be present per pH unit for generation of a stepless pH course? In a real system, this might turn out to be an impossible task. However, for an ideal system, comprising monovalent ampholytes with equal diffusion coefficients and electric mobilities, equal relative concentrations and evenly spaced pI values on the pH scale and exhibiting the same difference between their two pK values, Almgren (1971) has been able to calculate that by computer simulation. Figures 1.9A, B and C give the results of such a computation. In order to solve this problem, we have to know the following parameters: ΔpI and ΔpK (both in pH units) and the width of the gaussian distribution curve at steady-state (expressed in Fig. 1.9A, B and C as the abscissa yM: because it is a column coordinate, but it has the dimensions of moles/litre). As shown in Fig. 1.9A, a system comprising 8 carrier ampholytes, distributed over 0.25 pH units (from pH 7.0 to pH 7.25), in which each individual species has a $\Delta pK = 2$, and spaced at $\Delta pI = 0.05$, is able to generate a smooth pH course. The conductivity course (H) is also constant — even over this pH interval. According to these data, it would appear that at least 30 carrier ampholytes/pH units are needed for production of a stepless pH gradient. It should be noticed, however, that the gaussian distribution of each ampholyte is very wide: each species penetrates within two adjacent zones of focused ampholytes. How can we ensure that since, given a $\Delta pK = 2$, these species should be considered as fairly good carrier ampholytes, exhibiting steep pH/mobility slopes in the proximity of their pI values? This can be obtained either by a fairly high initial concentration (the peak value here being 10 mM) or by a rather low voltage gradient in the separation column, or by a combination of the two.

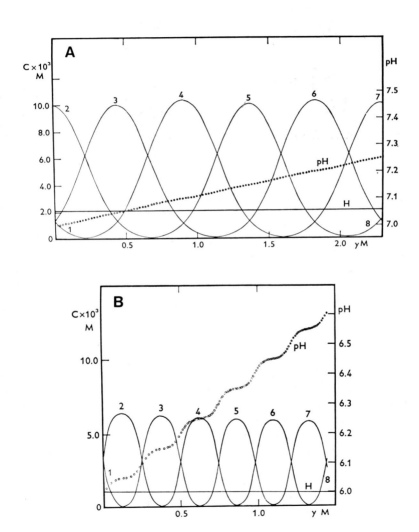

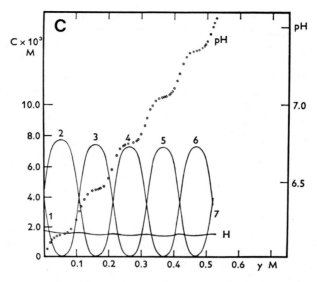

Fig. 1.9. Computer simulation of focusing of ideal, mono-monovalent ampholytes with equal diffusion coefficients and electric mobilities. The abscissa (yM) has the dimensions of moles/liter; the left ordinate is in moles, H is the conductivity and the dotted line is the pH gradient. (A) Focusing of 8 carrier ampholytes over 0.25 pH units, having $\Delta pK = 2$ and spaced at $\Delta pI = 0.05$. (B) Focusing of 8 carrier ampholytes, over 0.5 pH units, having $\Delta pK = 2$ and spaced at $\Delta pI = 0.1$. (C) Focusing of 7 carrier ampholytes having $\Delta pK = 4$ and $\Delta pI = 0.3$. Only in case A is a stepless pH gradient generated (from Almgren, 1971).

If we now take the same system of ampholytes, with $\Delta pK = 2$, but spaced at $\Delta pI = 0.1$, there will be only 15 species per pH unit (Fig. 1.9B), the conductivity course (H) will still be even and constant, but the pH gradient will increase by steps. The steps (or plateaus) are centred in the middle of each peak, while the pH grows linearly in the boundary between each pair of ampholytes. Probably, however, the pH increases stepwise not only because there are half as many amphoteric compounds, but also because each species has a narrow gaussian distribution over the column length, occupying 1/3 the width as compared with Fig. 1.9A. As a consequence, each ampholyte penetrates only within a single, adjacent species and its concentration falls to zero in the centre of the neighboring ampholyte. Since the ΔpK here is the same as in Fig. 1.9A, this narrower distribution can be obtained by a decreased ampholyte concentration (the peak value here being 6 mM) or by a higher voltage gradient, or by a combination of the two.

As a third hypothesis, if we take a rather poor system of carrier ampholytes, with $\Delta pK = 4$ and spaced at $\Delta pI = 0.3$, two phenomena will be apparent (Fig. 1.9C): not only will the pH gradient grow by steps, but also the conductivity course (H) will not be linear any longer: there will be a conductivity minimum at the centre of each peak (corresponding to the pH plateau) and a maximum at the boundary between each pair of ampholytes. Here again, as in Fig. 1.9B, each ampholyte penetrates only within a single, adjacent species and its concentration falls to zero in the centre of the neighboring ampholyte. However, since each species has a very narrow gaussian distribution over the column length, occupying only 1/2 the width of the species in Fig. 1.9B, they must have been focused at a very high voltage gradient, since they have a quite shallow pH/mobility slope about their pI values. These theoretical considerations have some very important practical applications. We have seen that they already specify the minimum number of ampholytes per pH unit (at least 30) needed to generate a stepless pH course. Moreover, they warn us of another risk: at extremely high voltage gradients (Pharmacia recommends as much as 250–300 V/cm; Kinzkofer and Radola (1981) in their miniature system go up to as much as 400–700 V/cm) even good carrier ampholytes, being driven into extremely sharp zones, might create conductivity gaps (or minima) in the system, which might generate local joule effects and damage heat-labile enzymes.

1.6. Synthesis of commercial carrier ampholytes

We have seen that in a series of articles entitled *Isoelectric fractionation, analysis and characterization of ampholytes in natural pH gradients*, Svensson (1961, 1962a,b; Rilbe, 1973a) laid the theoretical foundations for present-day IEF. However, a long search through chemical catalogs for commercially available ampholytes exhibiting the hallmark of a good carrier ampholyte, i.e., $(pI - pK) < 1.5$, which is to say good conductivity and good buffering capacity in the isoprotic state, was practically unsuccessful. Table 1.2 gives a list of such compounds (Svensson, 1962a).

TABLE 1.2

Svensson's (1962a) table of possible carrier ampholytes arranged in the order of increasing isoionic points

Ampholyte	pI	$pI - pK_1$
Aspartic acid	2.77	0.89
Glutathione	2.82	0.70
Aspartyl-tyrosine	2.85	0.72
o-Aminophenylarsonic acid	3.00 (?)	0.77 (?)
Aspartyl-aspartic acid	3.04	0.34
p-Aminophenylarsonic acid	3.15 (?)	0.92 (?)
Picolinic acid	3.16	2.15
L-Glutamic acid	3.22	1.03
β-Hydroxyglutamic acid	3.29	0.96
Aspartyl-glycine	3.31	1.21
Isonicotinic acid	3.35	1.51
Nicotinic acid	3.44	1.37
Anthranilic acid	3.51	1.47
p-Aminobenzoic acid	3.62	1.30
Glycyl-aspartic acid	3.63	0.82
m-Aminobenzoic acid	3.93	0.81
Diiodotyrosine	4.29	2.17
Cystinyl-diglycine	4.74	1.62
α-Hydroxyasparagine	4.74	2.43
α-Aspartyl-histidine	4.92	1.90
β-Aspartyl-histidine	4.94	2.00
Cysteinyl-cysteine	4.96	2.31
Pentaglycine	5.32	2.27
Tetraglycine	5.40	2.35
Triglycine	5.59	2.33
Tyrosyl-tyrosine	5.60	2.08
Isoglutamine	5.85	2.04
Lysyl-glutamic acid	6.10	1.65
Histidyl-glycine	6.81	1.00
Histidyl-histidine	7.30	0.50
Histidine	7.47	1.50
L-Methylhistidine	7.67	1.19
Carnosine	8.17	1.34
α,β-Diaminopropionic acid	8.20	1.40
Anserine	8.27	1.23
Tyrosyl-arginine	8.38 (?)	1.00 (?)
	8.68 (?)	1.13 (?)
L-Ornithine	9.70	1.05
Lysine	9.74	0.79
Lysyl-lysine	10.04	0.59
Arginine	10.76	1.72

In this list, compounds with (pI − pK) > 2.5 have been discarded as useless, while species covering the interval 1.5 < pI − pK < 2.5 are included but are classified as poor carrier ampholytes. Thus, between pH 5 and pH 7, there is a huge gap with practically no species satisfying Svensson's requirements. I have incorporated this table, however, because it gives an idea of how extremely difficult and laborious the gestation period of IEF has been, both theoretically and practically. Moreover, this table forms the basis of what, years later, Caspers et al. (1977) have invented as 'separator IEF' and Beccaria et al. (1978) renamed 'pH-gradient modifier IEF'. While the chemicals listed here could hardly be used for a proper IEF separation, they could be added individually, against a background of Vesterberg-type carrier ampholytes, to flatten out a certain portion of the pH gradient, to maximize separation in that region (separator IEF) (see also §4.11).

Back in Stockholm, now (the search condensed in Table 1.2 had been performed at the Gates and Crellin Laboratory of Chemistry, at Caltech, in L. Pauling's laboratory), the research took another turn. It was considered that peptides could be used as a proper buffering species for IEF separation of proteins. Initially, casein hydrolysates were used (Svensson, 1962b), but then hemoglobin or whole blood were preferred, due to the high hystidine content of hemoglobin. After proteolysis, the peptide mixture was desalted and fractionated by stationary electrolysis in a 20-chamber apparatus and proved useful for IEF separation of proteins, with a resolving power of about 0.1 pH unit (Vesterberg and Svensson, 1966). However, even this approach was beset with difficulties. It was impossible to obtain colorless peptide preparations; moreover oligopeptides focusing between pH 5.0 and 6.5 proved to possess poor carrier ampholyte properties.

A real breakthrough came in August 26, 1964, when Olof Alfred Yngve Vesterberg (for his friends, Olof tout court) filed a Swedish patent on the synthesis of carrier ampholytes suitable for IEF (a more comfortable English version can be read as the USA extension granted on December 23, 1969 as patent No. 3,485,736). With that, present day IEF was coming of age.

1.6.1. Synthesis of Ampholine (LKB)

The basic references in this field are the following: Vesterberg (1969a,b), Haglund (1975), Davies (1970) and Gasparic and Rosengren (1974, 1975). Albeit never clearly stated in the literature, it is understood that commercial Ampholine's are indeed the carrier ampholytes first devised and patented by Vesterberg. Thus, according to the patent, they are 'polyprotic amino carboxylic acids, each containing at least four weak protolytic groups, at least one being a carboxyl group and at least one a basic nitrogen atom, but no peptide bonds'. How did Vesterberg succeed in such a remarkable yet, by today's standards, simple process? Well, Svensson's theory was at hand: carrier ampholytes had to have an even distribution of pI values, and thus pK values, in the pH interval 3−10. Vesterberg realized that all monovalent carboxylic groups dissociate between pH 4 and 5 and all monovalent aliphatic amines between pH 9 and 10. But if we took oligo and polyamines, especially the series of (linear or branched) species in which each nitrogen group is spaced from the adjacent one by an ethylene bridge, the pK values were spaced at graded intervals, the more so the more complex was the oligo-amine. Table 1.3 gives the pK values of such compounds, up to TEPA (tetra-ethylene pentamine). More recent work (Righetti et al., 1975a) indicates that the highest oligo-amine commercially available (pentaethylene hexamine, PEHA) might have a pK_6 as low as 1.1. Thus, Vesterberg realized that suitable amines for the synthesis of carrier ampholytes would be: ethylene diamine, diethylene triamine, triethylene tetramine (TETA), TEPA and PEHA, especially when used in a mixture. As acids to be coupled to the basic compounds, Vesterberg used α,β unsaturated

TABLE 1.3

Dissociation constants for ethylene amines[a]

Amine	pK_1	pK_2	pK_3	pK_4	pK_5	Temp. (°C)
Ethylenediamine	10.1	7.0				20
Diethylenetriamine	9.9	9.1	4.3			20
Triethylenetetramine	9.9	9.2	6.7	3.3		20
Tetraethylenepentamine	9.9	9.1	7.9	4.3	2.7	25

[a] From Vesterberg (1969b).

compounds, preferably with ten carbon atoms at most. He indicated the following compounds: acrylic, methacrylic and crotonic acids, but also suggested the use of unsaturated dicarboxylic species, such as maleic and itaconic acids. Here too, the acids could be used either singly or in a mixture, albeit it is believed that the commercial product contains only acrylic acid derivatives. The synthetic process was at hand too, and had been described by Engel in 1888: by dripping acrylic acid in a proper mixture of polyamines (the best one being PEHA), at appropriate ratio (usually 2 N atoms/1 carboxyl group) at 70 °C and letting the reaction proceed until all acrylic acid has reacted (usually a minimum of 5 h, with yield of nearly 100%) carrier ampholytes with evenly and closely spaced pI and pK values in the pH 3–10 range are obtained. The synthesis goes via an anti-Markovnikov addition after the mechanism of the Michael reaction and therefore β-amino acids (β-amino propionic residues) are obtained. Figure 1.10 reports a hypothetical general structural formula of these ampholytes together with the narrow pH-cuts available for IEF fractionation in narrow ranges. The latter, except for the very extreme pH 2.5–4 and pH 9–11 ranges, are obtained by sub-fractionating the 'wide pH-range' mixture in multi-compartment electrolysers.

The ampholytes thus synthesized are only able to buffer between pH 3.5 and pH 10, due to lack of suitable pK values below and above these limit pH values. Subsequently, Vesterberg (1973c,d) has investigated means of extending the fractionation range beyond these limits. In the alkaline range, compounds containing guanidine groups would generate higher pI species, but were abandoned because they are unstable in alkaline solutions. Vesterberg (1976) found that aliphatic oligo-amines having amino groups more than three methylene groups apart had higher pK values than the TETA, TEPA and PEHA series and also higher pI values upon coupling with a carboxyl group. Thus, 1,6-diaminohexane has pK_1 = 10.6 and pK_2 = 11.7 and its carboxylic acid derivative a pI = 11.15 (all values measured at 4 °C); a similar behavior is exhibited by homologous diamines such as 1,4-diaminobutane and 1,5-diamino-pentane. Thus it is quite possible that the extreme alkaline range pH 9–11 is made of mixture of similar amines, properly coupled with a carboxyl group. A similar, ad hoc synthetic principle must also apply to the extreme acidic range pH 2.5–4. In this last case, stronger acids had to be used: as suggested by Vesterberg (1976) some dicarboxylic

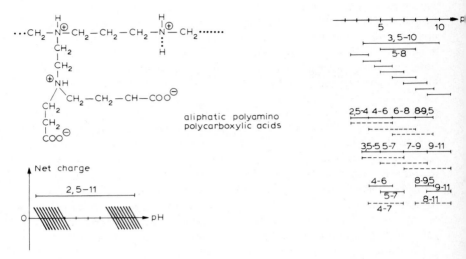

Fig. 1.10. Composition of Ampholine. On the upper left side a representative chemical formula is shown (aliphatic oligo amino-oligocarboxylic-acids). On the lower left side, portions of hypothetical titration curves of Ampholines are depicted. Right: different pH cuts for wide and narrow range carrier ampholytes (courtesy of LKB Produkter AB).

acids would appear to be suitable, since, due to mutual interaction, one pK will be quite acidic, the other rather weak. Thus malonic acid ($HOOC-CH_2-COOH$) has $pK_1 = 1.8$ and $pK_2 = 5.7$. The various commercial pH ranges of Ampholine are sold as 25 ml bottles, containing 40% solids, in sterile vials. This does not apply to the two extreme ranges, pH 2.5–4 and pH 9–11, which, having about twice the conductivity and buffering capacity, are sold as 20% solutions. Although never stated by the manufacturer, their physico-chemical behavior has suggested to Gelsema et al. (1979) and to Bianchi Bosisio et al. (1981) that these two extreme ranges do not belong to the same 'Vesterberg' family' (i.e., a single synthetic step generating a wide pH 3.5–10 mixture which is then fractionated into narrow pH cuts) but are synthesized independently with other oligoamines and/or acids.

As stated by Davies (1970), LKB Ampholine had been reinforced at the acid end with glutamic and aspartic acids and at the alkaline extreme with lysine and arginine, since in the synthetic process the yield of ampholytes at both ends was lower than in the middle ranges. The presence of amino acids had in fact been reported by Earland and Ramsden (1969) and suggested also by Righetti et al. (1975b) on the basis of optical rotatory data. However, according to Haglund (1975) only Ampholine ranges produced from 1967 up to and including 1972 contained amino acids. With batches produced from 1973 onward, these additives have been excluded. For a period of time also (mostly during 1974 and 1975) ^{14}C-labeled carrier ampholytes, for binding studies, and small molecular mass species, for preparative peptide separation (Gasparic and Rosengren, 1974), have been available from LKB. To my knowledge, however, they have unfortunately become a collector's item. During that period, LKB had also marketed extremely narrow pH cuts (encompassing only 0.5 pH units, instead of the customary 2 pH units) such as pH 7–7.5 for hemoglobin fractionation, or pH 4–4.5 for α_1-antitrypsin typing, but they are no longer available.

1.6.2. Synthesis of Servalyte (Serva)

Their synthesis and properties have been described by Pogacar and Jarecki (1974) and by Grubhofer and Borja (1977). Originally, for the synthesis, TEPA, PEHA and a low molecular mass polyethylenimine were used. More recently, the polyamine mixture is obtained by condensing ethylene imine with propylene diamine and distilling the product in order to eliminate compounds with $\overline{M}_r$ greater than 400. The reaction is illustrated as follows:

$$
\begin{array}{c}
\underset{\displaystyle\overset{\displaystyle N}{\underset{\displaystyle H}{|}}}{\overset{\displaystyle CH_2-CH_2}{\diagdown\diagup}} \qquad + \qquad H_2N-CH_2-CH_2-CH_2-NH_2 \\[2mm]
\downarrow \; \text{catalyst} \\[2mm]
H_2N-CH_2-CH_2-\underset{\displaystyle\overset{\displaystyle|}{CH_2}}{\overset{\displaystyle|}{N}}-CH_2-CH_2-CH_2-\underset{\displaystyle\overset{\displaystyle|}{CH_2}}{\overset{\displaystyle|}{N}}-CH_2-CH_2-NH_2 \\
CH_2 \qquad\qquad\qquad CH_2 \\
NH_2 \qquad\qquad\qquad NH_2
\end{array}
$$

For introduction of the acidic groups, propansultone and Na-vinylsulfonate for the sulfonic, and Na-chloromethyl phosphanate for the phosphonic acid, are used, yielding products with the following formula as a representative example:

$$H_2N-CH_2-CH_2-N-CH_2-CH_2-CH_2-N-CH_2-CH_2-NH-CH_2-PO(OH)_2$$

$$
\begin{array}{ll}
\quad\quad\quad\quad | & \quad\quad\quad\quad\quad | \\
\quad\quad\quad CH_2 & \quad\quad\quad\quad CH_2 \\
\quad\quad\quad | & \quad\quad\quad\quad\quad | \\
\quad\quad\quad CH_2 & \quad\quad\quad\quad CH_2 \\
\quad\quad\quad | & \quad\quad\quad\quad\quad | \\
\quad\quad\quad NH_2 & \quad NH-CH_2-CH_2-CH_2-SO_3H
\end{array}
$$

Actually, it was found that propansultone in a basic medium gave the best results for the synthesis of the polyamino-polysulfonic acid mixture, since it reacts strongly with primary, secondary and tertiary amino groups (the last ones giving betaine-type compounds). However, these compounds produced ampholytes with pI values between pH 2–3.5 and pH 5.8–9.5, with a big gap between pH 3.5 and 5.8. Previous quaternarization of the amino groups with dimethylsulphate in order to increase their dissociation constant and their basicity did not eliminate this pH gap. Only through the introduction of weakly acid carboxyl groups with acrylic acid was it possible to generate ampholytes with pI values in the pH range 3.5–5.8. Thus, Servalyte contains, in addition to the polyamino backbone, the following acidic groups: carboxyl, sulphate and phosphate. Servalyte (sold through Bio Rad as Biolyte) is available in the following pH ranges: 2–4; 3–5; 4–6; 5–7; 6–8; 7–9; 8–10; 9–11 and the broad range 2–11. As with Ampholine, they are available as 40% solutions (except the pH cuts 2–4, 3–5 and 9–11, supplied at 20% concentration) in serum vials of 25 ml. Serva also supplies a technical grade, pH 3–10 range, which is strongly yellowish-brownish, but works just as well as the colorless pH ranges, at considerably reduced price. Needless to say, the narrow pH cuts are produced by isoelectric fractionation of the broad pH range mixture: Grubhofer and Borja (1977) use a 60 l chamber, divided into 21 compartments by porous ceramic diaphragms, into which a 2% ampholyte pH 2–11 solution is focused in ca. 20 h.

1.6.3. Synthesis of Pharmalyte (Pharmacia)

In the summer of 1978, a new generation of buffering amphoteres was made available, the Söderberg-type CA's (Williams and Söderberg, 1979; Söderberg et al., 1980) which have appeared on the market under the trade name Pharmalyte. Figure 1.11 gives a general, hypothetical structure of a Pharmalyte constituent ampholyte containing six amines. The basic synthetic process involves the copolymerization of amines, amino acids and dipeptides with epichlorohydrin. By a suitable choice of amines and amino acids, five narrow pH intervals are directly generated. Thus, the wide-range Pharmalyte pH 3–10 is prepared by blending the five narrow intervals, just the opposite of what is usually done with traditional carrier ampholytes. An even buffer capacity and a low and even conductivity, are claimed by the manufacturer, and substantiated by independent work by Gelsema et al. (1979).

The only way by which even conductivity and buffering capacity courses could be obtained during IEF, would be to start with an enormously heterogeneous mixture of synthetic carrier ampholytes, with closely spaced pK and pI values over a given pH interval. Since, within each narrow pH interval, seven different amines are cross-linked with epichlorohydrin, and the average degree of polymerization is six, logic suggests that no more than a handful of different amphoteres could be

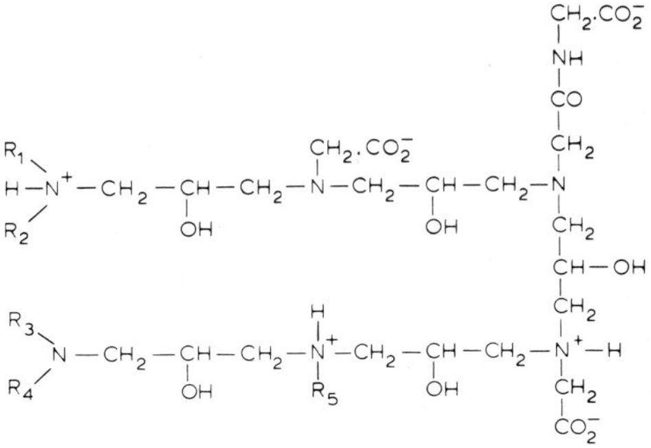

Fig. 1.11. Hypothetical structure of a Pharmalyte constituent ampholyte containing six amines. R corresponds to hydrophilic groups. For details see text. (from Williams and Söderberg, 1979).

generated. The new idea here is the introduction of the concept of sterical isomerism. By using D,L-epichlorohydrin, D,L-amines and D,L-amino acids, the possible number of amphoteric species could be increased enormously, as all the diastereoisomeres of the ampholytes, in contrast with the mirror-image compounds, will have different (even if very slightly) isoelectric points. The number is further increased by the fact that penta- and heptapolymers are generated, together with branched and cyclic forms. It has also been hinted that there is a minimum of 1,000 different amphoteres per pH unit. Since it has been demonstrated (Radola et al., 1977) that about 100 carrier ampholytes per pH unit are present in Vesterberg-type CA's this would represent an amazing increase of one order of magnitude in the population of species generated during the synthesis.

Structurally, the Pharmalyte ions contain mostly tertiary amino groups, most of them linked to at least one β-hydroxyl group. Based on the average degree of substitution (six amines and five epichlorohydrins) their molecular mass ($\overline{M}_r$) should be around 750–800, substantially higher than the average $\overline{M}_r$ derived by gel filtration (ca. 450 daltons). Notwithstanding the rather high $\overline{M}_r$, Pharmalyte appears to have a quite compact structure (6–8 Å), possibly due to intramolecular hydrogen bonding. The main buffering groups, with respective buffering pH ranges, are α-aminocarboxylic (pH 2–3); glycyl-glycine residues (pH 3–4); β-hydroxylamines (pH 4–9) and dialkylaminopropyl (pH 9–11).

1.7. Laboratory synthesis of carrier ampholytes

As we have seen, there are now three commercial sources of carrier ampholytes: LKB's Ampholine, Serva's Servalyte and Pharmacia's Pharmalytes. Even though somewhat chemically different, they all follow the same basic idea of Vesterberg: the synthesis of amphoteric compounds from polyamines (or condensed amino acids) and organic and/or inorganic acids in such a way as to obtain the most heterogeneous synthetic product. This mixture is then subfractionated into narrow ranges by an IEF step, or the synthesis is made ad hoc to produce a narrow, specified pH cut. As companies over the years have vied to outdo each other with, presumably, better and better carrier ampholytes, so have scientists put their wits to the production of these compounds. This has been beneficial since, while marketing companies have a strong tendency to keep tight-lipped on their products, scientists are inclined to talk and spread their knowledge as much as they can. We will see how, over the years, several laboratories have put forward some different and ingenious synthetic procedures which, albeit basically in line with Vesterberg's thoughts (there seems to be no way to circumvent it), have greatly increased our knowledge on these chemicals.

1.7.1. The Vinogradov–Righetti procedure

Vinogradov et al. (1973) showed that it was possible to reproduce Vesterberg's synthetic procedure with simple laboratory equipment. They obtained suitable carrier ampholytes in the pH range 4–8 by coupling pentaethylenehexamine (PEHA) with acrylic acid. These preparations were satisfactory for many purposes but their conductivity profile was not uniform over the whole pH range. These experiments, nevertheless, indicate the relative ease with which suitable ampholytes could be prepared. These workers also described optical methods to identify focused ampholytes and thereby allow some much needed characterization of ampholyte banding patterns.

On the basis of these observations, Righetti et al. (1975a) used a combination of different polyamines with acrylic acid to produce ampholytes which formed smooth and stable pH gradients over the pH range 3–9. The method involves the coupling of hexamethylenetetramine (HMTA), triethylenetetramine (TETA), tetraethylenepentamine (TEPA) and PEHA either singly, or in a mixture, to acrylic acid. Experimental details for the synthesis are given below.

All oligoamines should be distilled under nitrogen and reduced pressure just before use. Originally, we had also suggested distillation of acrylic acid (it contains 200 ppm p-methoxyphenol) under nitrogen, but today we prefer to do it in an air stream: in fact, as the product distills under N_2, it might, due to the high temperature, start polymerizing with itself, producing oligo- and poly-acrylic acid, while the oxygen in the air should inhibit self-polymerization. The distillation process for TETA, TEPA and acrylic acid is quite simple, since these substances have low boiling points under reduced pressure. Perhaps the most difficult task is the distillation of PEHA which distils over the range 200–290 °C under a pressure of 500 μm Hg. A suitable distillation apparatus is shown in Fig.

A

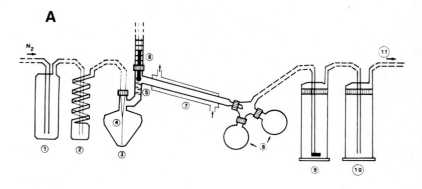

1.12A. High quality nitrogen should be used. Less pure grades should be freed from oxygen either by passage through a catalytic burner or an appropriate filter. The nitrogen is bubbled into the solution via a very thin capillary reaching the floor of the distillation chamber. Because of the high viscosity of PEHA at low temperatures, the condenser should only be air cooled. The whole system is maintained under reduced pressure by a vacuum pump protected by a Fresenius tower containing concentrated H_2SO_4, to absorb polyamine vapors, and by a second tower containing drierite (alternatively, a dry-ice trap can be used). The distillation of reagent and subsequent coupling should be performed within the same day. Acrylic acid should be distilled last since it might polymerize once the inhibitor is removed. Excess acrylic acid may be stored frozen, preferably at $-80\,^\circ$C or in liquid nitrogen. The distilled amines and acrylic acid are then combined in appropriate amounts. A suitable reaction chamber (Fig. 1.12B) consists of a two-necked flask, equipped with a thin capillary in one arm for nitrogen flushing, and a burette for acrylic acid addition. The burette has a built-in side arm to provide for nitrogen escape from the flask, and for an inert N_2 atmosphere in the burette itself. The mixture is stirred with a bar magnet. The acrylic acid is added dropwise to a 1.5 M solution of the polyamine

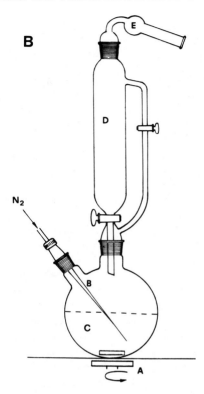

Fig. 1.12. (A) Scheme of the distillation apparatus: (1) copper wire catalyst kept in an oven at 450 °C; (2) cooling serpentine; (3) distillation flask; (4) capillary for nitrogen flushing; (5) Vigreux column; (6) thermometer; (7) condenser; (8) collection flasks; (9) Fresenius tower with conc. H_2SO_4; (10) $CaCl_2$ trap; (11) connection to the vacuum pump. The shaded regions represent ground-glass joints. (From Righetti et al., 1975a.) (B) Reaction chamber for carrier ampholyte production. A) magnetic stirrer; B) capillary for nitrogen flushing; C) polyamine solution; D) acrylic acid in a burette with a side arm; E) $CaCl_2$ trap for gas escape. (Gianazza, E. and Righetti, P.G., unpublished.)

in water under continuous stirring, over a period of 60 min, to a final N/COOH ratio of 2:1. The flask is then stoppered and held for 16–20 h at 70 °C in a Dubnoff shaker. The reaction can be summarized as follows:

$$\text{HMTA + TETA + TEPA + PEHA + acrylic acid} \xrightarrow[16-20\,h]{H_2O,\ 70\,^\circ C} \text{carrier ampholytes}$$

The ampholytes thus obtained are diluted to 40% (w/v) and stored frozen in dark bottles. Yellow compounds formed during the synthesis can be removed by repeated charcoal treatment, as reported by Vinogradov et al. (1973). This may not be necessary, however, since the chromophores are also ampholytes and serve a useful function as 'built-in' markers for the pH gradient. While 'chromophoric' ampholytes need not interfere with protein separation and detection in analytical systems, they might interfere with protein detection in preparative systems, especially in sucrose density gradients, where proteins are often analyzed in presence of carrier ampholytes.

By several criteria, the product described above is equivalent to commercially available ampholytes, the main difference relates to the strong absorption at 315 and 368 nm (Righetti et al. 1975a). The synthesis is simple, the starting reagents are readily available, and large amounts can be produced in a short time. However, I should like to mention that, recently, many oligoamines (notably PEHA, which produces the best carrier ampholytes in the oligoamine series) are no longer available from most commercial sources, since they have been discontinued due to their carcinogenic properties. In a recent paper (Binion and Rodkey, 1981) a source of PEHA is given as Columbia Organic Chemicals, Columbia, SC, USA. Most of the carrier ampholytes thus synthesized exhibit a conductivity minimum centred around pH 6 (range 5.5–6.5). This leads to a minimum of buffering capacity, which in turn means that, whenever these pH ranges are used, the highest potential drop will be located in the pH interval 5.5–6.5, while other parts of the gradient will be underfocused. This high field strength in the pH zone 5.5–6.5 can generate considerable joule heat, with deleterious effects, especially when using high-voltage techniques. In order to circumvent that, Righetti et al. (1977a) have coupled oligoamines with a mixture of acrylic and itaconic acids (pK_1 3.85; pK_2 5.45). The latter seemed to be an ideal compound for bridging the conductivity gap. In fact, once reacted, only the carboxyl group in the β-position to the amino group will lower its pK from 3.85 to ca. 3.35, while the γ-carboxyl pK will be virtually unaffected or, at most, be lowered by 0.1 pH unit. Itaconic acid carrier ampholytes have improved the conductivity in the pH zone 5.5–6.5 by as much as 400–500% as compared with conventional carrier ampholytes.

1.7.2. The Charlionet et al., synthesis

An interesting variant of the basic synthetic approach of Vesterberg (1969a) has been reported by Charlionet et al. (1979, 1981). It is known that, in classical acrylic acid-type ampholytes, most of the buffering power in the pH range 2–11 will have to be found in the different pK values of the amino groups of the polyamino backbone. Actually, a judicious blend of α-carboxyls (as in Pharmalyte), β-carboxyls (acrylic acid) and of β- and γ-carboxyls of itaconic acid (Righetti et al., 1977) in carrier ampholytes should provide a good buffering capacity in the pH range 2–6 but still the overall quality will be dictated by the properties of the polyamine. By increasing the heterogeneity of the starting polyamines, ampholytes of improved qualities are synthesized. In an effort to improve the quality of available polyamines, Charlionet et al. (1979), before coupling them to acrylic acid, react them with the following compounds: acrylamide, N,N'-methylene bisacrylamide (Bis), 2,3-epoxypropanol-1 and 1,2–7,8-diepoxyoctane. By these reactions, primary and secondary amino groups in polyamines are transformed into secondary and tertiary amines, respectively. Moreover, while the reaction with monofunctional compounds (acrylamide and epoxypropanol) is used solely for amino group modification, the reaction with the bifunctional agents (Bis and diepoxyoctane) leads to cross-linking of two different polyamines together. While the reaction of acrylamide, Bis and acrylic acid follows the well-known addition of nucleophilic groups of polyamines to the β-carbon of α-β-unsaturated compounds, the reaction of an epoxide group with a polyamine is expected to follow the reaction sequence:

$$R-NH_2 + \underset{CH_2-CH_2}{\overset{O}{\diagup\diagdown}} \rightarrow R-NH-CH_2-\overset{OH}{\underset{|}{CH}}-$$

$$\underset{R}{\overset{R}{\diagdown}}NH + \underset{CH_2-CH_2}{\overset{O}{\diagup\diagdown}} \rightarrow \underset{R}{\overset{R}{\diagdown}}N-CH_2-\overset{OH}{\underset{|}{CH}}-$$

thus leading to secondary hydroxyl groups and secondary and tertiary amines. By using empirically-found stoichiometric amounts, a highly heterogeneous mixture of modified polyamines is obtained which, upon reaction with acrylic acid, leads to carrier ampholytes with good conductivities and buffering capacities in the pH range 3–10. These ampholytes, for use in α_1-antitrypsin typing (α_1-AT), could be further fractionated into very narrow ranges, encompassing less than 0.5 pH units (e.g., pH 4.4–4.7 range). In these very shallow pH gradients, α_1-AT phenotypes differing in pI values by as little as 0.001 pH units could be resolved. Since the resolution limit was previously given as 0.02 pH units (Vesterberg and Svensson, 1966), this represents an amazing 20-fold increase in resolving power. It must be emphasized that a previous claim by Allen et al. (1974) of a resolution of 0.0025 pH units has not been substantiated by the present authors. In a more recent article, Charlionet et al. (1981) have investigated the limiting factors for the resolving power of IEF in natural pH gradients. As we have seen (§ 1.5.7) the resolving power (i.e., the smallest difference of pI values between proteins which can be separated) is a function, among other variables, of the slope of the pH gradient over the separation distance $[(\mathrm{dpH})/\mathrm{d}x]$. How shallow can a pH gradient be? It appears that the minimum pH over the separation column length cannot be lower than 0.3 pH units. Attempts at improving this lower limit by the use of high $\bar{M}_r$ ampholytes in the form of succinylated bovine albumin have failed: proteins in general have too few protolytic groups at pH = pI, so that they do not have enough conductivity and buffering capacity and are useless as carrier ampholytes (also, if one were to increase their concentration, there would be substantial aggregation and flocculation at their pI). However, if one uses as buffering ions the Vesterberg type polyamino-polycarboxylic acids, the limiting factor to resolving power in IEF appears to be only the $\bar{M}_r$ of the ampholytes: the higher their $\bar{M}_r$, the better the resolving power expected. It must be stated that, in the system of Charlionet et al. (1979), ampholytes with $\bar{M}_r > 5,000$ daltons are already generated. We will see further on (§ 1.7.4) that this assumption, in real practice, is fallacious and that there is a limit to the $\bar{M}_r$ of carrier ampholytes.

1.7.3. The Just approach

The year 1980 gave birth to yet another variant of Vesterberg's synthesis, the Just-type carrier ampholytes (Just, 1980), which we nicknamed the 'mysterious strain' (Righetti et al., 1980), since their synthesis had been announced at every international meeting since 1976, but details were withheld. Now, however, the plot has been unravelled and once again the synthetic approach involves the addition of α-β-unsaturated compounds to oligoethylene oligoamines, notably PEHA (although the source of this chemical is being kept secret). However, several variants make the Just synthesis a very interesting approach. First of all, instead of acrylic acid, its methyl ester is used. Secondly, the reactants are brought together not batchwise, but in a flow-through system, by mixing them in appropriate ratios with the aid of the Ultrograd gradient former. Typically, 1 M PEHA and 4 M acrylic acid methyl ester in methanol are poured in each reservoir of the gradient former. The reaction is started with a percentage concentration ratio of 80:20 (PEHA:acrylic derivative; basic carrier ampholytes) and is completed with a ratio of 35:65 (acidic carrier ampholytes). Figure 1.13A shows the pH distribution of the carrier ampholytes obtained with PEHA and other oligoamines and Fig. 1.13B the type of template used in the Ultrograd for continuously varying the PEHA: methyl acrylate ratio. The mixed reactants travel in a capillary column at 40 °C for 1.5 h and are collected in a fraction collector, thus being automatically divided into narrow pH ranges (in Fig. 1.13A seven narrow ranges are represented by the dots in the curve). In order to maintain the number of ampholyte molecules synthesized per time unit fairly constant, and depending on the initial PEHA concentration, the amount of =NH equivalents, going from basic to acidic carrier ampholytes, is kept constant by proportionally increasing the time available for the synthesis of one pH unit (this is seen in Fig. 1.13B as a progressive widening of the pH axis). After evaporating the methanol from the reaction products, the methyl ester is hydrolyzed for 2 h at 120 °C. The advantages of using methyl acrylate are two-fold: its reaction rate is enormously higher as compared with acrylic acid (half-life of 3.5 min vs. > 30 h at room temperature); in addition, since the ester is uncharged and volatile, its excess is easily removed by vacuum distillation.

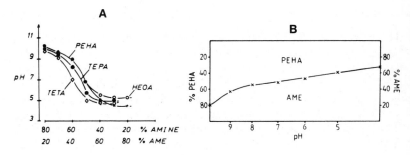

Fig. 1.13. (A) pH of the hydrolyzed reaction products obtained by coupling various percentage ratios of 4 M acrylic acid methyl ester (AME) with various 1 M amine solutions in methanol (HEOA = heptaethylene octamine). (B) Template used for the synthesis of PEHA ampholytes covering the pI range 3.5–10. Each one-unit pH interval is made to contain the same amount of PEHA by progressively widening the pH axis towards acidic pH values. The two solutions were mixed with the LKB Ultrograd gradient former. Both figures from Just (1980).

1.7.4. The Righetti–Hjertén high $\overline{M}_r$ species

From the above, whilst it would appear that all possible synthetic variants of carrier ampholytes have been tried, this is not quite the case. My group has been interested over the years in IEF of peptides (see §5.1.): while analytically this technique has advanced considerably, preparatively it does not quite work, since it is almost impossible, even by advanced techniques such as hydrophobic interaction chromatography (Gelsema et al., 1980b), to separate small peptides from contaminant carrier ampholytes. One way to achieve that would be to use small $\overline{M}_r$ ampholytes (Gasparic and Rosengren, 1974) but these are no longer commercially available and it is doubtful that they could be properly separated from peptides; moreover, according to Charlionet et al. (1981), the smaller the ampholyte $\overline{M}_r$, the worse is the pH gradient generated. That this is so, has also been demonstrated by Gelsema et al. (1979) who have calculated that the ratio between molar buffer capacity and conductivity is proportional to the cube root of molecular mass. Thus, Righetti and Hjertén (1981) have set out to solve this problem at the opposite extreme of the $\overline{M}_r$ scale, i.e., by using high $\overline{M}_r$ ampholytes for preparative separation of peptides. Two approaches were first tried: oligoamines were coupled to soluble dextrans (in the $\bar{M}_r$ range 10,000–20,000) and then reacted with acrylic acid; alternatively, albumin and hemoglobin were carbamylated or maleylated to generate 'charge-shift trains' which would encompass as much as 2–3 pH units in pI differences. Neither approach worked because, as also demonstrated by Charlionet et al. (1981), the ratio buffering groups/molecular mass is too unfavorable. Thus, we started out with a giant polyethylene imine (PEI, $\bar{M}_r$ 40,000–60,000) and coupled it to acrylic acid. This synthesis cannot be performed by a batch procedure, as in the conventional Vesterberg method (see §1.7.1) since only a few, neutral pI species are generated. Thus, we resorted to a modified Just approach. As shown in Fig. 1.14 the PEI solution is mixed, in a flow-through chamber, with a linear gradient of acrylic acid, the solution segmented, every 1 ml, by a small nitrogen bubble and loaded in a capillary reaction coil. In a typical experiment, the Erlenmeyer contained 50 ml of 10% PEI, while the gradient mixer had 25 ml of 18% (v/v) acrylic acid in the reservoir (R) and 25 ml of 2%

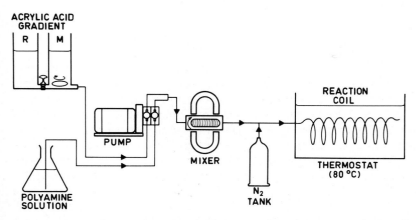

Fig. 1.14. Synthesis of giant carrier ampholytes. A 10% polyethylenimine solution is mixed with a linear 2–18% (v/v) acrylic acid gradient with the aid of an ultrograd mixing chamber (in this chamber the magnet bar, immersed in the magnetic field, rotates over its long axis). The flowing gradient is segmented, every 1 ml, with a small nitrogen bubble. Once the reaction coil is filled, the reaction is allowed to proceed for 48 h at 80 °C (from Righetti and Hjertén, 1981).

(v/v) acrylic acid in the mixing chamber (M). The acrylic acid gradient had been calculated to give about 10% substitution in alkaline carrier ampholytes and ca. 100% substitution at the acidic extreme. Once the reaction coil is filled with the gradient, it is sealed and the reaction allowed to proceed for 48 h at 80 °C.

These giant carrier ampholytes have proved useful in the preparative fractionation of model dipeptides, allowing recoveries up to 85% and no contamination of the purified peptide. However, there are several draw-backs with these high $\bar{M}_r$ species: they have an enormous tendency to aggregate by ionic bonds (multipoint attachment) giving heavy precipitates, especially among acidic and alkaline compounds; moreover, they still exhibit the conductivity minimum centred around pH 6, just as do regular size ampholytes. This shows why the Charlionet et al. (1981) hypothesis (progressive increase in resolution with increasing $\bar{M}_r$ of ampholytes) is fallacious: above a given $\bar{M}_r$ (possibly already at 5,000–6,000) no further improvement in the quality of the pH gradient is obtained. In fact, according to Murel (personal communication) no further improvement in the quality of ampholytes is obtained above a length of 12–15 nitrogens in the polyamine.

1.8. Detection of carrier ampholytes

A method for revealing focused ampholyte zones (they should form numerous, discrete bands which are parallel to each other and perpendicular to the electric field) has at least a two-fold value: it can be used to inspect the quality of the commercial product as well as the quality of the IEF separation (e.g., strongly distorted or wavy ampholyte patterns suggest anomalies in the sample under fractionation, such as the presence of high salt or buffer levels). The first observations on focused distributions of carrier ampholytes came from Jonsson et al. (1973) and Rilbe (1973b) who demonstrated refractive index gradients generated upon IEF of Ampholine and suggested that these zones consisted of well-focused ampholytes. According to Fries (1976), there are two ways for observing focused zones: the shadow method (formation of a shadow-gram) which depends on the second derivative of the refractive index ($d^2n/dl^2 = 0$) and the Toeplen–Schlieren method (formation of a schlieren image) which just shows the refractive index gradient ($dn/dl = 0$).

Righetti et al. (1975a) have used the shadow method for inspecting the quality of 'home-made' carrier ampholytes synthesized with different oligoamines. Figure 1.15 shows the patterns given by ampholytes prepared from coupling TETA and acrylic acid at a nitrogen–carboxyl ratio of 2:1. Several major bands of ampholytes are revealed by this method. These bands probably represent clusters of ampholytes rather than individual components. This preparation does not give smooth pH gradients between pH 3 and 10 and its conductivity is very uneven all along the pH gradient. Inspection of the focused plate revealed several rather large gaps corresponding to regions of low ampholyte concentration.

Better results were obtained with ampholytes produced with TEPA. TEPA ampholytes generated many more species, especially in the acidic region. The ampholyte clusters were tighter and more closely spaced, with fewer regions of low ampholyte concentration. Although the basic region contained more ampholyte species than the TETA ampholytes, there appeared to be fewer basic ampholytes than acidic ones (Fig. 1.16). In addition, the conductance varied considerably throughout the gel, being lowest between pH 5–7.

The best results with a single amine were given with ampholytes prepared by coupling PEHA to acrylic acid. Since, in this case, both the acidic and basic regions appear to contain many, well-focused ampholyte species, the pH gradient was smoother than with TEPA and also the conductivity profile was higher and more uniform than had either TETA or TEPA ampholytes. Conceivably, the number of ampholyte species could be increased by a judicious mix of ampholytes prepared by coupling different amines to acrylic acid. Recently, Edwards and Anderson (1981) have used a combination of Fries–Schlieren method with the Ronchi grid of Kolin (1958) for inspection of focused ampholyte patterns in horizontal gel slabs. These authors suggest that the ampholyte ridges observed in the gel slab are due to the osmotic pressure exerted by the focused ampholyte species, which results in a net transfer of water from adjacent regions where the ampholyte concentration is lower.

Other methods have been described to detect focused Ampholine patterns. Thus Felgenhauer and Pak (1973) reveal them by the glucose caramelization technique. After focusing in a Sephadex thin layer, by the method of Radola (1973a,b), they roll on to the Sephadex gel a sheet of filter paper soaked in 5% glucose. Upon incubation at $110\,^{\circ}C$, for 8 min,

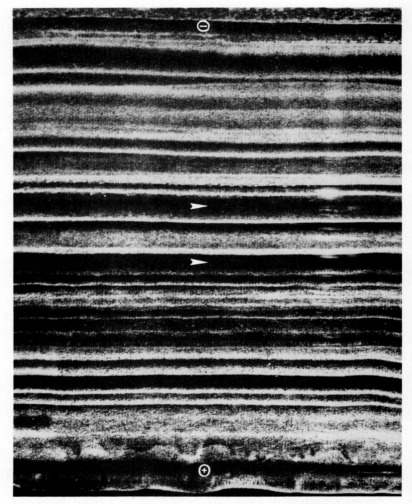

Fig. 1.15. TETA ampholytes focused in a gel slab. At equilibrium, the gel was photographed against a black background with side illumination. The rope-like structures are clusters of focused ampholytes. This is a picture of a transparent, unstained gel and the ampholytes are detected on the basis of different refractive indices. The anode (+) and the cathode (−) are marked (from Righetti et al., 1975a).

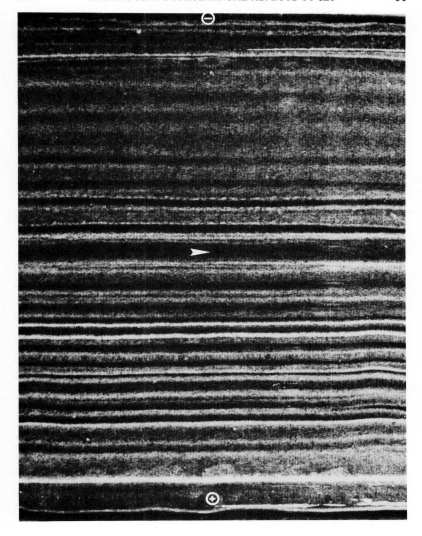

Fig. 1.16. TEPA ampholytes focused in a gel slab. The direction of the pH gradient is indicated by + (anode) and − (cathode). All other conditions as in Fig. 1.15 (from Righetti et al., 1975a).

the Ampholine-glucose caramel products can be evaluated by their fluorescence at 350 nm or in daylight. This detection technique has revealed 38 distinct bands in commercial Ampholine pH range 3.5–10.

Another detection method is based on the formation of insoluble complexes between Ampholine and Coomassie Violet in half saturated picric acid containing 5% acetic acid. This method reveals acidic and neutral Ampholine but fails to detect basic ampholytes since they form soluble complexes (Frater, 1970). LKB Ampholines have also been analyzed by ion exchange chromatography on a sulfonated polystyrene resin, by using the amino acid auto analyzer (Brown et al., 1976). 60 to 70 ninhydrin positive components were resolved in the wide pH range 3.5–10. To ensure complete elution of Ampholine from the ion exchange bed, a pH gradient over the range 3.8 to 11.0 had to be used. From the elution behavior of acidic and alkaline pH ranges, it was concluded that other factors, such as hydrophobicity and molecular size, than the pI of the ampholyte components play significant roles in their ion-exchange behavior.

Bonitati (1980, 1981) has described another method for visualization of carrier ampholyte patterns in granular gel slabs or on paper prints, by using ultraviolet light (UV). The fluorescence of ampholyte species decreases in the order: Servalyte > Ampholine > Pharmalyte. When viewed under short- (254 nm) and long-wave (350 nm) light, focused Servalyte give a series of sharply defined dark and bright bands, purple and yellow-green in color, respectively. As seen in Fig. 1.17 the short-

---

Fig. 1.17. Classification of the fluorescent isoelectric pattern of Servalyte. Individual fluorescent bands are identified according to the classification scheme of Bonitati (1980). Arrows indicate demarcation bands which divide the pattern into sections A–E. The three bright bands, C-(8,9,10)-b(254 nm), are shaded in the diagram to indicate that they frequently appear slightly darker than the other bright bands in the pattern. The main features of the fluorescent patterns depicted here are highly consistent from run to run; the following variations (described in terms of the *shortwave* pattern) have been observed, however: (a) Section A is sometimes compressed in size and the first two or three bands are missing. (b) One batch of Servalyte gave a pattern in which section A was very dark, and individual bands were not clearly visible. (c) In a few runs, a thin dark band of moderate intensity appeared between B-4-d and B-5-d. (d) Bands C-(11, 12)-d, and the lower three dark bands of section D were sometimes very faint. These variations in the shortwave pattern were accompanied by parallel changes in the longwave pattern (from Bonitati, 1980).

wave pattern contains a total of 56 bands: 29 dark and 27 bright. The long-wave pattern contains 50 bands: 25 of each kind. The two patterns coincide across most, but not all, of the bands. The short- and long-wave patterns are, for the most part, complementary with respect to color of the individual bands: bright bands in one pattern correspond to dark bands in the alternate pattern. These spectra of zones can be used analytically in a number of ways: as a map for locating and identifying

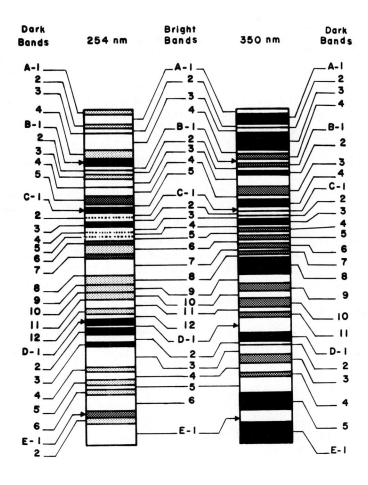

specific protein bands in gel slabs and on prints; as a means for assigning pI values to protein bands on prints; and as a tool for monitoring the quality of different batches of carrier ampholytes. At a preparative level, the fluorescent banding pattern of Servalyte is utilized as a map to locate the proteins indirectly and non-destructively, and to guide their excision for recovery from the gel slab.

1.9. Fractionation of carrier ampholytes

Two methods may be used to prepare ampholytes that buffer in a narrow pH range. The first is to alter the relative amounts of amine and acid in the coupling reaction. Although feasible, the method is difficult to control and excess free amines or acids contaminate the product (Righetti et al., 1975a) and need to be removed. A better method is to fractionate the wide pH range ampholytes by IEF, as is done commercially. Narrow pH ranges, encompassing only one or two pH units (Gianazza et al., 1975) may be obtained by a simple method based on the continuous-flow IEF principle of Fawcett (1973). This method has advantages over multi-compartment electrolyzers, as described by Rilbe (1970) and Vesterberg (1969), since it avoids problems of osmotic pressure and polarization at membranes, and also reduces anodic oxidation of ampholytes. In practice, both methods are used in the fractionation of commercial products. In fact, while LKB and Serva use multicompartment electrolyzers for preparation of narrow pH cuts of Ampholine and Servalyte, Pharmacia prepares directly narrow pH ranges of Pharmalyte by ad hoc synthetic procedures. Carrier ampholytes can also be separated in narrow ranges by using the zone convection IEF apparatus of Valmet (1969). By this technique, this author has separated Ampholine encompassing the pH range 3.5−10 into a number of fractions, each covering about 0.2 of a pH unit. In practice, however, according to Charlionet et al. (1981) the minimum pH interval which can be successfully used in an IEF fractionation spans 0.3 pH units.

For preparation of small quantities of limited pH range ampholytes it is sufficient to fractionate broader pH intervals either in the Rilbe density gradient columns (110 ml and 440 ml) (Vesterberg et al., 1967) or in Radola's (1973a, 1975) granulated flat beds. Otavsky et al. (1977) have also used horizontal troughs filled with a mixture of Sephadex and

Pevikon C870 (plastic beads made of polyvinyl chloride and polyvinyl acetate): after IEF, the recovery of the desired ampholyte pH cut can simply be achieved by centrifuging the cake through a sintered glass).

Particularly attractive also are the flow-through systems described by Just (1980) and by Righetti and Hjertèn (1981): since the carrier ampholytes are synthesized along the length of a capillary tubing, fractionation of its content gives automatically narrow pH cuts. In principle, any apparatus used for preparative IEF can be utilized for preparation of narrow pH ranges. From this point of view, equipment which does not utilize capillary systems as anticonvective media is of interest, since the fractionated carrier ampholytes do not have to be separated from the inert support (usually a granulated gel). Thus, other chambers that can be exploited for this purpose are the polyethylene coil tubing of Macko and Stegemann (1970) and the thin-layer, continuous free-flow apparatus of Just et al. (1975a and b).

1.10. General properties of carrier ampholytes

I shall attempt to review the data now available on the physico-chemical properties of carrier ampholytes. The reader should bear in mind that most of these data refer to LKB Ampholine, since these chemicals have been available since 1967: however, some properties of Servalyte and

TABLE 1.4
Properties of carrier ampholytes and Ampholine (from Haglund, 1975)

Fundamental 'classical' properties
    (a) buffering ion has a mobility of zero at isoelectric pH
    (b) good conductance
    (c) good buffering capacity

Performance properties
    (a) good solubility
    (b) no influence on detection systems
    (c) no influence on sample
    (d) separable from sample

'Phenomena' properties
    (a) 'plateau' effect, drift of the pH gradient
    (b) chemical change in sample
    (c) complex formation

Pharmalyte have also been described and mention will be made in the text of data which is pertinent to these particular commercial chemicals. The general properties of carrier ampholytes have been summarized by Haglund (1975) and are given here in Table 1.4.

1.10.1. Molarity and ionic strength of isoprotic carrier ampholytes

One of the major problems in IEF experiments has been the total lack of knowledge of the molarity and ionic strength (I) of focused carrier ampholytes. The only parameter measurable with certainty after an IEF experiment is the pH, whose course is easily mapped, both in sucrose density gradients and in gels. By way of contrast, when performing electrophoresis, these three physico-chemical parameters which define the buffer medium, are rigorously known by the experimenter. This has made it impossible, up to now, for any comparison to be made between electrophoresis and IEF data and has generated uncertainty in data, especially when performing IEF of cells, since their measured pI is a strong function of the environmental ionic strength (Sherbet, 1978). Recently, however, McGuire et al. (1980) have been able to measure accurately the molarity of focused Ampholine. By focusing 1% Ampholine in a free liquid curtain, in the Hannig free-flow apparatus, and measuring the osmolarity of the 48 eluted fractions, they have derived a molarity of 9–10 mM in the pH range 3.5–10. This is in fairly good agreement with the value 15 mM given by Gelsema et al. (1978), by assuming an average molecular mass ($\overline{M}_r$) for Ampholine of 700 daltons. The ionic strength, however, which in IEF is vanishingly small, has been a much more elusive parameter to measure. It has been known for a long time that the classical definition of ionic strength of Lewis and Randall (1921):

$$I = \tfrac{1}{2} \Sigma c_i z_i^2 \tag{35}$$

could not be applicable to such conditions as found in IEF. On the other hand, direct measurement of I during an IEF experiment has so far baffled any attempt. Righetti (1980) has circumvented that by using red blood cells (RBCs) as a probe and exploiting their well known pI dependence from the buffer ionic strength (Seaman, 1975). By plotting data obtained from electrophoresis and IEF of RBCs, it was possible to extrapolate a value of $I = 0.5$ mg ion/l for focused 1% Ampholine (see Fig. 1.18). Also theoretical considerations and practical measurements of conductivity point to a value of $I = 0.5$ to 1 mg ion/l. The following equations have been proposed:

$$I = 1/20\ \bar{C}_{amph} + C_H \text{ in the pH range } 2.5-7 \text{ and} \qquad (36)$$

$$I = 1/20\ \bar{C}_{amph} + C_{OH} \text{ in the pH range } 7-11, \qquad (37)$$

where $\bar{C}_{amph}$ is the molarity of focused carrier ampholytes and C_H and C_{OH} the molarities of protons and oxydryl ions, respectively, at a given pH.

Actually, Gelsema et al. (1978), on the basis of Rilbe's (1976) theoretical considerations, have derived, for the ionic strength of focused Ampholine, the following inequality:

$$I \leqslant 1/3\ \bar{C}_{amph} + 1/2\ C_H \qquad (38)$$

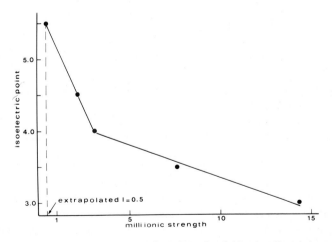

Fig. 1.18 Dependence of the isoelectric point of red blood cells on the medium ionic strength. The first point to the left has been obtained from IEF experiments by Just et al. (1979) and McGuire et al. (1980), while points 2–5 are from electrophoretic data tabulated by Seaman (1975). The ionic strength in IEF is the missing coordinate value of the first point of the curve (from Righetti, 1980).

Assuming an average $\overline{M}_r = 700$ and a molarity of 15 mM for focused 1% (w/v) Ampholine, these authors have derived an upper limit for I of ca. 5 mg ion/l, which is about 10 times the value I have reported. In a letter to me (dated Utrecht, May 13, 1980) Gelsema has argued that perhaps a distinction should be made between some kind of effective I pertaining at equilibrium during an IEF experiment and the I of (defocused) post-IEF fractions. He has made the following calculations: taking the minimum value of conductivity (χ) of focused Ampholine as $\chi = 0.5 \times 10^{-4}$ Ω^{-1} cm^{-1} (from his data together with my own) and considering that:

$$\chi = F \Sigma \, c_i z_i u_i \qquad (39)$$

where F is the faraday (96,500 coulombs) and u_i the mobility of the ion with the concentration c_i and charge z_i, this means that in the pH region 5–8, where neither the solvent nor catholyte or anolyte ions contribute significantly to the conductivity, the value of $\Sigma \, c_i z_i u_i$ will be

$$\frac{\chi}{F} \simeq 0.5 \times 10^{-9} \text{ (equiv. } \Omega^{-1} \text{ cm}^{-1} \text{ coul}^{-1}) \qquad (40)$$

In the narrow pH intervals of post-IEF fractions, z_i of ampholytes can only be 0, +1 or −1 and therefore $\Sigma \, c_i z_i u_i = \Sigma \, c_i u_i$. If we now suppose the mobilities u_i of the univalently charged ampholyte species to be nearly equal ($=\bar{u}$) we obtain:

$$\Sigma c_i z_i u_i = \Sigma c_i u_i = \bar{u} \, \Sigma c_i \simeq 0.5 \times 10^{-9} \text{ (equiv. } \Omega^{-1} \text{ cm}^{-1} \text{ coul}^{-1}) \quad (41)$$

By introducing in Eq. (41) my value of $I = 0.5$ mM $= 0.5 \times 10^{-6}$ (M/ cm^3) this would give a value of average mobility of: $\bar{u} = 10^{-3}$ (cm^2 V^{-1} s^{-1}) which appears to be too high since the limit mobility of acetate and benzoate ions is about 0.4×10^{-3} and 0.3×10^{-3} cm^2 V^{-1} s^{-1}. Therefore Gelsema thinks that a value of $I = 1$–2 mg ion/l in focused 1% Ampholine would be a more realistic estimate.

1.10.2. Isoelectric and isoionic (isoprotic) points

An ampholyte is characterized by having more than one protolytic group, and by having dissociating groups of both acidic and basic nature in each molecule. There are two concepts which are intimately related to ampholytes: the isoelectric point (pI) and the isoionic point. The pI is the pH where the electric mobility is zero, while the isoionic point of an ampholyte is the pH where anions and cations are present in equal concentrations, i.e., the ampholyte has zero net charge. Experimentally, the isoionic point is defined as the pH of an ampholyte solution which will not change when the concentration is increased by dissolving more of the pure ampholyte; in mathematical terms:

$$\left[\frac{\partial (pH)}{\partial C} \right]_c = 0 \tag{42}$$

where C = concentration of ampholyte and c = concentration of strong acid or base. Since the term 'isoionic point' does not conform to this equation, Rilbe (1973a) has suggested instead the term 'isoprotic point'. In the IEF practice, it has been found that isoelectric and isoionic (isoprotic) points lie very close together, so that the difference between them is insignificant. Therefore, in the present treatise, no distinction will be made between 'isoelectric' and 'isoionic' ('isoprotic') points, and the three terms will be used interchangeably. These concepts also apply to proteins as separated by IEF, since in the vast majority of cases no interaction between the focused protein zone and the surrounding buffering ions (carrier ampholytes; see also §4.11) has been demonstrated. However, in conventional electrophoresis, the protein pI, as measured by extrapolation to zero mobility and zero I, might greatly diverge from its true isoprotic point in the cases in which the protein macroion binds substantial amounts of the buffering ions. Many examples of this have been reported (Cann, 1966) thus rendering IEF uniquely suited for a proper assessment of a protein pI (Righetti and Caravaggio, 1976; Righetti et al., 1981).

1.10.3. Molecular size

Another highly controversial aspect of carrier ampholyte chemistry is the definition of their actual molecular mass. Several research groups have proposed all possible ranges of $\bar{M}_r$ values: 5,000–7,000 daltons (Gierthy et al. 1979); 800–1,200 daltons for Servalyte, 1,000–6,000 daltons for Ampholine and 1,000–15,000 daltons for Pharmalyte (Radola, 1980b; Goerth and Radola, 1980); even an upper limit of 20,000 daltons has been suggested (Baumann and Chrambach, 1975). These high $\bar{M}_r$ values, apparently, are not just exhibited by a tiny fraction of the population of carrier ampholytes, but by a substantial proportion of them, expecially in the alkaline pH ranges. On the other hand, direct measurements made by gel filtration and osmometry have given, for Ampholine, a $\bar{M}_r = 700$, with only 0.7% of the species above 1,000 daltons and 0.03% in the proximity of 4,000 daltons (Haglund, 1975). As scientists are never inclined to believe manufacturers, it is comforting that these data have been fully substantiated by independent work by Gelsema et al. (1979) who have found that, in almost all pH intervals, $\bar{M}_r$ [Pharmalyte] $> \bar{M}_r$ [Ampholine] $> \bar{M}_r$ [Servalyte]. Moreover, in all three cases, the highest $\bar{M}_r$ values were found in the acidic, not in the basic fractions of carrier ampholytes. This is immediately evident also from theoretical considerations (Gianazza et al., 1979a). The actual $\bar{M}_r$ values calculated were: $\bar{M}_r = 710$ for Ampholine and $\bar{M}_r = 870$ for Pharmalyte. I feel it is important to critically evaluate these data, since a proper assessment of carrier ampholyte $\bar{M}_r$ values is fundamental to all protein chemists using IEF. The supporters of the high $\bar{M}_r$ hypothesis fail to account for the following experimental observations: (a) precipitation of proteins with up to 100% saturation with ammonium sulphate removes more than 99.99% carrier ampholytes (Nilsson et al., 1970). High $\bar{M}_r$ species would have been most certainly precipitated; (b) IEF of peptides (Righetti and Chillemi, 1978) is made possible by the fact that, above a minimum critical length (ca. 12–14 amino acids, i.e., ca. 1,500 daltons), they are fixed and stained by simultaneous exposure to a dye–trichloroacetic acid (TCA) mixture. Under these conditions, carrier ampholytes are fully soluble.

The $\bar{M}_r$ determinations of Gierthy et al. (1979) rely on electrophoretic migration in sodium dodecyl sulphate (SDS) gels. In this system (except for Servalytes, which perhaps do not stain), the most alkaline pH ranges (pH 9—11) exhibit the highest $\bar{M}_r$ values (5,000—7,000) while, if anything, the opposite should be true. I feel that these authors have failed to realize that SDS-electrophoresis of basic proteins is unreliable for their $\bar{M}_r$ assessment (Panyim and Chalkley, 1971). Moreover, Poduslo and Rodbard (1980) have demonstrated that the error in $\bar{M}_r$ determination increases as the actual $\bar{M}_r$ of the basic protein decreases. Thus, in the case of two basic proteins from myelin of rat central nervous system, having $\bar{M}_r$ values of 18,400 and 14,300, the apparent $\bar{M}_r$ values found in SDS-gels were 21,000 and 19,000, respectively. Moreover, if the carrier ampholytes bound little or no SDS, as could be suspected if they were indeed small molecules, the negative charge density of basic carrier ampholytes would be much lower than that of acidic ones, resulting in a much higher apparent $\bar{M}_r$ in SDS-electrophoresis. The $\bar{M}_r$ data of Goerth and Radola (1980), obtained by thin-layer gel chromatography on Bio-Gel P-10 in 25 mM phosphate, pH 7.2, and 2 M urea, are less susceptible to criticism. I can only offer as an explanation that Servalyte, Ampholine and Pharmalyte interact to a different extent among themselves, giving aggregates of different composition and stoichiometries, which then exhibit a much higher apparent $\bar{M}_r$. In fact Gianazza et al. (1979a) have recently demonstrated that Ampholine aggregates exist even during the IEF process, and that 8M urea is required to split them apart. Moreover, they have found basic carrier ampholytes to be more hydrophobic than acidic ones, as they bind substantially higher amounts of neutral detergents, such as Nonidet P-40 (NP-40). That this is so has also been demonstrated by Gelsema et al. (1980a) using gel filtration: especially LKB Ampholines aggregate considerably when their concentration is raised from 0.6 to 4% (w/v). Moreover, from their chromatographic behavior in molecular sieves, they have inferred that normal range Ampholines (3.5 < pI < 10) contain a large proportion of species derived from only a few polyamines, probably the isomers of pentaethylenehexamine. In the case of Pharmalytes, these authors have found that early batches (produced till the end of 1979) contained high $\bar{M}_r$ species, especially the pH 5—8 range, in which a substantial amount of material was eluted in the void volume of a Bio-Gel P-4 resin. Apparently, in more recent batches (possibly from 1980 onward), these high $\bar{M}_r$ compounds have been eliminated.

Up to now, however, no direct value for the $\bar{M}_r$ of any carrier ampholyte pH range has been given. Recently, Bianchi Bosisio et al. (1981), by considering that osmolarity measurements in dilute solutions should still be the best approach to $\bar{M}_r$ determination, have measured the relative average $\bar{M}_r$ of each narrow pH range Ampholine. The data are given in Fig. 1.19 and show that indeed the masses are distributed from a minimum of 625 daltons for the alkaline species (pH 8–10) up to a maximum of 910 daltons for the acidic ranges (pH 3.5–5). The two extreme ones, though, (pH 2.5–4 and presumably pH 9–11) appear to have much smaller size ($\bar{M}_r = 325$) suggesting, as also postulated by Gelsema et al. (1980a), that they do not belong to the PEHA × Ac (acrylic acid) family. Thus, from these data, it would appear that, at least for Ampholine, no species with $\bar{M}_r > 1000$ daltons exist. Unfortunately,

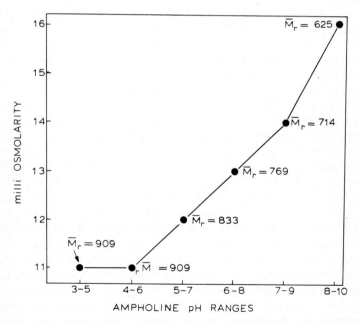

Fig. 1.19. Measurement of average molecular mass ($\bar{M}_r$) of narrow Ampholine pH ranges by osmolarity data. Values obtained on 1% Ampholine solutions in the pH ranges indicated in the abscissa (from Bianchi Bosisio et al., 1981).

similar experiments with Pharmalytes cannot be performed, since their actual concentration in solution is not known by the manufacturer (only the conductivity values of each pH range are given).

1.10.4. Optical properties

The optical properties of Ampholine are also very favorable for protein fractionations, since most do not absorb appreciably around 280 nm. Consequently, carrier ampholytes do not usually interfere with protein detection at this wavelength. The UV absorbance of Ampholine is generally low, and is usually confined to components of low pI. Although this is generally true for analyses of IEF experiments in sucrose density gradients where fairly large fractions are collected, it is not always true when scanning gels or columns under voltage where much sharper resolution is obtained and the absorbance of individual ampholytes may be considerable (Righetti and Drysdale, 1971; Catsimpoolas, 1973c).

UV spectra of different Ampholine ranges, by Righetti et al. (1975b) have shown that most Ampholine ranges have a peak absorbance around 365 nm. The pH range 9—11 also exhibits a peak at 290 nm and the pH range 3.5—5 has a strong chromophore at 310 nm. This acidic pH range has a marked UV absorbance and is also distinctly yellow in color compared to other pH ranges. Some of these chromophores are pH-dependent, and they also exhibit pH-dependent fluorescence spectra (especially in the acidic pH ranges). Vinogradov et al. (1973) and Righetti and Drysdale (1974) attributed the characteristic spectra to heterocyclic nitrogen structures among the population of amphoteric molecules, but this has not been substantiated by NMR analysis (Righetti et al., 1975a).

Recent batches of Ampholine do not exhibit any ORD spectrum, and this is consistent with their known structure. However, in batches prepared before 1970, the two extreme pH ranges were found to rotate the plane of polarized light (Righetti et al., 1975b). This could be due to the fact that the extreme pH ranges were reinforced with Glu and Asp (in the acidic side) and with Lys and Arg in the basic side (as also pointed out by Davies (1970)). Similar evidence for the supplementation of ampholytes was also obtained by amino acid analysis (Earland and Ramsden, 1969). As stated by Haglund (1975), though, after the end of 1972 Ampholines were no longer supplemented with amino acids.

Not much has been published in the literature on the optical properties of Servalyte and Pharmalyte, albeit it is generally known that their UV (280 nm) absorbance is very low and thus favorable for direct protein scans. Williams and Söderberg (1979) have given the UV (280 nm) profile of focused Pharmalyte in the pH 4–6.5, 5–8 and 3–10 ranges. There too we can see that the A (280 nm) is localized mostly at the acidic end of each pH interval with peaks that, for the pH 4–6.5, can be rather high (0.8 A). From Bonitati's work (1980, 1981) it is now also known that, in addition to Ampholine, Servalyte and Pharmalyte fluoresce when excited at 254 and 365 nm. The strongest fluorescence emission intensity is actually given by Servalyte, and this property has been exploited for mapping protein zones and ampholyte distribution profiles after IEF (see also § 1.8).

1.10.5. Buffering capacity and conductivity of Ampholine, Servalyte and Pharmalyte

A wealth of information is available on LKB Ampholine (Davies, 1970; Haglund, 1975; Fawcett, 1975a) and on home-made products (Righetti et al., 1975a, 1977). Comparative studies between Ampholine and Servalyte (Fredriksson, 1977a) and among the three commercial products (Gelsema et al., 1978) have been published. According to Fredriksson, below pH 8 the buffer capacity of Servalyte pH 2–11 is 1.5–2.5 times lower than that of Ampholine pH gradient 3.5–10, whereas above pH 8.5, the buffer power of the two products is equal. For both gradients, the buffer capacity is at a minimum around pH 6.2, this minimum being 5.2 μequiv. pH^{-1} ml^{-1} for Ampholine and as low as 1.8 μequiv. pH^{-1} ml^{-1} for Servalyte (these values refer to a 1% solution focused in a sucrose density gradient). They are in good agreement with the data of Davies (1970) who found, for a 1% Ampholine solution, in the pH interval 5–7, a buffering power of 6 μequiv. pH^{-1} ml^{-1}.

Gelsema et al. (1978) have made a thorough comparison of the buffer capacities and conductivities of the three commercial species. Since their data have been obtained after focusing in a free liquid phase (no sucrose added as an anticonvective medium), in a coiled polyethylene tube (Bours, 1976), they must be regarded as the most reliable values. Their results are plotted in Fig. 1.20. We can see that, in the pH region 5.5–8.2, Pharmalytes have the best and most uniform while Servalytes have the

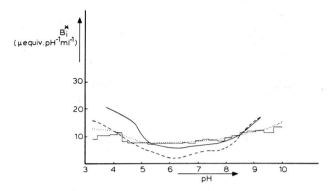

Fig. 1.20. Buffering capacity versus pH of focused Ampholines (——), Servalytes (— — —) and Pharmalytes (·····); for Pharmalytes the data supplied by the manufacturer are plotted in a stepwise graph. (From Gelsema et al., 1978.)

worst buffering power. However, between pH 3.5 and 5.5 Ampholine contains the best buffering species and, above pH 8.5, both Ampholine and Servalyte buffer better than Pharmalyte. If we now plot the ratio between molar buffer capacity (B_i) and conductivity (χ_i) vs. pH of the focused fraction, we obtain the graph in Fig. 1.21. Now, it is known that:

$$\frac{B_i}{\chi_i} \simeq \sqrt[3]{\overline{M_r}} \tag{43}$$

and inspection of this figure shows that practically throughout the pH gradient, and especially between pH 5.5 and 8.5, Pharmalyte have the highest value of this ratio, followed closely by Ampholine and then by Servalyte. Thus, these three curve profiles follow the relative distribution of their respective $\overline{M_r}$ values known to be: $\overline{M_r}$ Servalyte $< \overline{M_r}$ Ampholine $< \overline{M_r}$ Pharmalyte. Therefore Pharmalyte have the best buffering power in the pH 5.5—8.5 region because they have the highest $\overline{M_r}$, in agreement with the concept of Charlionet et al. (1981) that the limiting factor for the resolving power in IEF is only the $\overline{M_r}$ of the ampholyte dictating the pH gradient. These comparisons are only valid if we assume that, in the three commercial species, the number of buffer-

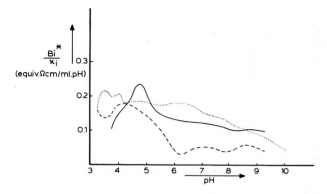

Fig. 1.21. Values of B_i/x_i versus pH of focused Ampholine (——), Servalytes (– – –) and Pharmalytes ($\cdots\cdots$). B_i = molar buffering capacity in the isoelectric state; x_i = conductivity in the isoelectric state; B_i/x_i is proportional to the cube root of molecular mass (From Gelsema et al., 1978.)

ing protolytic groups per mass unit are roughly constant (see also §1.7.2 and §1.10.3). One more observation comes to my mind. We have seen that Pharmalyte has the highest and most even buffering power in the pH 5–8 range, with no conductivity minimum at pH 6.2. The pH 5–8 species also have a quite high $\overline{M}_r$, since they are eluted in the void volume of Bio-Gel P-4 (Gelsema et al., 1980a). On the other hand, as we have seen in §1.10.3, after 1980 Pharmacia has produced lower $\overline{M}_r$ species and recent data have also shown that Pharmalytes also appear to have a minimum of conductivity (and thus buffering capacity) in the pH 6–6.5 range. I am tempted to suggest that the removal of these high $\overline{M}_r$ species has brought this as a consequence but, unfortunately, the one who should know best, the manufacturer, has kept quite silent about it.

1.10.6. Distribution of carrier ampholytes upon focusing. Total number of individual species

Starting from the differential equation of IEF (Eq. 1), which represents the balance between electrophoretic and diffusional mass flow in the steady-state, Svensson (1961) has derived for the concentration C of a focused ampholyte in a linear pH gradient of constant specific conductivity χ_0 the equation:

$$\left(\frac{C}{C_0}\right)_{\text{Gaussian}} = \exp\left(-\frac{P\,i\,x^2}{2\,q\,\chi_0\,D}\right) \tag{44}$$

which is the symmetrical Gaussian function, with i = electric current, q = cross-sectional area of the focusing medium; x = distance along the direction of the current ($x = 0$ at the concentration maximum C_0 of the ampholyte and $x > 0$ towards the cathode); D = diffusion coefficient of the ampholyte and P as defined by Eq. (17). However, in the case in which the specific conductivity is not constant, but is a linear function of x, the solution of Eq. (1) becomes:

$$\left(\frac{C}{C_0}\right)_{\text{skew}} = \exp\left[\frac{P\,i\,\chi}{D\,q\,r^2}\cdot\ln\left(1+\frac{rx}{\chi_0}\right)-\frac{P\,i\,\chi}{D\,q\,r}\right] \tag{45}$$

which is a skew distribution function (r = zone width). Thus, in IEF, the carrier ampholytes can have both, a Gaussian or a skew concentration profile at their pI values. In fact, Brown et al. (1977) have found asymmetric distributions of quite a few carrier ampholytes upon IEF. On the other hand, Gelsema and De Ligny (1979) have argued that, in all practical cases, the conductivity variations along a focusing column, albeit substantial, have a negligible effect on the concentration distribution of an ampholyte about its pI, thus they cannot be responsible for any skewness. Therefore, the origin of experimentally obtained skew distributions should not be sought in this effect, but rather in ampholyte inhomogeneity or in interaction of the ampholyte with other components in the focusing system, i.e., with other (carrier) ampholytes or with the anticonvective medium. For instance, Gianazza et al. (1979a) have reported not only the interaction of carrier ampholytes among themselves, but also with other components added in the system, especially neutral detergents used to solubilize membrane proteins (such as Nonidet P-40 and Tween 80).

Another controversial aspect of this subject is the actual width of a single, focused carrier ampholyte along the separation column length. Brown et al. (1977) have concluded that, upon focusing, each ampholyte component is distributed over a wide area (7–15% of the column length, typically above 10%) in some cases extending over as much as 40% of the total separation distance. This position does not seem tenable, also in view of the fact that it is based on ampholyte patterns resolved by ion-exchange chromatography (which possibly resolves only 10–15% of the total species) and on the IEF distribution of His which, at higher loads, precipitates and aggregates all over the focusing column. Considerable amount of evidence, on IEF of plant pigments (Jonsson and Petterson, 1968), of dyes (Righetti et al., 1977b), of free amino acids (Catsimpoolas, 1973c), of peptides (Righetti and Chillemi, 1978) and detection of focused carrier ampholyte patterns by refractive index gradients (Righetti et al., 1975a; Rilbe, 1973b) and by specific stains (Radola et al., 1977), suggests that indeed even small $\bar{M}_r$ ampholytes focus as sharp peaks, notwithstanding their higher diffusion coefficients, as compared to proteins, typically extending along 0.1 to 0.5% of the column length, especially at loads of only a few micrograms (Righetti, 1977). However, things might be different in alkaline pH ranges, since it has been demonstrated (Gianazza et al., 1979a) that these contain a relatively higher proportion of 'poor' carrier ampholytes, as compared with acidic pH ranges. Thus their distribution about the pI could be broader.

One last aspect relates to the total number of individual species present in any of the three commercial products. Vesterberg (1973c) had calculated that, by his synthetic procedure, more than 360 isomers and homologues are obtainable. This number could still be increased by adding some methyl or ethyl groups on the amino groups. However, by ion-exchange chromatography, only a total of 62 different species could be resolved in LKB Ampholine pH 3.5–10 (Brown et al., 1976). On the other hand Radola et al. (1977) by developing ampholyte patterns, upon focusing in a Sephadex G-75 bed, with formaldehyde, lactose or ninhydrin, were able to count, in a two pH unit range, as many as 150 different Ampholine components. In narrow cuts of only 0.5 pH units, as many as 45 peaks were resolved by ion-exchange chromatography. On the basis of these data, more than 500 ampholyte species should be present in the wide Ampholine pH range. Actually according to other authors and manufacturers, the total number of individual ampholytes in a mixture could be enormously high. Söderberg et al. (1980) suggest that there is a minimum of 1,000 different ampholytes per pH unit in Pharmalytes. This suggests that broad pH range Pharmalyte should contain more than 6,000 individual species, an amazing number indeed. Were it not enough, then I should like to mention that Charlionet et al. (1981) have calculated that, by their synthetic procedure, just in the pH zone 4.3—5.0 (barely a 0.7 pH unit span) more than 4,000 ampholyte species with different pI and pK values could have been created. I am afraid I should rather stop here, before these numbers grow up to galactic dimensions. . .

1.10.7. Chelating properties

It has been known for a long time that carrier ampholytes chelate metals (Davies, 1970) but little quantitative information has been available on this subject. Galante et al. (1975) have found that metal chelation by Ampholine usually involves two adjacent nitrogen functions, giving rise to 5-membered rings. Chelation maxima occur in the pH region 8.5— 10.5, where most of the nitrogen functions are unprotonated. Mixed type chelation on a nitrogen and carboxyl function is a possible though less likely event, since it generates less stable seven-membered rings. Little, if any, chelation occurs solely on carboxyl functions since this would lead to twelve-membered rings.

As compared to EDTA, the chelating power of Ampholine is generally weak. With the exception of Cu^{2+} a given amount of divalent cation (Mg^{2+}, Ca^{2+}, Zn^{2+}) requires a 100 to 500 molar excess of ampholytes for chelation in contrast to EDTA, which usually binds metals in the pH range 3—11 in a 1:1 molar ratio. Carrier ampholytes are, however, as powerful chelators of Cu^{2+} as EDTA and, consequently, many copper containing proteins may be rapidly inactivated on exposure to ampholytes in IEF. Albeit at the time of this study we were not aware of it, I have now found data in the chemical literature which fully support our findings. According to Christensen et al. (1971) a polyamino backbone, whether cyclic, branched or linear, is indeed a powerful complexing agent for Cu^{2+}. Table 1.5 lists the log of binding constants for a series of tetramines: in the case of the cyclic structures, the ion is believed to fit rather snugly in the central hole, so that four more-or-less equivalent nitrogen atoms are coordinated in a single plane about the metal ion. In any event, in all cases, the complexes exhibit a stoichiometry of 1:1 molar ratio. It is of interest to notice that Vesterberg-type carrier ampholytes (in which the nitrogens are usually spaced two methylene groups apart) are particularly suited for Cu binding: as seen in Table 1.5, they have a much stronger affinity for the metal than the same oligoamine

TABLE 1.5

Binding constants for the 1:1 complexes of tetra-amines with copper(II)[a]

Ligand		Log of binding constant (liter/mole)	Temperature (°C)
	Macrocyclic		
(cyclic tetra-amine structure)	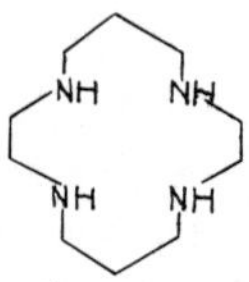 blue complex	20	25
	red complex	28	25
	Noncyclic		
$N((CH_2)_3NH_2)_3$		13.1	25
$NH_2(CH_2)_3NH(CH_2)_2NH(CH_2)_3NH_2$		17.3	20
$N((CH_2)_2NH_2)_3$		18.8	20
$NH_2(CH_2)_2NH(CH_2)_2NH(CH_2)_2NH_2$		20.1	25
$NH_2(CH_2)_2NH(CH_2)_3NH(CH_2)_2NH_2$		23.9	25

[a] From Christensen et al. (1971).

in which the nitrogens are separated by a propionic chain. At present, there are no studies available on the chelating properties of Servalyte and Pharmalyte.

In order to minimize metal chelation, proteins should be run to equilibrium from the anodic side. It is also advisable to apply samples to pre-focused gradients to minimize their contact time with ampholytes. Ideally the samples should be introduced as close as possible to their expected pI in prefocused gradients. Fortunately, loss of activity of many enzymes by metal chelation is seldom irreversible. Often, full enzymatic activity may be restored by incubating the focused enzyme with appropriate amounts of sequestered cofactor. The main problem with metal chelation is the generation of multiple banding patterns, due to multiple forms containing differing amounts of metal.

1.10.8. Biological toxicity

Studies by Wadström et al. (1974) have indicated that Ampholines are fairly innocuous as judged by both animal and in vitro cytotoxicity experiments. Ampholines do not appear to interfere in immunodiffusion tests or in the hemagglutination inhibition test and have been used as pH stabilizers in tissue culture (Walther and Schubert, 1974). Since Ampholine is easily removed from a purified protein, and no clear evidence of stable complexes between a protein and Ampholine exists, it should be possible to use Ampholine for preparative purposes either in IEF or in isotachophoresis for preparation of proteins for human or animal use.

However, in tissue culture, things might be different. According to Gierthy et al. (1979) some ampholytes are mitogenic on quiescent human diploid lung fibroblasts. The mitogenic activity is particularly strong with LKB Ampholine pH 9–11, whose stimulatory effect amounts to as much as 30–50% of the maximum response obtained by stimulation with human platelet factor. None of the other ampholytes tested were mitogenic to a significant extent, including Phisolyte (a brand sold by Brinkmann), Servalytes and Pharmalytes. However, all carrier ampholytes, from any commercial source, when used at high concentrations (usually above 2–3 mg/ml) have an inhibitory effect and are cytotoxic.

While their data on the biological activity of carrier ampholytes appear to be sound, their conclusions that the mitogenic material (Ampholine pH 9–11) and also the non mitogenic products (Phisolytes pH 9–11, Pharmalytes pH 8–10.5 and Servalytes pH 4–9) contain high $\bar{M}_r$ components ($\bar{M}_r$ 5,000–7,000 daltons) do not seem to be tenable. As we have seen (§ 1.10.3), and as demonstrated for the extreme pH ranges of Ampholine (pH 2.5–4 and pH 9–11), they have in fact the smallest molecular size of all the species buffering between pH 4 and 10 ($\bar{M}_r$ of ca. 300 to 400 daltons) (Bianchi Bosisio et al., 1981; Gelsema et al., 1979).

1.10.9. Solubility of carrier ampholytes in different solvents

Carrier ampholytes appear to be soluble in the most common protein precipitating agents, such as sulphosalicylic acid and TCA (in this last case, at least up to 50% TCA) (Radola, 1980b). However, all the three commercial species, in all pH ranges, are precipitated at room temperature by 0.6% picric acid, suggesting that they form complexes with it (unpublished experiments with A. Görg, W. Postel and R. Westermeier). LKB Ampholines are also soluble in 100% saturated solutions of ammonium sulphate, i.e., under conditions which precipitate reversibly most proteins (Nilsson et al., 1970). They are soluble too, at least up to 2%, in

100% chloral hydrate, a solvent which has been used for electrophoresis of proteins (Ballou et al., 1974), but appear to be modified by this aldehyde, probably by formation of a Schiff base (Brenna et al., 1982). Unfortunately, IEF cannot be performed in this solvent, because above pH 5 proteins will start reacting strongly with the aldehyde. LKB Ampholines are also fully soluble in 100% formamide, a classical solvent for electrophoresis of DNA and RNA (Rickwood and Hames, 1981). Here too, however, we have been unable to perform reproducible IEF experiments, since formamide is strongly hydrolyzed at the cathodic end, producing formic acid and free ammonia, thus disrupting the pH gradient. This hydrolytic process is not abolished even when using weak catholytes, such as dilute solutions of Tris (unpublished experiments with E. Gianazza). We have recently started an investigation on the solubility limits of Ampholine in different organic solvents (mostly alcohols) and in typical disaggregating agents, such as dimethyl sulphoxide (DMSO), dimethyl formamide (DMF) and tetramethyl urea (TMU). This is because often it is highly desirable to substitute, during IEF, 8 M or 9 M urea solutions with these organic solvents, since urea is known to carbamylate proteins in a pH-dependent fashion. For instance, storage proteins in cereals (prolamines) are well soluble in alcohols (isopropanol) which could be an ideal substitute for urea during IEF fractionation.

In Table 1.6 I have listed the solubility limits of Ampholine (pH 3.5−10 range) in hydro-organic mixtures: these data refer to a 2% (w/v) solution of Ampholine, at room temperature (22 °C), under non-focusing conditions. Presumably, upon focusing, the solubility will be somewhat less, since the carrier ampholytes have a minimum of charge.

1.10.10. Recycling of carrier ampholytes

The possibility of recycling ampholytes, especially after a preparative step, would be of important practical use. For this purpose, conventional and hollow fiber dialysis, gel chromatography, salting out, ion-exchange chromatography and electrophoresis were proposed by Vesterberg (1976) and by Callahan et al. (1976). As another approach, Goerth and Radola (1980) have suggested ultrafiltration. To this purpose, they have used Amicon and Kalle membranes either of the cellulose acetate type, or of polyamide or of other synthetic polymer. Their procedure is based on two steps: in the first, the ultrafiltration is performed with mem-

TABLE 1.6

Solubility of carrier ampholytes in different solvents

Solvent	Solubility limit (22 °C, 2% Ampholine pH 3.5−10)
Sulphosalicylic acid	fully soluble up to 5% (w/v) [a]
Trichloroacetic acid	fully soluble up to 50% (w/v) [a]
Picric acid	heavy precipitate in 0.6% (w/v)
Ammonium sulphate	fully soluble up to 100% (w/v)
Chloral hydrate	fully soluble up to 100% (v/v)
Formamide	fully soluble up to 100% (v/v)
Sulfolane	soluble up to 85% (v/v)
Ethanol	soluble up to 70% (v/v)
Dimethyl sulphoxide	soluble up to 65% (v/v)
Dimethyl formamide	soluble up to 55% (v/v)
Isopropanol	soluble up to 50% (v/v)
Tetramethyl urea	soluble up to 45% (v/v)

[a] Solubility above this level not tried.

branes with $\overline{M}_r$ cut-offs ranging from 5,000 to 20,000 daltons; in the second, recycling is done with membranes of cut-off levels of 500−5,000 daltons. It appears that commercially available ampholytes can be recycled in their two-step procedure at reasonable flow rates. However, complete elimination of ampholytes from the protein zone is only obtained with the most porous membranes (20,000 daltons and only of the polyamide type (PA 20 from Kalle)). The retention behavior increases in the following order: Servalyte < Ampholine < Pharmalyte and this has lead Goerth and Radola (1980) to conclude that the last two chemicals have rather high $\overline{M}_r$ values while, in the light of recent data (Bianchi Bosisio et al., 1981; Gelsema et al., 1980a) it appears that they have a rather strong tendency to aggregate (see § 1.10.3). Interestingly, it seems that the properties of recycled carrier ampholytes are better than those of the commercial material: the cathodic drift is reduced and their size distribution is more uniform.

1.11. pH gradients generated by other means

I will review here some other methods for generating pH gradients for possible use in IEF. Some of them have more theoretical interest than practical applications; others have been claimed to be useful for IEF fractionations.

1.11.1. Bacto peptone carrier ampholytes

Blanicky and Pihar (1972) have prepared ampholytes from bactopeptone by removing proteins by precipitation in 60% ethanol. After ethanol elimination, they obtained a mixture of possibly several dozen peptides, which appear to be satisfactory for IEF. The conductivity and buffering capacity in the neutral region was improved by adding His. This type of carrier ampholyte is similar to those described by Vesterberg and Svensson (1966) and may therefore suffer from the same drawbacks. A similar approach has been used by Molnarova and Sova (1974) who used casein hydrolysates to obtain good pH gradients down to pH 2.5.

1.11.2. Thermal pH gradients

A radically different approach to the generation of pH gradients which does not rely on carrier ampholytes or on prolonged electrolysis has been made by taking advantage of the temperature coefficient of the pH of a buffer as well as of the pI of the proteins to be separated (Kolin, 1970; Luner and Kolin, 1970, 1972; Lundhal and Hjertén, 1973). A pH gradient can be established in a buffer solution within seconds by taking advantage of the temperature dependence of the pK. By establishing a temperature gradient within the buffer, pH gradients can be obtained that span a pH range of about 1 pH unit. The slope of the pH gradient can be controlled by simply varying the temperature limits. This method would have the advantage of eliminating the long process of establishing the pH gradient by stationary electrolysis as well as the use of additives such as the carrier ampholytes which have to be separated from the proteins after their IEF fractionation.

As an example, the pH of a Tris–HCl buffer is determined by the equilibrium relation:

$$K = [H^+] \, [Tris] / [Tris \, H^+] \tag{46}$$

The temperature dependence of the H^+ activity (α_H) of a buffer solution can be expressed as:

$$(\partial p\,\alpha_{\mathrm{H}}/\partial T) \simeq -(\partial \log K/\partial T) - (2z+1)(\partial \log \gamma/\partial T) \qquad (47)$$

where γ is the activity coefficient of an average ion of valence $z = 1$. In the example of the Tris buffer, the second term of Eq. (47) is less than 10% of the first term, and thus can be neglected in the present approximation. The first term $(\partial \log K/\partial T)$ can be obtained from the following relation involving the gas constant R and the molar heat of dissociation of an acid ΔH^0:

$$(\partial \ln K/\partial T) = (\Delta H^0/RT^2) \qquad (48)$$

thus:

$$(\partial p\,\alpha_{\mathrm{H}}/\partial T) \simeq (\Delta H^0/RT^2) \qquad (49)$$

However, it must be remembered that the temperature affects not only the pH of the buffer (buf) but may also affect the pI of the protein (pro) to be focused. According to Lundahl and Hjertén (1973) and Kolin (1976) focusing can only occur if $(\partial \mathrm{pH}/\partial T)_{\mathrm{buf}} - (\partial \mathrm{pI}/\partial T)_{\mathrm{pro}} \neq 0$. In practice, however, these pH gradients are difficult to utilize for IEF because the pH of ordinary buffers and the pI of proteins do not exhibit sufficiently large differences in temperature coefficients. Moreover, many buffers have too small $\partial \mathrm{pH}/\partial T$ values to be of any practical use. Thus, in phosphate, $\partial \mathrm{pH}/\partial T = -0.003$, which means that, by establishing on the separation column a temperature gradient of as much as $100\,^\circ\mathrm{C}$ (enough to denature most, if not all, proteins) the pH gradient would be barely 0.3 pH units. Among the few useful ones is Tris, which exhibits a $\partial \mathrm{pH}/\partial T = -0.028$ (no ordinary buffer has a larger temperature coefficient than this).

1.11.3. Dielectric constant and boundary pH gradients

Troitzki et al. (1974, 1975) have formed pH gradients by using common buffers in gradients of organic solvents, such as ethanol, dioxane, glycerol, or in polyol gradients such as mannitol, sucrose, sorbitol. Taking advantage of the pH variations of these buffers in different concentrations of these solvents, they were able to generate gradients of approximately 1.5 pH units in different regions of the pH scale. These pH gradients were stable up to 12 days of IEF under voltages up to 1,000 V. They have shown separations of rabbit hemoglobin in pH gradients 7–8.6 formed with borate buffers in a mixed gradient of 0.5% glycerol and 0–30% sucrose, and of human serum albumin in a pH gradient 4.5–5.8 formed with acetate buffer in a 0–90% glycerol gradient.

Actually, exploitation of such gradients was suggested by Lundahl and Hjertén (1973) who remarked that pH gradients are formed by passage of current through solutions containing gradients of neutral solutes such as sucrose or even acrylamide (e.g., pore-gradient electrophoresis). These authors also reported formation of steep pH gradients at the boundary of buffer solution and acrylamide gel (probably due to variation in transference number of the ions at the liquid/gel boundary). An interesting suggestion from them concerns the δ and ϵ boundaries well known from moving boundary electrophoresis. They point out that these boundaries are stationary and that they represent jumps in pH and conductivity. They suggest the possibility of exploiting these phenomena for generation of stationary pH gradients for IEF. However, if work does not progress beyond these initial stages, it is doubtful that both, dielectric constant and boundary methods, would be of any practical use. In my experience when working in polyacrylamide gels (unpublished data with B. Brost and R.S. Snyder) or in free solution (unpublished results with M. Bier and N. Egen) these pH gradients (which, at best, are semi-stationary) are either too steep or too shallow to be possibly utilized in IEF.

1.11.4. pH gradients generated by buffer diffusion

These gradients have already been described briefly in § 1.1 and § 1.5.1. They were termed 'artificial' by Svensson (1961) since they are obtained by diffusion of buffers of different pH and rapidly decay under an electric field. The pH gradient was generated by placing the substance to be separated at the interface between an acidic and basic buffer, in a Tiselius-like apparatus, and allowing diffusion to proceed under an electric field (Kolin, 1954, 1955a and b, 1958). In these 'artificial' pH gradients Kolin was able to obtain 'isoelectric line spectra' of dyes, proteins, cells, microorganisms and viruses on a time scale ranging from 40 s to a few minutes, a rapidity still unmatched in the field of electrophoresis.

This technique was also used by Hoch and Barr (1955) for separating serum proteins and by Tuttle (1956) and Maher et al. (1956) for separating hemoglobins. Due to poor reproducibility and instability of the pH gradients, this approach was abandoned until 1979, when Pollack described a modified version of it. This author has gone a step further: not only has he abandoned carrier ampholytes, but also buffers and performs focusing in a gel containing only deionized water between a weak catholyte (usually 0.1 M Tris, pH 9.5) and a weak anolyte (0.1 M acetic acid). Is this a focusing experiment? Notwithstanding the claims of the author, it seems to me that separations achieved represent only protein zones stacked within huge conductivity gaps. In fact the pH gradient (if any) only spans 10% of the separation column length, huge pH plateaux existing at both cathodic and anodic ends. Thus, there must be, close to the middle of the gel column, an ion vacuum, a sort of a Royal Gorge (Colorado) into which the protein ions are trapped and unable to move (see also 'apparent focusing' of cytochrome c in Kolin experiments (1954) correctly interpreted by him as a conductivity trap). With these experiments, we are thrown back to the early forties, to the electrolysis experiments of Tiselius and Svensson (Rilbe, 1976), where a pH step was generated in the middle of the electrolyzer as a result of collection of H_2SO_4 at the anode and NaOH at the cathode upon electrophoretic separation of Na_2SO_4. In fact, according to Pollack (1979) as little as 1 mM salt added to anolyte or catholyte or both is enough to prevent 'apparent focusing' of any protein ion, suggesting that even a minute increase in ionic strength is enough to let the proteins escape from the conductivity traps.

1.11.5. pH gradients obtained with mixtures of acids, bases and ampholytes

It all started in 1969, when Pettersson demonstrated that acidic pH gradients between pH 1 and 3 could be formed by electrolysis of a mixture of acids and ampholytes. He used 6 acids (Nos. 1, 2, 4, 6, 8 and 11 in Table 1.7) and two ampholytes. , glutamic (pI 3.22) and aspartic (pI 2.98) acids, and found that they distribute themselves in order of increasing pK values under an electric field. The strongest acid was closest to the anode, while the weaker acid was almost completely non-ionized in the region of its zone closer to the anodic side. By this

TABLE 1.7

Acidic and basic buffers and ampholytes used in BEF (from Righetti, 1979)

Acids	pK	Bases	pK	Ampholytes	pK (of amino group)
Dichloroacetic	1.48	Pyridine	5.5	MES	6.4
Phosphoric	2.12	4-Picoline	6.2	ADA	6.6
Malonic	2.18	BisTris	6.9	PIPES	6.8
Monochloroacetic	2.84	Lutidine	7.0	BES	7.15
Lactic	3.08	HEM	7.2	MOPS	7.2
Citric	3.08	Imidazole	7.5	ACES	7.3
Malic	3.45	NEM	8.0	HEPES	7.55
Formic	3.75	TEA	8.4	TES	8.0
Succinic	4.16	Tris	8.8	Tricine	8.6
Glutaric	4.34	MEA	10.4	Bicine	8.7
Acetic	4.75			Glycine	9.5
Propionic	4.85			Taurine	9.7
				GABA	11.3

Abbreviations: MES = Morpholino ethane sulfonic acid; MOPS = Morpholino-3-propane sulfonic acid; ACES = (N-2-acetamido-2-amino ethane) sulfonic acid; TES = N-tris-(hydroxymethyl) methyl-2-amino propane sulfonate; Tricine = N-tris (hydroxymethyl) methyl glycine; Bicine = N,N'-bis (2-hydroxyethyl) glycine; GABA = γ-amino butyric acid; ADA = N(2-acetamino iminodiacetic acid; BES = N,N-bis(2-hydroxyethyl)-2-amino ethane sulfonic acid; HEPES = N-2-hydroxyethyl piperazine-N'-2-ethane sulfonic acid; PIPES = Piperazine-N,N'-bis(2-ethane) sulfonic acid; BisTris = Bis-(2-hydroxyethyl-amino-tris-hydroxymethyl) methane; HEM = hydroxyethyl morpholine; NEM = N-ethyl morpholine; TEA = Triethanol amine; Tris = Tris (hydroxymethyl) amino methane; MEA = monoethanolamine.

method, pepsin (pI 2.5) and seven colored components from red beet sap, with pI values between pH 1.3 and 3.1, were separated. These pH gradients, however, increased stepwise, rather than as a smooth function. A similar approach was used by Stenman and Grasbeck (1972) for the fractionation of the R-type vitamin B_{12}-binding protein. They use a mixture of nine acids (Nos. 3, 5, 7–12 in Table 1.7) and the same two ampholytes. Upon focusing for 5 days at 400 V, in glycerol density gradients, under quasi-equilibrium conditions, they obtained a smooth pH-gradient from pH 1.5 to 4.5.

Recently Chrambach's group, in a series of papers (Nguyen and Chrambach, 1976; Nguyen and Chrambach, 1977a,b,c; Nguyen et al., 1977a,b; Caspers and Chrambach, 1977) has demonstrated that the same focusing results can be obtained in the neutral and basic pH region by using mixtures of non amphoteric, as well as amphoteric, buffers, including mixtures of amino acids. They have used this technique, called buffer focusing (BEF), for the study of pH gradient formation and decay in IEF and for demonstrating a possible parallelism between IEF and ITP. Their results can be summarized as follows.

(A) The finding that non-amphoteric buffers, placed in an electric field between acid anolyte and basic catholyte can produce stable pH gradients suggests that focusing is non-isoelectric.

(B) This hypothesis, in turn, suggests that focusing is steady-state stacking (ITP) (in order of pK values for non-amphoteric and in order of pI values for amphoteric compounds) under conditions such that the buffers used as catholytes and anolytes in ITP are replaced by base and acid, thereby providing a 'pH cage' which prevents the stack from migrating out of the separation column.

The spin-offs of their studies are two-fold:
(1) it is possible to stabilize pH gradients in IEF against decay (cathodic drift) by equalization of the anolyte pH with the pH of the most acidic amphoteric component in a given Ampholine pH range. In the case of Ampholine pH 6–8, this is achieved by using Thr at the anode and His at the cathode (Nguyen and Chrambach, 1977b);
(2) an ITP system with multiple trailing buffer constituents (cascade stack) can be converted into a cascade IEF when the leading and trailing constituents in ITP are replaced with strong acid and bases. Conversely, an IEF system can be transposed into a cascade stack, when the electrolytes are changed into the appropriate leading and terminating buffers of the corresponding ITP system (Nguyen et al., 1977a). This is very useful when performing preparative IEF in gel media in electrophoretic apparatus provided with elution chambers.

It seems to me that these kinds of gradients are only useful for a theoretical, rather than practical, approach to IEF. Moreover, it remains to be seen whether or not IEF is indeed ITP. Since this parallelism has been drawn mostly by buffer focusing, it must be stressed that, according to Rilbe's (1976) theory, non-amphoteric BEF can never reach equilibrium conditions, as in IEF, because weak acids and bases, which stop migrating electrophoretically due to total loss of charges, still keep diffusing in strong acids or bases lying on their anodic or cathodic side, respectively. It also remains to be seen whether or not abolition of pH gradient decay is generally applicable to all pH ranges. However amphoteric buffers different from Ampholine can be useful additives for the formation of non-linear pH gradients in the new technique 'pH gradient modifier IEF' (see §4.4).

1.11.6. pH gradients generated by two-three ampholytes: computer simulation

All of the above approaches have been purely empirical and have suffered from the absence of an explicit model demonstrating the contribution of individual components to gradient formation. An excellent theoretical and experimental study has finally appeared from Bier's group (Bier et al., 1981; Palusinski et al., 1981). They have generated pH gradients with mixtures of only two to three ampholytes, of known electrochemical properties, under a set of known physical parameters. The electrochemical properties of an ampholyte, which have to be known to define the system, are:

(a) the mobility coefficient (m) of all the components of the mixture. Not only the m values of the neutral, but also of the positive and negative subspecies should be known, since the cation can have a mobility up to 5% higher than the anion, due to the greater hydration of the carboxylate group (Edward and Waldron-Edward, 1965);

(b) the dissociation constants of all the components (pK). For polyampholytes, only the constants nearest the pI need to be used;

(c) the isoelectric points (pI) of all components. From the latter two, ΔpK and ΔpI values will be derived, which will be the quality marks of the IEF separation (see also Almgren, 1971).

The physical parameters which define the IEF separation are:

(a) the column dimensions (length and cross-section);

(b) the applied current density (A/cm^2);

(c) the initial (boundary) concentrations of the ampholytes used. Once all these experimental variables are known, the computer will generate the following output:

(a) the concentration profiles of all subspecies of both ampholytes along the column axis;

(b) the pH, conductivity and electrical potential profiles along the column axis, as well as other complementary data.

Bier et al. (1981) have listed in a Table (1.8) binary systems which could lead to successful IEF fractionation over a range of 1 to 2 pH units. I am reproducing in Fig. 1.22 a computer simulation of concentration and pH profiles of a mixture of aspartic and *m*-amino benzoic acids, which shows that quite symmetrical concentration and nearly linear pH profiles can be obtained in this system. Paradoxically, the pair His-His/His, which is expected to be an excellent combination, generates useless concentration and pH gradients. The most striking result of their simulation is the following: in simple mixtures of ampholytes (as well as, presumably, in buffer focusing, § 1.11.5) useful pH gradients can only be obtained at extremely low voltages, which in turn means extremely long focusing times. For example, hemoglobin could be focused in a Gly-Gly-Gly/His system but in 6 days and with a voltage gradient of 0.5 V/cm (this should be compared with the 400 V/cm of Radola et al., 1981). Already at 2 V/cm or higher the separation would be impaired since the distribution of the ampholytes in the system will no longer be symmetrical, the interpenetration of the two zones will be less pronounced and conductivity gaps will develop which will prevent passage of the protein ions. This study is fundamental for understanding the focusing process. According to these authors, this type of focusing (as well as buffer focusing, BEF) is definitely unpromising for analytical purposes, and even for preparative runs might create severe problems.

1.11.7. Immobilized pH gradients

As yet there are no data on grafted pH gradients, but I have the feeling this could be the newest example of a natural evolution process. My group has been working on this over the last few years, mainly on preparative aspects (Righetti and Gianazza, unpublished). On the other hand, in a recent discussion, it was brought to my attention that Rosengren et al. (1978, 1981) had indeed patented this process. So, after following the evolution of the species, from Kolin's 'artificial' pH gradients, through Rilbe's 'natural' IEF, up to Bier's computer simulations, it occurred to me that immobilized pH gradients could be made in a rather simple way. We have seen from Bier et al. (1981) that IEF with only two ampholytes is feasible, but unpractical because, even at minimal voltage gradients, the two ampholytes are swept away towards anode and cath-

TABLE 1.8

Input parameters and results of IEF simulations of real ampholytes. From Bier et al. (1981)

System components	pK_1	pK_2	ΔpK	ΔpI	Mobility coefficient $\times 10^4$	Current (mA/cm^2)	Results
β-Ala-His	6.83	9.51	2.68	2.59	2.30	0.098	Good
Arg	9.04	12.48	3.44		2.71		
β-Ala-His	6.83	9.51	2.68	1.53	2.30	0.108	Good
Orn	8.65	10.76	2.11		3.11		
α-OH-Asn	2.31	7.17	4.86	2.68	2.95	0.08	Good
His-Gly	6.27	8.57	2.30		2.40		
Asp	1.88	3.65	1.90	0.98	2.97	0.40	Good
m-ABA	3.12	4.74	1.62		3.01		
α-Asp-His	3.02	6.82	3.80	2.67	2.11	0.079	Good
His	6.0	9.17	3.17		2.85		
α-Asp-His	3.02	6.82	3.80	0.93	2.11	0.25	Fair
Isogln	3.81	7.88	4.07		2.96		
β-Ala-His	6.83	9.51	2.68	1.53	2.30	0.096	Good
Orn	8.65	10.76	2.11		3.11		
Gly	2.35	9.78	7.43	3.63	4.11	0.10	Good
Orn	8.65	10.76	2.11		3.11		
Gly-Gly	3.15	8.25	5.10	2.47	3.08	0.08	Good
β-Ala-His	6.83	9.51	2.68		2.30		
His	6.0	9.17	3.17	0.58	2.85	0.18	Good
β-Ala-His	6.83	9.51	2.68		2.3		
His	6.0	9.17	3.17	0.30	2.85	0.384	Poor
His-His	6.8	7.8	1.0		1.49		
His	6.0	9.17	3.17	1.09	2.85	0.096	Good
Tyr-Arg	7.55	9.80	2.25		1.58		
His-Gly	5.8	7.82	2.02	1.38	2.4	0.20	Good
β-Ala-His	6.83	9.51	2.68		2.3		
His-His	6.8	7.8	1.0	0.89	2.40	0.30	Fair
β-Ala-His	6.83	9.51	2.68		2.30		
Isogln	3.02	6.82	4.07	1.74	2.96	0.09	Good
His	6.0	9.17	3.17		2.85		
Tri-Gly	3.23	8.09	4.86	1.94	2.59	0.076	Good
His	6.0	9.17	3.17		2.85		

Table 1.8 continued

System components	pK_1	pK_2	ΔpK	ΔpI	Mobility coefficient $\times 10^4$	Current (mA/cm²)	Results
Tyr-Arg	7.55	9.80	2.25	2.08	1.58	0.079	Good
Arg	9.04	12.48	3.44		2.71		
Tyr-Arg	7.55	9.80	2.25	1.02	1.58	0.108	Good
Orn	8.65	10.76	2.11		3.11		
Tyr-Tyr	3.52	7.68	4.16	1.99	1.56	0.06	Good
His	6.0	9.17	3.17		2.85		

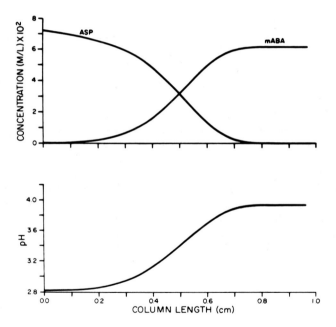

Fig. 1.22. Computer generated concentration and pH profiles for the focusing of aspartic acid (pI = 2.95, ΔpK = 1.91) and *m*-aminobenzoic acid (pI = 3.93, ΔpK = 1.62) at a current density of 0.40 mA/cm². The computer calculated voltage across the column was 0.59 V (from Bier et al., 1981).

ode, leaving in the middle a huge conductivity gap and a very steep pH step, which prevents proper IEF separations. Chemically grafted pH gradients are the answer to that, and might represent the last evolutionary step, like the Mule in Asimov (1952) classical Foundation trilogy. Let us take a gradient former: in the mixing chamber we will have, e.g., a 5% T, 5% C acrylamide solution, 30% sucrose, 10 mM acrylic acid (pK 4.5) and 5 mM vinyl imidazole (pK 6.5). In the reservoir we will pour 5% T, 5% C acrylamide, 5% sucrose, 5 mM acrylic acid and 10 mM vinyl imidazole. After filling the gel mould with a linear density gradient, upon polymerization the gel should contain, chemically grafted to the matrix, a linear gradient from 10 to 5 mM of acrylic acid (from anode to cathode) and a reciprocal linear gradient from 10 to 5 mM vinyl imidazole (from cathode to anode). These cross-gradients should automatically generate a linear pH gradient from pH 4.5 to 6.5, but with a minimum of conductivity and buffering capacity at midpoint between the two pK values. In this system, since the buffering species are immobilized, the proteins can be driven to equilibrium with the very high voltages typical of IEF. This could be the new revolution of focusing: first of all, we should be able to perform IEF in a chemically defined medium, of known molarity and ionic strength; secondly, by a proper selection of reactive, weak acids and bases, we could immobilize any pH gradient of 1 to 2 pH units for tailor-made separations; thirdly, since the buffers are grafted to the matrix (which could be a polyacrylamide, or a Sephadex, or an agarose gel), the focused proteins could be eluted isoionically (i.e., freed of macro-molecular and micromolecular contaminants!). I am taking bets that this approach should work, even though it sounds too good to be true. If it does not, this book will be subjected to Bradbury's drastic (1953) treatment: *Fahrenheit 451*.

CHAPTER 2

Preparative IEF

Over the past years, IEF has been shown to have many favorable attributes as a preparative method. Ideally, a good preparative method should provide high resolution separations of large amounts of material and give good recoveries, without altering the physico-chemical or biological properties of samples. IEF meets many of these criteria and also has additional advantages. For instance, unlike other methods, IEF has the almost unique advantage of affording improved resolution at increased load levels. I have already discussed how a high sample load is compatible with increased resolution and how both are favored by focusing in shallow pH gradients where a broad zone may accommodate more protein than the same pH increment in a wider pH range (see §1.5.9). Another advantage is that, by judicious selection of the pH range, many unwanted substances can be eliminated from the separation zone. Finally, since IEF is an equilibrium method, dilute samples are concentrated as they are separated.

Several approaches have been used to achieve the inherent potential of IEF, have confirmed theoretical predictions of load capacity and zone resolution and have given good recoveries of sample and of biological activity. Among the successful variations for preparative IEF are vertical columns with density gradient stabilization, horizontal troughs with self-generating density gradients, zone convection systems, continuously polymerized gels such as polyacrylamide or granulated gels such as Sephadex either in flat beds or in columns. Each system has its own particular advantages. Some capitalize on phenomena, such as isoelectric precipitation, which are anathema to many others. This section will describe many of the presently available techniques for preparative IEF, with particular reference to capacity, resolution, sample detection and recovery. Three basic systems will be discussed: (a) IEF in liquid media; (b) IEF in Sephadex beds; (c) IEF in polyacrylamide gel. They allow protein separation in the milligram to gram scale. As general reviews in the field of preparative IEF, the following articles should be consulted: Rilbe (1970); Winters and Karlsson (1976); Fawcett (1975b, 1976a); Bours (1976); Righetti (1975a); Radola (1976); Quast (1979) and Just and Werner (1979).

2.1. Preparative IEF in liquid media

2.1.1. The LKB columns

One of the earliest methods for preparative IEF is the use of columns in which the pH gradient is stabilized against convection by a vertical sucrose density gradient. LKB adapted a design by Svensson (1962b) and produced two columns of 110 ml (LKB 8100-1) and 440 ml (LKB 8100-2) capacity. The design of these columns is shown in Fig. 2.1. Approximately 5 to 25 mg protein per zone can be focused in the smaller column and approximately 20 to 80 mg per zone in the larger. The upper load limit depends on the number of components, their solubility at their pI and the shape of the pH gradient. For most proteins, the load should not exceed 5 mg/cm^2 of cross sectional area. These columns require a considerable amount of carrier ampholytes. For most purposes a minimum concentration of 1% (w/v) Ampholine is recommended, corresponding to

---

Fig. 2.1. Electrolysis column of 110 ml capacity equipped with platinum electrodes for isoelectric focusing and separation of proteins. This is an improved version of the one described earlier by Svensson (1962b). Isoelectric focusing takes place in compartment B. The cooling jacket A has an inlet at 7a and an outlet at 7b. One electrode 5 is at the top of the column in contact with the liquid in compartment B. The gas formed at this electrode escapes at 6. The central tube C, containing the other electrode, is connected during focusing to compartment B. The valve 3 is connected by a rigid teflon bar with the top 1a. The central tube is open when the stopper 3 is kept in the down position with the aid of the hook 2. The gas formed at electrode 1, rises through the central tube, escaping at 4. Before draining the column the central tube is closed with the valve 3, actuated by the hook 2. At the bottom of the column there is a plug 8 with an attachment for a capillary tube to enable fractions to be taken (from Vesterberg et al., 1967).

about 1.1 g and 4.4 g Ampholine, respectively, for each column. However, Ampholine ranges 2.5—4 and 9—11 are used at a concentration of 0.5% (w/v), since their conductivity is twice that of the other pH ranges.

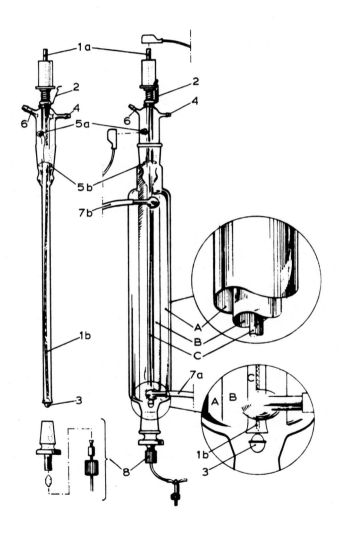

The density gradient. The density gradient is usually prepared with sucrose solutions, although other solutes may also be used. Ideally, the solute should have a high-solubility but low-viscosity in water. It should be non-ionic and have a high-density, a minimum density gradient of 0.12 g/cm^3. Finally, the material in the gradient should not react with the sample nor interfere with its subsequent detection. Analytical reagent grade sucrose is recommended for the preparation of the sucrose density gradient. The gradient can be formed by mixing a light solution containing only ampholytes and sample with a dense solution containing ampholytes, sample and sucrose. These solutions may be mixed in certain proportions to produce a series of solutions of increasing densities. These are then layered sequentially on top of the dense electrolyte solution in the lower chamber of the electrofocusing column. The layers are allowed to diffuse into one another to produce a fairly smooth gradient. Alternatively, more reproducible density gradients may be prepared with a gradient mixer such as LKB 8121 and LKB 8122 Ampholine gradient mixers, for use with the 110 ml and 440 ml columns, respectively. These mixers (Svensson and Pettersson, 1968) provide linear density gradients from two equal volumes of light and dense solutions having a density difference of 0.2 g/cm^3. This density difference corresponds to light and dense solutions containing 5 and 50% sucrose, respectively. Other solutes, such as sorbitol (Vesterberg, 1971) or glycerol (Vesterberg and Berggren, 1966) may be used as anticonvective media, especially when performing IEF in alkaline regions, where sucrose ionizes. Ethylene glycol (Ahlgren et al., 1967; Kostner et al., 1969) can be used at concentrations of 60–70% in the dense solution, because of its lower density. Other useful chemicals include mannitol and polyglycans, such as Dextran (Leise and LeSane, 1974) and Ficoll (Sherbet et al., 1972; Sherbet and Lakshmi, 1973) from Pharmacia (Uppsala, Sweden). Mixed sucrose-Ficoll gradients are also used for cell separations (Boltz and Todd, 1979) since they can be made isoosmotic (303 mOsm) in a range of densities from 1.035 to 1.055 g/cm^3. A 1 to 10% Ficoll gradient corresponds to a reciprocal 5.0 to 7.0% sucrose density gradient. Fredriksson and Pettersson (1974) and Fredriksson (1975) have described the use of deuterium oxide as a medium for density gradients. D$_2$O gradients have so far been used only in the small (1.5 ml) spectrophotometric cell of Jonsson et al. (1973). These gradients are formed in situ by free interdiffusion of three D$_2$O solutions for 3 min

(Rilbe and Pettersson, 1968). While the density increment obtainable with D_2O is rather small (0.1 g/cm^3), its viscosity at 5 °C is, however, about 50% lower than the viscosity of an aqueous sucrose solution of similar density. This means that a more favorable distribution of the field strength along the column and shorter focusing times can be obtained. However, pH readings in D_2O gradients are higher than in sucrose gradients, due to shifts of the asymmetry potential of the glass electrode in D_2O (compared with H_2O). It remains to be seen whether or not D_2O gradients will be of general applicability also in preparative columns.

The density gradient with ampholyte and sample is layered on top of a dense cushion containing an electrolyte at the bottom of the column. The second electrolyte is now floated on top of this gradient (see Fig. 2.2). Table 2.1 lists three types of density gradient stabilization (sucrose,

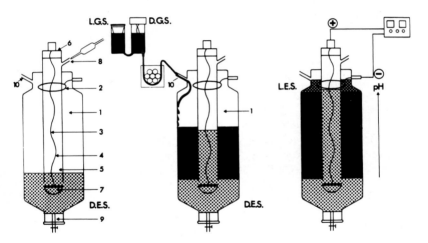

Fig. 2.2. Setting up the Ampholine Electrofocusing column. Left: the dense electrode solution (D.E.S.) is layered at the bottom of the column with the valve (7) of the central electrode compartment (5) open. This solution is pipetted in via the gas escape (8), to avoid contamination of the separation chamber. Center: the density gradient is introduced in the column via the sample inlet (10) with the aid of a peristaltic pump. The density gradient is best produced with a gradient mixer, shown schematically, L.G.S. (light gradient solution) and D.G.S. (dense gradient solution). Right: the light electrode solution (L.E.S.) is now floated on top of the gradient, until the platinum loop (2) is submerged. The electrodes are then connected to a power supply. (Courtesy of LKB Produkter AB.)

TABLE 2.1

Density gradient solutions for the 110 ml and 440 ml LKB columns for three types
of stabilizing media (sucrose, glycerol and ethylene glycol)

Density gradient solution	110 ml Column LKB 8100-1	440 ml Column LKB 8100-2
Sucrose stabilization		
Dense gradient solution		
Sucrose	27 g	107.5 g
Volume of H_2O + Ampholine (+ sample if added)	37 ml	150 ml
Total volume	54 ml	215 ml
Concentration of sucrose	50% (w/v)	50% (w/v)
Light gradient solution		
Sucrose	2.7 g	10.75 g
Volume of H_2O + Ampholine (+ sample if added)	53 ml	207 ml
Total volume	54 ml	215 ml
Concentration of sucrose	5% (w/v)	5% (w/v)
Glycerol stabilization		
Dense gradient solution		
Glycerol − 87% (v/v)	30 ml	118 ml
Volume of H_2O + Ampholine (+ sample if added)	24 ml	97 ml
Total volume	54 ml	215 ml
Concentration of glycerol	60% (w/v)	60% (w/v)
Light gradient solution		
Total volume of H_2O + Ampholine (+ sample if added)	54 ml	215 ml
Concentration of glycerol	0%	0%
Ethylene glycol stabilization		
Dense gradient solution		
Ethylene glycol − 100% (v/v)	40.5 ml	161 ml
Volume of H_2O + Ampholine (+ sample if added)	13.5 ml	54 ml
Total volume	54 ml	215 ml
Concentration of ethylene glycol	75% (v/v)	75% (v/v)
Light gradient solution		
Total volume of H_2O + Ampholine (+ sample if added)	54 ml	215 ml
Concentration of ethylene glycol	0%	0%

TABLE 2.2

Electrode solutions for the 110 ml and 440 ml LKB columns to be used either with the cathode or with anode uppermost

Electrode solution	110 ml Column LKB 8100-1	440 ml Column LKB 8100-2
Anode at top of column		
Cathode solution (pH − 11.7 approx.)	15 g sucrose + 10 ml H_2O + 6 ml 1 M NaOH	48 g sucrose + 30 ml H_2O + 20 ml 1 M NaOH
Total volume	25 ml	80 ml
Concentration of sucrose	60% (w/v)	60% (w/v)
Anode solution	1.5 ml 1 M H_3PO_4 + 8.5 ml H_2O	6 ml 1 M H_3PO_4 + 34 ml H_2O
Total volume	10.0 ml	40 ml
Cathode at top of column		
Cathode solution	25 ml 1 M NaOH + 7.5 ml H_2O	10 ml 1 M NaOH + 30 ml H_2O
Total volume	10.0 ml	40 ml
Anode solution (pH − 1.2 approx.)	15 g sucrose + 12 ml H_2O + 4 ml 1 M H_3PO_4	48 g sucrose + 38 ml H_2O + 12 ml 1 M H_3PO_4
Total volume	25 ml	80 ml
Concentration of sucrose	60% (w/v)	60% (w/v)

glycerol or ethylene glycol) for the 110 ml and 440 ml columns. Table 2.2 gives suggestions for electrode solutions to be used in both columns either when the anode or when the cathode is uppermost.

Electrode solutions. Choice of electrode polarity. In order to protect the carrier ampholytes from anodic oxidation and cathodic reduction the electrodes are surrounded by an acid and an alkaline solution, respectively. During IEF, the acid and the base are drawn to the respective electrodes, while the ampholyte buffers at the two ends of the column are repelled from the electrolyte chamber. In general, dilute strong acids (1% phosphoric or sulfuric acids) are used as anolyte, while dilute bases (2% ethanol amine, ethylendiamine or 8% NaOH) are used as catholyte. However, when narrow pH ranges are used (e.g., 1 pH unit span) it is suggested that progressively weaker acids and bases are used, so that their pH is not too distant from the pH interval of the IEF fractionation, for prevention of big pH and conductivity jumps at the interface between electrolytes and separation column. Thus, aspartic, glutamic, acetic and citric acids can be used at the anode, and arginine, lysine, Tris or histidine at the cathode.

For the choice of electrode polarity, it is preferable to have the anode at the bottom of the column to minimize disturbances due to ionization of sucrose by the strong base at the cathode (Flatmark and Vesterberg, 1966). The pI of the components of interest should be considered too: electrodes should be selected so that they focus in the lower half of the column, close to the elution port. This will reduce diffusion during subsequent elution which is performed in the absence of the electric field. However, the choice of electrode polarity depends also on conductance considerations. It is known that the conductivity in a fully established pH gradient shows a minimum between pH 6 and 7 (see § 1.5.5). Another factor that influences the conductivity distribution is the viscosity change along the density gradient, in that increasing viscosity leads to decreasing conductivity. Normally, the polarity is chosen so as to have these two effects counteracting each other, by placing the conductivity minimum at the upper end of the column. Thus when running pH ranges lower than pH 6, the cathode should be at the top of the column and, when using ranges higher than pH 6, the anode should be at the top. In particular, when using Ampholine pH 9−11, it is a must to have the anode at the top of the column. At the opposite extreme, when using Ampholine pH 2.5−4, the cathode must be at the top. However if, in preparative IEF of crude materials, a precipitate is known to occur, the polarity should be chosen so as to have the precipitate forming at the lower end of the column. Since, in this last case, the elution capillary might be clogged, stepwise elution from the top should be used (see § 4.1).

Sample application.   Because IEF is an equilibrium method, the way the sample is applied is not as critical as in other forms of electrophoresis where a thin starting zone is necessary. The sample may be incorporated into either the dense or the light sucrose−ampholyte solution, or into both. Since samples are concentrated during the experiment, the volume of sample is not critical and may be as much as 80% of the final column volume. Alternatively, the sample may be introduced as a narrower zone close to the place where it is expected to focus. In this way the run will be shortened, and exposure to pH extremes avoided. As has been discussed in the previous chapter, the upper load for the sample depends on several factors. These include the geometry of the column, the number and distribution of components in the sample, their solubility at their pI, the pH range and level of carrier ampholytes and the strength of the electrical field. Since large cross-sectional areas and narrow pH ranges tolerate higher loads, the larger column which has the same height as the smaller column usually carries proportionally more sample than might be indicated by the four-fold difference in volume. The amount of salt and buffer in the sample should be low and should not exceed 0.5 mM for the 110 ml column and 1.5 mM for the 440 ml column. High salt concentrations will create high current densities at the beginning of the experiment and also generate acid at the anode and base at the cathode. A good solution is to dialyze the sample against 1% Gly (Vesterberg et al., 1967; Wadström, 1967) or 1% Ampholine solutions at appropriate pH. Due to its property of 'poor carrier ampholyte', Gly is essentially uncharged in the pH 4–8 range, but contributes to an increase of dipole moment of the solvent, thus stabilizing proteins. Prior to dialysis, the tubing should be washed in hot, 10% acetic acid followed by thorough rinsing in distilled water and soaking in the solution intended for dialysis. When dialyzing against Ampholine, since most proteins are generally more stable in weak alkaline solutions, the pH 6–8, 7–9 and 8–10 ranges are often appropriate.

Electrolysis conditions.   The total power applied to the column should not exceed its capacity to dissipate the heat generated. Since the separation column is a hollow cylinder, with cooling on all sides, heat dissipation in the LKB columns is very efficient. With a temperature of circulating coolant of 2–4 °C, it is customary to run the small column (110 ml capacity) at about 3–4 W and the large column (440 ml) at about 8–10 W. However, with the introduction of constant wattage power supplies, these upper limits have been increased (to a maximum of 15 W [1,600 V_{max}] for the 110 ml column and 30 W [2,000 V_{max}] for the 440 ml column) (Winter and Karlsson, 1976). It is advisable not to thermostat the column below 2–4 °C (Haglund, 1971). The temperature coefficient of the viscosity and conductivity of the sucrose–Ampholine mixture is nearly constant down to around 1 °C but rises steeply below this point.

Consequently, small temperature differences across columns maintained below 2 °C may cause wavy zones and impair resolution.

The current is highest at the beginning of the experiment when the ampholytes are randomly distributed throughout the column; consequently the initial voltage must be carefully controlled to avoid excessive heat. Under normal conditions with a level of 1% (w/v) Ampholine, an initial voltage of 400–500 V may be appropriate. Later, when the conductivity decreases as the ampholytes migrate to their pI, the voltage may be increased up to 1,600–2,000 V to improve resolution. This is done automatically with constant wattage power supplies. The system reaches equilibrium when the current stabilizes at a low plateau. The actual time required to achieve equilibrium will, of course, depend on the actual electrophoretic conditions and the temperature of the circulating coolant, but it is usually between 24 and 72 h with both columns when run with conventional power supplies. This time may be reduced by using constant wattage power supplies.

Columns with narrow pH ranges or more than 1% ampholyte usually require considerably longer focusing periods. The correct electrolysis period can only be determined experimentally. It is emphasized that the final movement to equilibrium of carrier ampholytes and proteins occurs so slowly that the final drop in conductivity may not be apparent. Consequently, it is advisable to err on the safe side and continue electrolysis for a few hours after the apparent plateau.

When the system has reached equilibrium the contents of the column are either drained from the bottom or displaced upwards with a denser solution (see Fig. 2.3). It is customary to collect fractions of 1–4 ml and to determine the amount of protein and the pH of each fraction. This may be done manually or with the aid of flow cells which monitor UV absorbancy and pH. Ideally, the pH of the fractions should be determined at the same temperature as that prevailing during IEF (see §4.5). Elution flow rates of 60 ml/h for the 110 ml column and 240 ml/h for the 440 ml column have proved to be suitable.

Although these columns have proved useful for many purposes, they are subject to some technical problems which limit the amount of sample that may be focused and impair the resolution actually obtained after fractionating the column contents. Most proteins tend to be least soluble at their pI and may precipitate from solution. Another form of precipita-

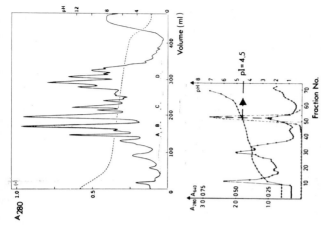

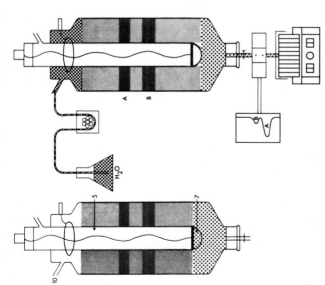

Fig. 2.3. Emptying the Ampholine Electrofocusing Column. Left: at the end of the experiment, the central valve (7) is closed and the light electrode solution removed via inlet (10). Center: a tubing connected to a peristaltic pump and to a distilled water reservoir is plugged into inlet (10). The upper part of the column is filled with H_2O and then inlet (10) is tightly stoppered. When possible, the elution capillary is connected to a UV flow cell and a recorder. Elution by a pump at the column top is superior to elution by gravity flow or by a pump connected with the elution capillary at the column bottom. Right: examples of elution profiles and pH gradients obtained by this method. The graph on the upper right side shows how to determine the pI from a UV peak. (Courtesy of LKB Produkter AB.)

tion can occur when the density of a focused band of a large amount of protein exceeds that of the supporting gradient. Other problems arise from diffusion during lengthy elution periods in the absence of the electric field and from remixing of zones in the lower chamber as the column contents merge into a continuous stream. Since additional mixing of zones can also occur in the parabolic flow profile of the capillary path between the column and fraction collector, this path should be kept as short as possible. If flow cells are used, the volume of their chambers should be kept to a minimum. Another source of diffusion is from peristalsis generated by pumps, especially when they operate on the eluting capillary. This situation may be avoided by emptying the column by gravity. Alternatively the pump tube may be inserted into the top of the column (inlet No. 10 in Fig. 2.2) so that the water displaces the column content downwards. The rate of outflow will thus be dictated by the speed of the pump.

Boddin et al. (1975) have devised a method by which the column content may be eluted under the electric field. They adapted a commercial polyacrylamide gel apparatus by closing the lower end of the column with a simple one hole stopper and separated the sucrose density gradient from the lower electrode with a membrane which allowed direct electrical contact.

2.1.2. The ISCO columns

Instrumentation Specialty Company (ISCO) has produced two columns for IEF, an analytical or small-scale preparative unit (model 210) and a preparative version (model 630). The former has a bore of 1 cm and is about 30 cm long. The design of this column is shown in Fig. 2.4. The maximum gradient volume is 23 ml and the maximum load capacity is about 10 mg of protein. When cooled at $2\,^{\circ}C$ an electrical load of up to 15 W can be applied to the column. The larger model is built on the same principle and can accommodate gradients up to 160 ml in volume and 32 cm in length. When cooled at $2\,^{\circ}C$ it can be run at ca. 20 W. An interesting feature of both systems is that the electrode chambers are built coaxially around the separation column. Two concentric membranes separate each electrode from the gradient to inhibit the introduction of electrolysis products and pH changes in the column. This membrane also prevents net transfer of gradient components into or out of the inner column.

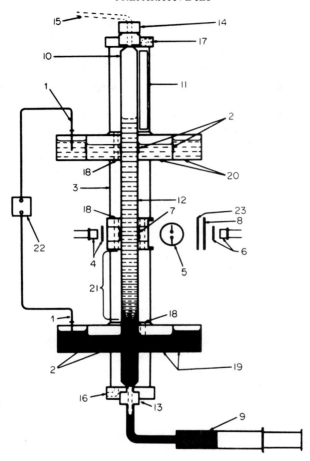

Fig. 2.4. Schematic diagram of the ISCO model 210 density gradient electro-phoresis column. Key to the drawing: 1. Platinum wire electrode; 2. Semi-permeable membranes; 3. Coolant jacket; 4. Filter and measuring photocell; 5. Low-pressure mercury vapor lamp; 6. Filter and reference electrode; 7. Quartz windows; 8. Adjustable shutter; 9. Motor-driven syringe; 10. Transparent central tube; 11. Centimeter scale; 12. Teflon central tube (1 cm. i.d.); 13. Lower column support; 14. Removable cap for insertion of gradient and sample; 15. To sample collector; 16. Coolant inlet; 17. Coolant outlet; 18. Coolant passage; 19. Lower electrode chamber; 20. Upper electrode chambers; 21. Gradient column; 22. Constant current power supply; 23. UV absorbance standards. (From Leaback and Wrigley, 1976.)

Another advantage is that the progress of the experiment may be monitored by scanning the column. To do this, the current is switched off and the entire gradient column is raised past a densitometer to produce an absorbance profile at one or more wavelengths. Thus progress on sample separation can be recorded at selected intervals and the attainment of equilibrium confirmed. In addition, a special applicator allows the sample to be introduced into the center of the column at any desired time. Thus a preformed pH gradient can be scanned before sample application to provide a baseline for Ampholine absorbance (Righetti and Drysdale, 1971). This is achieved by injecting a dense sucrose solution into the bottom of the column to force the gradient through a flow-cell and into a reservoir. Since scanning and unloading procedures are performed with pulseless syringe pumps, and the apparatus has no mixing chambers, sharp zone recoveries are possible without undue problems of convection, turbulence or laminar flow associated with other systems. The principle of these columns has been described by Brakke et al. (1968). This type of intermittent scanning is different from the method of Catsimpoolas (1973c) for in situ scanning of columns while still under voltage. This technique, called transient state IEF, can provide much useful information on the kinetics of IEF (Catsimpoolas, 1973c–f). The use of the ISCO columns for IEF fractionation was first described by Grant and Leaback (1970) and its methodological aspects extensively reviewed by Leaback and Wrigley (1976).

Fawcett (1975b) has pointed out that similar experiments can be conducted with much simpler apparatus. For example, he has built modified U-tubes, with a quartz column in one limb (separation arm) and a three-way tap at the bottom bend. This tap is used for filling and emptying the column. The separation limb is fitted with a cooling jacket. The column is emptied by removing the upper electrode and pumping dense sucrose solution into the base of the quartz tube to displace the gradient upwards. A Uvicord II detector head is fitted to the upper part of the quartz tube, so that a UV scan is obtained during elution. An adapter moulded of silicone rubber attached to the column top allows fractions to be collected. Fawcett (1975b) has also built multiple U-tube units (such as two and three U-tube apparatus) to be used when precipitates occur or when high concentrations of impurities spread into focused zones. These units are fitted with stop cocks in appropriate places, to facilitate removal of focused bands. These columns are usually cooled by immersion in a tank.

2.1.3. The Poly-Prep 200

Catsimpoolas et al. (1980) have adopted the Poly-Prep 200 from Buchler Instruments (Fort Lee, New Jersey) for IEF fractionation. A diagram of

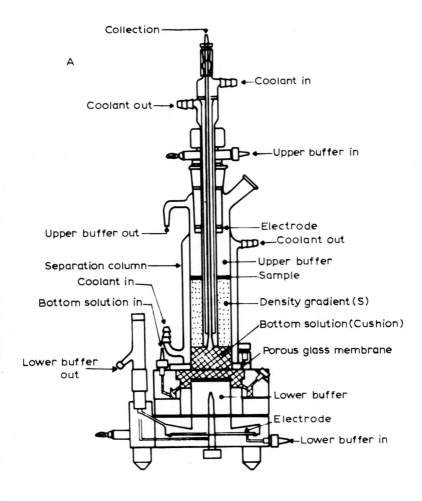

DENSITY GRADIENT ISOELECTRIC FOCUSING

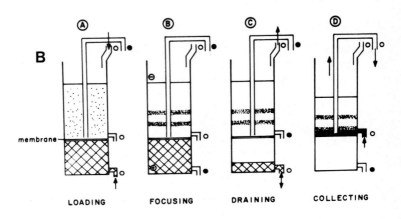

Fig. 2.5. Drawing of Poly-Prep 200 (Buchler Instruments). (A) Schematic diagram of the assembled apparatus. (B) Experimental stages of isoelectric focusing in a density gradient. The arrows indicate the direction of liquid flow and the circles the open (o) and closed (●) position of the ports. (From Catsimpoolas et al., 1980.)

the assembled apparatus is shown in Fig. 2.5A. It consists of four parts: the column assembly, the internal glass-cooling tube and two lucite blocks for the lower buffer electrode section. When assembled, the apparatus has a glass separation chamber with outer and inner cooling. An annulus (hollow cylinder) is thus produced having a cross-sectional area of 17.6 cm. A rigid porous glass membrane separates the lower buffer reservoir from the separation chamber. Inlet ports are available at the bottom of the separation column and a narrow-bore outlet is provided in the center of the inner cooling piece. Thus, the bottom of the gradient can be pumped upwards through this narrow channel and into the fraction collector tubes by a dense chase solution. This apparatus was originally designed for preparative gel electrophoresis by Jovin et al. (1964). Its operation in IEF involves four stages, schematically shown in Fig. 2.5B: (A) generation of the density gradient and loading of the sample and basic and acidic electrolytes; (B) focusing; (C) draining of the acidic (lower) electrolyte and washing of the lower chamber and (D) collection of the separated components using a chasing solution. Separation of erythrocyte nucleoside phosphorylase has been reported.

2.1.4. Rilbe's columns

In addition to the type of vertical columns produced for LKB (Svensson, 1962b), Rilbe and his colleagues have developed new vertical columns and have also improved the design of horizontal multi-compartment electrolysers. In 1973 in Glasgow, Rilbe described a system for preparative IEF in short density gradient columns with vertical cooling. Following Philpot's (1940) early suggestion of using short and thick columns with vertical heat dissipation, Rilbe and Pettersson (1975) built two types of columns which differ essentially in the method of cooling. Both are extremely short and thick. In one, the distance between electrodes is only 1.55 cm and the cross-sectional area 283.5 cm^2 with a column volume of 440 ml. The electrodes are sheets of an alloy of 75% palladium and 25% silver and are soldered to brass plates at the column extremities. Since the electrodes are also the major cooling surfaces, a vertical temperature gradient is created in the separation column. Equilibrium is achieved in a very short time, 30 min, in a final potential gradient of 75 V/cm. Because of its geometry, the column must be filled and emptied from a lateral position. Drawbacks of this type of cell include possible contamination from the metal electrodes by anodic oxidation and rapid diffusion of focused bands during the time required to turn the column through 90° prior to fractionation.

The second type of column is similar in principle and has a capacity of 110 ml and smaller electrodes of platinum wire. This column is vertically cooled through cellophane membranes by the electrolytes which are chilled and stirred. Focusing is usually achieved in 90 min at a final potential difference of 125 V. This column must also be filled and emptied from a lateral position. In this type of cell, Rilbe and Pettersson (1975) fractionated more than 1 g of sperm whale myoglobin in a pH 7–9 gradient in 3 h. The main band (Mbl, pI 7.6 in the ferrous form or pI 8.3 in the ferric form) contained about 800 mg protein, an appreciable amount to be carried by a density gradient. The great advantage of this column type is therefore its high load capacity and the relatively short focusing time.

Rilbe's group in Gothenburg has continued along this line and described recently a new version of a density gradient column based on the principle of horizontal separation, followed by a vertical fractionation (Jonsson et al., 1980). A schematic drawing of this system can be seen in Fig. 2.6. The column has a volume of 112 ml and allows a rapid, low-resolving pre-focusing over its shorter dimension (20 mm), followed by a high-resolving fine-focusing over the longer dimension (112 mm) when the column is turned through 90°. This is possible due to its rectangular cross-sectional shape and to a special design of the electrode vessels. The entire IEF separation, including the fine-focusing in the vertical position, is performed in just four hours, with a total capacity, in terms of maximum sample load, of at least 1 g.

2.1.5. Zone convection IEF according to Valmet

In 1969 Valmet described a new method for preparative IEF based on a new electrophoretic principle, called 'zone convection IEF'. A unique

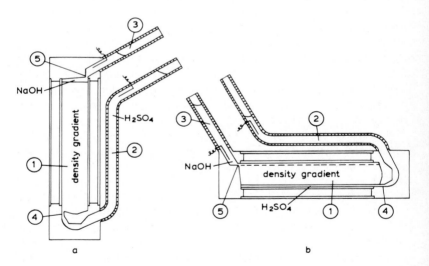

Fig. 2.6. Schematic drawing of a parallelepipedic density-gradient column, connected for isoelectric focusing either in the vertical (a) or in the horizontal (b) position. 1 = Separation compartment; 2 and 3 = inclined tubes containing the electrodes; 4 and 5 = slits facilitating filling and emptying of the separation compartment. (From Jonsson et al., 1980.)

feature of this method is that it does not require any anti-convective medium. Separations are conducted in a series of linked U tubes. The separation cell (shown in Fig. 2.7) is made of plastic with a wall thickness of 1 mm. It resembles a series of Tiselius cells with a cross section of 3 × 40 mm. Each U-tube unit is about 10 mm high. The channels are

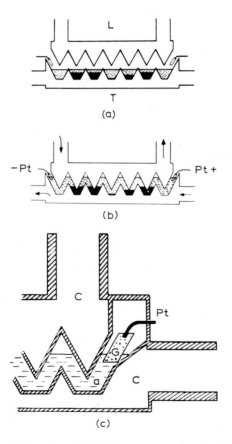

Fig. 2.7. Schematic drawing of the Valmet zone convection electrofocusing apparatus (a) the lid (L) and the trough (T) before or after the experiment; (b) lid and trough put together, as during the experiment; (c) detail of one of the electrode compartments. (From Valmet, 1969, courtesy of LKB Produkter AB.)

spaced 10 mm apart. The cell consists of two separate units, the trough and the lid. The trough is a shallow rectangular box with a corrugated bottom. The lid has similar indentations so that when it is lowered on top of the trough, the interdigitating projections force the liquid in each pocket to overflow, thus closing the electric circuit on the cell. The electrodes are attached to both ends of the apparatus. Band stabilization is achieved by the Ludwig (1856) Soret (1879) effect. As a protein focuses into a pocket, it concentrates at the bottom against the cold wall, thereby generating a vertical density gradient within the channel, perpendicular to the current flow.

At the end of the run, the lid is removed thus causing the liquid to flow back into each U-tube so that the trough now acts as a fraction collector. The concentrated protein fractions can be recovered from the bottom of the U-tube with the aid of a peristaltic pump, allowing the remaining carrier ampholytes to be reused for other separations. In the case of the Valmet trough, isoelectric precipitation is an advantage rather than a disadvantage, since zone detection is simplified and samples may be recovered by centrifugation. On the other hand, non-isoelectric precipitation can result in serious contamination.

Valmet has also used this apparatus for fractionating ampholytes into narrow pH ranges. By using a 30-pocket cell, he was able to separate the pH 3–10 Ampholine into a number of fractions each covering about 0.2–0.3 of a pH unit. The sample load in Valmet's horizontal trough is much higher than in conventional vertical density gradient apparatus. He has reported separation of 350 mg serum protein in 50 ml of 1% Ampholine. The high sample load, the economically favorable Ampholine protein ratio, and other unique features make this apparatus particularly attractive for large-scale IEF. Unfortunately, it has not become available commercially because of technical difficulties in construction.

This method has been used by Bodwell and Creed (1973) for separating a wide variety of proteins. Fawcett (1975b) has used a similar approach by modifying an earlier design of Kalous and Vacik (1959). Although he achieved discrete fractionation of ovalbumin in a wide pH range, he did not obtain good resolution with myoglobin components in alkaline pH ranges, apparently because of poor cooling and electro-osmotic effects. An apparatus built on the same principle of the Valmet (1969, 1970) trough has also been described by Rose and Harboe (1970).

It consists of a series of U-shaped compartments with outlets in the bottom, linked in a wavy form resembling a sea-serpent. Kiryukhin (1972) has built a Valmet apparatus of extended length, by increasing the number of channels from 30 to 48. This modified unit has been used by Troitski et al. (1974a) and Tolkacho (1974) for isolation of pure albumin from rabbit serum.

2.1.6. Zone convection IEF according to Talbot

Fortunately, Valmet's concept of zone convection IEF was not forgotten. Talbot (1975) devised a more practical apparatus embodying Valmet's idea. This cell, besides eliminating difficulties associated with non-iso-

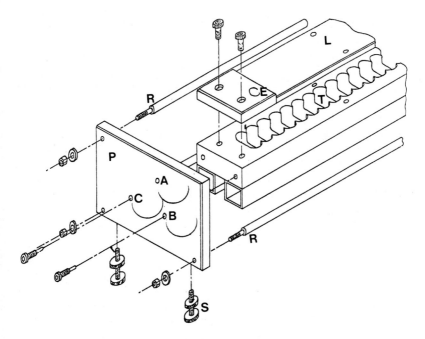

Fig. 2.8. Exploded diagram of part of the horizontal trough apparatus of Talbot for isoelectric focusing. (T) separation trough; (L) lid; (E) electrode – locating plate; (P) end plate; (R) stainless steel tie rods; (S) levelling screws; (C) pivot pin; (A,B) locking screws. (From Talbot and Caie, 1975.)

electric precipitation of material, provides easy access to the pH gradient and also incorporates a facility for auto-fractionation of the gradient at the end of the experiment. The separation trough is fashioned from a block of aluminium and is 1 inch high, 2½ inches wide and 24 inches long. The trough has a corrugated base and sides which form 52 pockets. The trough is coated with a varnish for electrical insulation and has cooling coils in its base for effective heat dissipation through direct metal contact. It is fitted with a lid to minimize evaporation and has electrode plates at each end. The block is held in a supporting framework, and can be pivoted about pins and locked either in a 45° or in a horizontal position. For operation, 30 ml of 1% (w/v) Ampholine is poured into the trough and distributed throughout the corrugations by tilting the whole apparatus from end to end. With the trough in the 45° position, the solution bridges the corrugations and forms a continuous electrical path along the length of the cell. 1% (v/v) phosphoric acid is used at the anode and 1% TEMED at the cathode. Fractionations are generally performed at 4 °C. The pH gradient is established before insertion of the sample. An initial voltage of 450 V at 1 mA is applied and stable pH gradients are formed in about 24 h. After the sample introduction, focusing is continued for a further 72 h. Figure 2.8 is an exploded diagram of the apparatus. The focused gradient is fractionated simply by returning the cell to the horizontal position so that the solution no longer bridges the trough corrugations and the focused fractions are separated in the bottom of each compartment. The pH of each fraction may be measured directly in the apparatus with a combination microelectrode. Working with labeled preparations from foot-and-mouth disease virus, Talbot and Caie (1975) have reported sample recoveries of 80 to 100%.

2.1.7. Zone convection IEF according to Denckla

An interesting variant of the Valmet apparatus has been patented by Denckla (1975, 1976). A drawing of this chamber is shown in Fig. 2.9. The cell is subdivided in a series of U-channels by short walls glued to the floor of the chamber (the alternate walls, labeled '9'). In turn, each U-compartment is separated into two limbs by a high wall (10), cemented to the side walls, which extends only proximate to the receptacle bottom (about 1 mm therefrom). Both series of walls are essentially hollow, having a channel for the passage of coolant. This system thus has two

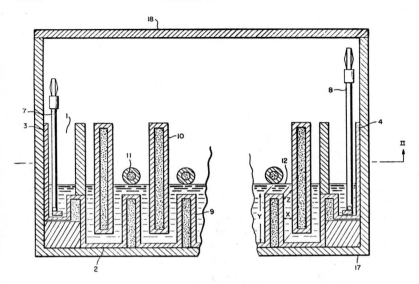

Fig. 2.9. Schematic drawing of the Denckla apparatus. (1) Rectangular receptacle for accomodating the ampholyte mixture to be separated; (2) bottom of the tank; (3,4) two opposite end walls; (7,8) electrodes (platinum foils or grids); (9) first alternate walls extending to the receptacle bottom; (10) remaining alternate walls extending proximate to the tank bottom (all these walls are hollow and accomodate circulating coolant through the entire tank); (11) additional cooling means arranged above first alternate walls 9; (12) surface of ampholyte solution during normal use; (17,18) non-conductive shield and cover, respectively; (y) height of the U-tube arms; (x,z) width and depth of the chamber, respectively. (From Denckla, 1975, 1976.)

basic characteristics: it is very well cooled and its design counteracts efficiently electroendoosmosis (the narrow gap between series of walls, labeled '10', and the floor is meant to block movement of the Helmholtz water layer). This device, as compared to the Valmet cell, allows an increase of ten-fold electric field strength per unit distance; three-fold resolving power; six-fold power dissipation per unit volume; eight-fold in speed and 500-fold in weight of ampholyte material which can be fractionated. This cell can be run at up to 60,000 V and can dissipate as much as 6,000 W (3 W/ml of solution). The key feature of the Denckla technique is also the concept of 'cascade separation', whereby a series of three successively smaller units having ten compartments each is utilized sequentially for purification of proteins in extremely shallow pH gradients. The three chambers are built so that the ratio of the volume of each unit to the next smaller one is approximately equal to the number of chambers in each unit. Thus, the content of one chamber in the bigger unit can be used to fill up all the ten channels in the next smaller unit, without substantial dilution of the carrier ampholytes. The resolution thus achieved (0.01 pH units) is comparable to that obtainable in a single Valmet cell composed of 700 U-compartments. Denckla (1974, 1977) has used this apparatus for purification of hormones from pituitary extracts.

2.1.8. Zone convection IEF in closed coils

An apparatus built on the principle of the Valmet (1969) column has been described by Macko and Stegeman (1970). IEF takes place in a coil of polyethylene tubing, 100 cm long, 4 mm inner diameter, 1 mm wall thickness. This tubing is coiled, with the aid of a double-face tape, around a copper tube of 20 mm diameter. The coiled tubing is held in place by a piece of wire screen fastened at the ends with strings. This is in fact an apparatus for zone convection IEF, where each turn of the spiral repre-

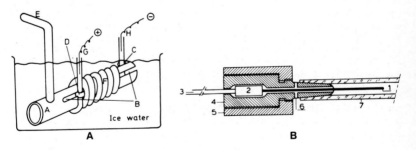

Fig. 2.10. Apparatus for free isoelectric focusing of proteins. A, brass tube, 25 × 170 mm; B, slit, 4 × 120 mm; C, fixed slot for tube F; D, movable slot for tube F; E, supporting bar; F, polyethylene tube (4 mm i.d., 6 mm o.d., length 1 m); G and H, platinum electrodes. (From Chilla et al., 1973.) (B) Schematic drawing of platinum electrode used in free isoelectric focusing. 1, platinum wire; 2, soldered connection; 3, cable to power supply; 4, Perspex holder; 5, Perspex protective collar (dimensions 18 × 25 mm); 6, gas outlet; 7, polyethylene tubing. (From Bours, 1973c.)

sents one compartment of the Valmet apparatus. The coil is filled with
10 ml of 10% sucrose and 1% Ampholine. The total sample load is in the
order of 10–15 mg. The two tube extremities, which are allowed to
stand up vertically, are filled with anolyte and catholyte. During IEF the
coil is immersed in a stirred ice bath or a 2 °C thermostat, with the loose
ends above the water level. After IEF, the coiled tubing is cooled down
quickly, preferably by immersing into liquid nitrogen, after which it is
cut into segments and the frozen contents emptied into small vials.

A drawing of the coiled IEF cell, as described by Chilla et al. (1973) is
shown in Fig. 2.10A. Figure 2.10B depicts the platinum electrode con-
nections, specially designed for safety measures by Bours (1973c). This
author uses a polyethylene tubing of 4 mm i.d., 6 mm o.d. and 2.5 mm
long: in this system, focusing is completed after 27 h at 3,800 V and 850
(initial) to 160 (final) μA. By free IEF in coils, separations of eye lens
crystallins (Bours, 1973a,b), glucose-6-phosphate dehydrogenase (Chilla
et al., 1973), ribose phosphate isomerase (Domagk et al., 1973, 1974)
and phosphoryl phosphatase (Zech and Zurcher, 1973, 1974) have been
reported.

2.1.9. Zone convection IEF in open coils

A drawing of the apparatus, constructed by Quast (1977), is shown in
Fig. 2.11. The principle is the same as the coiled cells described by
Macko and Stegeman (1970) and by Bours (1973c) but it differs in
several aspects, namely reusability of the helix, direct access to each of
the separated fractions and sample capacity in the range of grams. During
operation, the helix is immersed in a thermostated bath of fluorocarbon
liquid (Teflon); sample, solvent and ampholytes fill the turns of the helix
completely up to the base of the straight tubes. Separation proceeds as
described by Valmet (1969); upon termination of the run, the liquid in
the upper part of the turns of the helix is withdrawn simultaneously
after the run by small plastic tubings fitted to syringes fixed to a frame.
Air bubbles thus separate the liquid of neighboring turns, and the helix
can be regarded as a series of discrete U-tubes similar to the sea-serpent
shaped apparatus of Rose and Harboe (1970). The helix built by Quast
consists of 37 turns, each of 4.5 ml volume, corresponding to a
resolution of 37 fractions. The apparatus is quite compact in size, since
the glass tube has a 6 mm i.d., 9 mm o.d. with a diameter of the turns

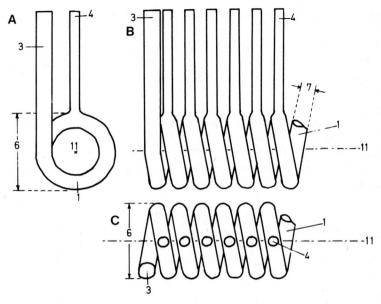

Fig. 2.11. Quast's open coil IEF apparatus. (A) End-on view of the glass helix; (B) Side-on view of the glass helix; (C) Top view of the glass coil. (1) turns of the helix, of equal diameter (6); (11) helix axis; (3) end of the helix, with electrode lodging; (4) vertical straight glass tubes on each turn of the helix; (7) diameter of the glass coil. (From Quast, 1977.)

of the helix of 4.5 cm. However, the total length of the tube between the electrodes, when extended, is 3.6 m which means that very high voltages must be applied (Quast has used as much as 25,000, i.e., 70 V/cm) or, if only standard power supplies are available, extremely long focusing times have to be used (e.g., 13 days at a potential differential of 7.5 V/cm). On the other hand, since the coils are freely accessible, the system can be run to equilibrium and the sample added toward the end of the run in the coils close to its pI. Interestingly, Quast has also investigated the feasibility of continuously changing the electrolyte solutions near both electrodes during the run, so as to achieve a constant, steady-state stream of ions of neutral salt through the gradient, to decrease protein coacervates at the pI. Basically, this is the same principle of steady-state rheoelectrolysis (see §2.1.16).

2.1.10. Free-zone IEF in Hjertén's rotating tube

This technique is based on the principle of free zone electrophoresis, as developed by Hjertén (1967). In carrier-free zone electrophoresis one of the main problems is how to stabilize separated zones or bands without the use of supporting media. This has been solved by Hjertén (1967) with the principle of rotational stabilization, as illustrated in Fig. 2.12. Lundahl and Hjertén (1973) and Hjertén (1976) have modified this apparatus for use in IEF. Here the separation chamber is a water-cooled horizontal quartz tube with an inner diameter of 3 mm, that rotates around its longitudinal axis at 40 rev./min to counteract convective disturbances. Coating of the tube with methyl cellulose eliminates electroosmosis. The tube is scanned in the UV light and the ratio of absorbancies at 320 and 280 nm is recorded. This minimizes disturbances due to irregularities in the rotating tube, dirt on its surface, dust in the cooling water, etc. At the two tube extremities, polyacrylamide beads are packed to avoid convective mixing of anolyte and catholyte with the Ampholine solution in the tube. A cellophane membrane at the tube ends prevents hydrodynamic liquid streaming. A review on the applications of both, free electrophoresis and free IEF has been published by Bours (1976).

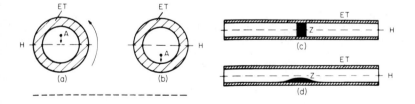

Fig. 2.12. Free zone electrophoresis. Drawing illustrating the principle involved in the stabilization of zones by rotation of the horizontal electrophoresis tube ET around its long axis. H represents the horizontal plane. A is a liquid element that has a higher density than the surrounding buffer solution. If we assume that gravity and frictional forces are the only forces acting on the liquid element, the direction of its movement must always be downward. Therefore, the liquid element moves away from the tube wall when it is in the upper half of the cross section (a) and toward the wall when in the lower half (b). According to these simplified diagrams the liquid element is forced by the rotation of the tube to move to-and-fro relative to the tube wall, which is equivalent to zone stabilization. If rotation is stopped, the sharp zone Z in (c) will spread rapidly, as shown in (d). (From Hjertén, 1967.)

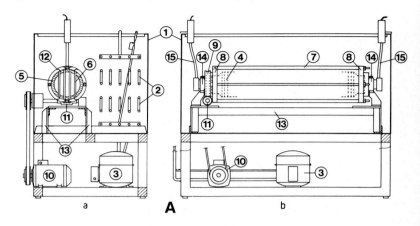

Fig. 2.13. (A) Longitudinal section of multicompartment electrolysis apparatus. The main parts are: 1 = cooling tank; 3 = refrigerating machine; 4 = electrolyzer; 7 = four pull rods, for assembling the electrolyzer, attached to a circular end piece (8); 10 = electrical motor for rotation of the electrolyzer via worm gearings (9) and (11); 13 = unit carrying the electrolyzer; 14 = slide bearings; 15 = gas-escape tubes. (B) Longitudinal section of one end of the electrolyzer. 1 = Gas-escape tube, connected to the electrode compartment (5) via a channel (2); 3 = platinum wire for connection to the electrode (4); 6 = membrane preventing the electrode gases from entering the electrolyzer; 7 = circular wall for uniformly spreading the electrode current into the separation compartments; 8 = rotary part, supported by the stationary part (9); 10 = O-rings tightening the bearings; 11 = rubber gasket sealing the electrode compartment; 12 = one of the 46 rings of the separation compartments, divided by membranes (13); 14 = hole for filling and emptying one compartment, fitted with a stopper (15) and O-rings (17) for proper sealing; 16 = set of concentric O-rings for sealing each compartment to the next one. (From Jonsson and Rilbe, 1980.)

2.1.11. Multi-compartment electrolyzers

Rilbe's group has also continued along the classical line of stationary electrolysis in efforts to improve multicompartment apparatus. For reviews see Svensson (1948) and Rilbe (1970). In 1974, in Milano, Rilbe presented the latest developments on three types of electrolysers: one with open cells, internal cooling, and no stirring; a second type with closed cells, external cooling and stirring. The last model embodies most

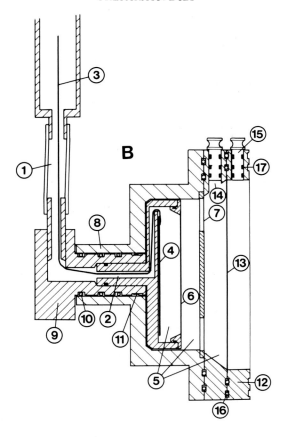

of the desired features of a good electrolyser. It solves most of the
technical difficulties connected with electrolysers, i.e., heat dissipation,
electroosmosis, homogeneity of solutions in each compartment, and
isolation of separated components without remixing. One of the
promising applications of this type of cell is in the very large-scale
fractionation of proteins. Stationary electrolysis can now be completed
in about 24 h. Since the volume in these 20-compartment electrolysers is
usually between 500 and 1,000 ml, and since rather high concentrations
of protein (about 4–5%) can be used, the load capacity is in the order of

several g/day. Electrolysers of this sort should offer a major improvement over many other preparative techniques using density gradients (Rilbe et al., 1975).

More recently, Jonsson and Rilbe (1980) have described an improved apparatus which allows sample fractionation in the gram range. The cell contains 46 separation compartments, its total volume is about 7.6 l and its length is 1 m. The compartments are closed, and internal cooling and stirring are affected by slow rotation of the whole apparatus in a tank filled with cold water. The apparatus can be run with an electric load of up to 5 kV, and an isoelectric focusing takes 2–3 days. Drawings of the entire assembly and of one end cell and electrode compartment can be seen in Fig. 2.13. Fourteen grams of whey protein could be completely separated into its main components, serum albumin (pI 4.60), α-lactalbumin (pI 5.01) and β-lactoglobulin (pI 5.13–5.23). Due to its mammoth size and the very high cost of the chemicals needed to operate it, I feel that, at present, this apparatus is more suitable for industrial type operations, rather than for research laboratories.

2.1.12. Free-flow, high voltage IEF according to Hannig

Continuous-flow techniques are particularly attractive for processing large quantities of protein mixtures, which would otherwise require a number of consecutive batch experiments in conventional, fixed bed apparatus. Continuous, free-flow electrophoresis (CFE) was first introduced by Barrollier et al. (1958) and developed by Hannig (1961, 1969, 1972). The chamber consists of two parallel glass plates 50 × 50 cm spaced only 0.3–0.5 mm apart, through which a laminar film of electrolytes flows. No anti-convectional packing is used to maintain stabilization and electrophoresis is conducted in free solution. The Hannig apparatus has been used mostly for electrophoresis of proteins, viable cells, cell organelles and membrane systems. Recently, McGuire et al. (1980) have reported IEF separation of red blood cells in this chamber. Hannig (1978) has also described a smaller flow through chamber with a built in scanning photometer for in situ detection of the separated zones and automatic mobility measurements. An exploded view of this chamber is shown in Fig. 2.14. In a 12 × 12 cm cell a laminar buffer film of 0.3 mm thickness flows from top to bottom. An electric field (up to 200 V/cm) is set up perpendicular to the buffer stream. The samples

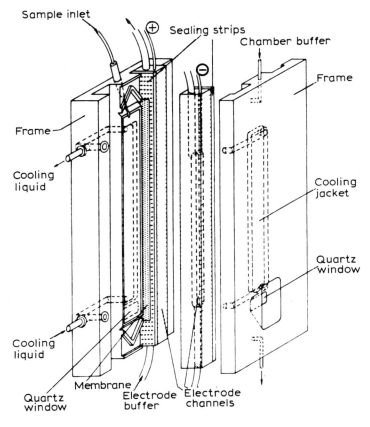

Fig. 2.14. Exploded view of the analytical separation chamber for free-flow electrophoresis and IEF (Hannig, 1978).

to be separated are injected continuously into the streaming medium. Hannig (1978) has used this apparatus also for IEF fractionation of sheep hemoglobins. Interestingly, since the flow-through of the liquid curtain would have been too fast to allow for an IEF separation, the liquid was kept recycling until the final stage of focusing was reached. This is similar to the principle developed by Bier's group (see §2.1.15).

2.1.13. Free-flow high voltage IEF according to Just and Werner

The apparatus, developed by Seiler et al. (1970a,b), has a non planar separation cell and is operated in the vertical position. The separation cell, illustrated in Fig. 2.15, is a U-shaped channel formed by fixing Lucite blocks (A and B) around three sides of a central Teflon-coated metal cooling plate (C). An electrode block (E) containing electrode chamber (F) is positioned at the back of the cooling plate. The complete assembly is attached together with tension bolts (G) and forms a compact unit. In the front block (A) and the electrode block (E) the bolt holes are enlarged, and with the aid of different spacer plates (M) it is possible to vary the thickness of the separating channel. The cell is 36 cm long and has an effective width of 11 cm; the thickness of the channel is normally 0.5 mm. This design affords easy assembly of the electrode compartments; the semipermeable membrane (L) is firmly held in position by the

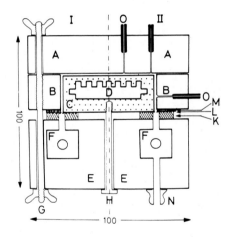

Fig. 2.15. Schematic drawing of two sections of the separation chamber for free-flow isoelectric focusing. (I) Vertical section 1 cm from the upper side. (II) Vertical section 2 cm from the upper side. (A) Front block of the separation chamber. (B) Side view of the separation chamber block. (C) Aluminium cooling block. (D) Channel for the cooling liquid. (E) Electrode block. (F) Electrode chamber with electrodes. (G) Holding screw. (H) Adjustable screw which regulates the tightening of blocks (E) against (K). (K) Silicon band. (L) Electrode membranes. (M) Leveling rubber band. (N) Inlet for electrode buffer. (O) Inlets for the buffer of the separation chamber. (From Seiler et al., 1970a.)

silicone gasket (K), forming a leak-free seal. As many as 48 fractions are collected from the bottom of the cell by capillary tubes connected to a peristaltic pump. Solutions of 5% acetic acid and 1.5% ethanolamine are used at the anode and cathode respectively, and are continuously pumped through the electrode compartments during IEF. The temperature within the separation cell is controlled by a thermistor. Samples are usually injected into the chamber by a micro pump connected to the cooling system. A typical run is performed at a field strength of 110 V/cm and at a liquid flow rate of 1 ml/min. Just et al. (1975a,b) and Just and Werner (1977a,b, 1979) have published extensive data on the use of this apparatus for the separation of cells and subcellular organelles. For example, they were able to separate mixtures of human, mouse and rabbit red blood cells in the pH ranges 3–10 and 5–7, at an injection rate of 3×10^7 cells/min. The flow-through time for the cells was only 7 min. Solutions of 1 mM EDTA were used to prevent cell clumping. Just et al. (1975a,b) have also reported preliminary data on the separation of the light mitochondrial fraction of rat liver into lysosomes and mitochondria. In this case, they used polyanionic substances such as heparin, chondroitin sulfate, polyvinyl sulfate, dextran sulfate and polyanethole-sulfonic acid to prevent aggregation. Notwithstanding this impressive amount of data, as well as Sherbert's (1978) work, IEF of cells has not been quite established (Righetti et al., 1980b). For instance, McGuire et al. (1980) have reported binding of carrier ampholytes to the cell surface during IEF while Catsimpoolas and Griffith (1980) have demonstrated lysis and loss of viability, upon IEF, of mouse spleen lymphocytes.

2.1.14. Free-flow high voltage IEF according to Prusik

The continuous free-flow apparatus used by Prusik (1974) for IEF is a more conventional design. This instrument, similar to that originally designed by Hannig (1961), but modified for IEF, is shown in Fig. 2.16. Carrier ampholyte solution is pumped through a horizontally oriented separation cell to give a liquid film 0.5 mm thick, 50 cm wide, and with an effective length of 44 cm. The apparatus is cooled by a current of cold air and is capable of dissipating 0.36 W/cm. Prusik first fractionates the ampholytes in a preliminary run through the apparatus. The six input reservoirs are then filled with the appropriate pooled fractions,

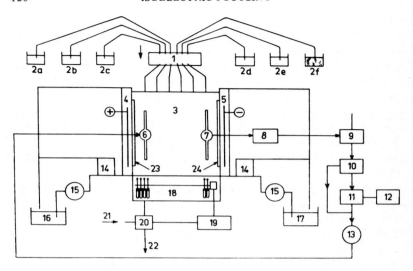

Fig. 2.16. Schematic diagram of continuous isoelectric focusing with prefractionated ampholyte carrier. 1 = Pump for electrolyte carrier; 2 = reservoirs of prefraction-ated electrolyte (Ampholine); $2f$ = solution of Ampholine fraction containing the protein solution; 3 = separating slit chamber; 4 = anodic electrode chamber; 5 = cathodic electrode chamber; 6 = air-cooling inlet; 7 = air-cooling outlet; 8 = temperature detector; 9 = pneumatic amplifier; 10 = proportional mixer; 11 = heat exchanger; 12 = cooling unit; 13 = fan; 14 = constant-level device; 15 = centrifugal pump; 16 = acid reservoir; 17 = base reservoir; 18 = fraction collector; 19 = high-frequency-level detector; 20 = air valve; 21 = atmospheric air inlet; 22 = vacuum; $23, 24$ = ion-exchange membranes. (From Prusik, 1974.)

one of which also contains the protein mixture. By introducing the prefractionated ampholytes in this way the pH gradient is established much more rapidly, thus reducing the residence time. Due to uneven development of Joule heat, the separation chamber must be located horizontally in order to limit thermal convections. Prusik uses a wide separation cell (50 cm); thus, because of the increased migration distance, a longer time is required to establish the focused condition. In experiments with pig pancreatic amylase using Ampholine pH 5–8 (0.3%) the residence time was 150 min and the applied potential 60 V/cm. In conclusion, apparatus for free-flow electrophoresis are available commercially in a variety of different models: a wide version for use in a horizontal plane and both wide and narrow versions for vertical use. It would seem that when used for IEF they require only minor alterations. In general, independent circulations for the acid and base electrode solutions will have to be provided and, with some units, it may be necessary to obtain a slower rate of pumping through the separation chamber in order to obtain optimal focusing.

2.1.15. Continuous-flow, recycling IEF

A new approach, called recycling isoelectric focusing, has been described by Egen et al. (1979), by Bier and Egen (1979) and by Bier et al. (1979). It is well known that continuous-flow techniques (Fawcett, 1973), which appear essential for large scale preparative work, are disturbed by parabolic and electroosmotic flows, as well as by convective flows due to thermal gradients. Egen et al. (1979) have improved this system by separating the actual flow-through focusing cell, which is miniaturized, from the sample and heat-exchange reservoir, which can be built up to any size. Minimization of parabolic flow, electroendoosmosis and convective liquid flow is achieved by flowing the sample to be separated through a thin focusing cell (the actual distance anode to cathode is only 3 cm) built of an array of closely spaced filter elements oriented parallel to the electrodes and parallel to the direction of flow. Increased sample load is achieved by recirculating the process fluid through external heat-exchange reservoirs, where the Joule heat is dissipated. During each passage through the focusing cell only small sample migrations toward their eventual pI is obtained, but through recycling a final steady-state is achieved. The IEF cell has ten input and output ports for sample flow-through, monitored by an array of ten miniaturized pH-electrodes and ten UV sensors. The entire system is controlled and operated by a computer. Schemes of the entire apparatus set-up and of the flow IEF cell can be seen in Fig. 2.17A, B. By activating pumps at the two extreme channels, the computer can alter the slope of the pH acting any effect of the cathodic drift, which results in a net migration of the sample zones towards the cathode. Today this appears to be among the most sophisticated instruments available for IEF fractionation. A commercial version of it is now marketed by Ionics Inc., Watertown Boston (Mass. USA).

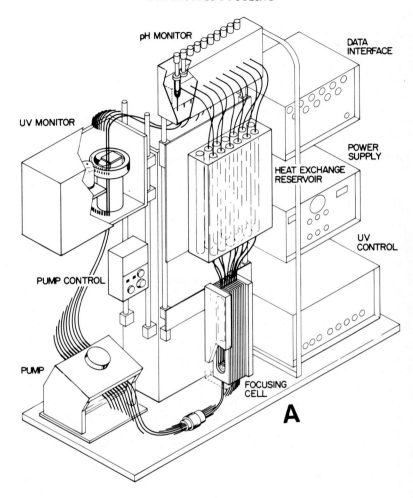

2.1.16. IEF in continuously flowing density gradients

A continuously flowing density gradient is another method for large scale preparative IEF. This idea was first adopted by Mel (1960, 1964) for preparative electrophoresis. Mel's apparatus consisted of a horizontal channel 30 cm long, 3 cm high and 0.7 cm wide with membranes separating the top and bottom of the channel from electrode compartments. Electrolyte solution in a range of sucrose concentrations is fed con-

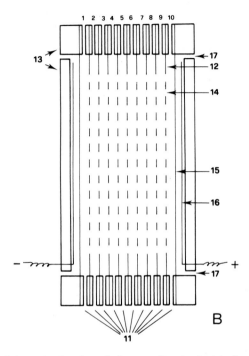

Fig. 2.17. (A) Schematic drawing of the recycling isoelectric focusing apparatus (RIEF). The 10 fractions from the reservoir are pumped through 10 flow-pH electrodes and then through 10 flow-UV cells to be cycled, via a multichannel pump, through a very compact focusing cell (the size of each compartment of the electrolyzer being 20 × 2 × 0.2 cm) and then back to the heat-exchange reservoir (from M. Bier, N.B. Egen, T.T. Allgyer, G.E. Twitty and R.A. Mosher, in E. Gross and J. Meienhofer (Editors), *Peptides: Structure and Biological Function*, Pierce Chemical Co., Rockford, IL, 1979, pp. 79–89). (B) Cross-sectional schematic representation of the RIEF cell. 1–10 = flow channels; 11 = connection to multichannel pump; 12 = plexiglass spacers; 13 = Perspex block (outer frame of the flow cell); 14 = filter elements (polyvinyl chloride) separating each compartment of the flow cell; 15 = dialysis membrane delimiting the electrode compartments; 16 = electrode platinum wires; 17 = input and output ports for recirculating electrolyte. (From Egen et al., 1979.)

tinuously into one end of the channel via twelve entrance tubes. It
flows horizontally through the apparatus at right angles to the applied
electric field and out through the exit tubes. Fawcett (1973, 1976), has
constructed two modified Mel's apparatus, one with a long and one with
a short migration path. The former consists of two cooling plates
assembled to form a separation cell 23 cm high, 25 cm long and 0.3 cm
wide. Fifty-four tubes evenly spaced along one end distribute the density
gradient; the horizontal flow is maintained by controlling both the
inflow and outflow with a 108-channel peristaltic pump. The latter,
shown diagramatically in Fig. 2.18, consists of a narrow Lucite frame (A)
20 cm long divided into three compartments by the strips (B). The lower
edge of (B) is tapered and is fixed 1.5 mm above the base of the apparatus
which is a thin plastic film (C). Efficient cooling is obtained by ensuring
good thermal contact between this plastic film base and a horizontal

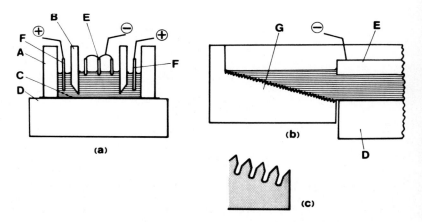

Fig. 2.18. Diagram of continuous-flow density gradient apparatus with short
migration path. (a) Transverse section of long horizontal trough (A) divided into
compartments by partition strips (B) and fitted with a thin plastic film base (C)
in contact with a cooling block (D). Long strips of stainless steel wire gauze form
the cathode electrode (E) and dip into the top layers of the density gradient contained in the central compartment. Short lengths of platinum wire gauze arranged
in pairs along the outer compartments form the anode electrodes (F). (b) Longitudinal section of end portion with grooved distribution block (G) for entry and
removal of the density gradient. (c) Enlargement of grooves in distribution block
(G). (From Fawcett, 1976.)

aluminium cooling block (D). The two outer compartments contain the dense electrode solution and the central one the density gradient. Both inflow and outflow of the density gradient are controlled by a multichannel pump connected by capillary tubing to a row of holes arranged along an inclined plane at each end of the trough. Each tube locates with the lower section of one of the grooves in the distribution block (G) which allows the smooth entry and removal of liquid layers 0.25 mm thick. Since this apparatus is cooled from the bottom surface only, as originally suggested by Philpot (1940), the vertical temperature gradient helps to strengthen the flowing density gradient. According to Fawcett (1976), this unit is capable of fractionating 1−2 g of protein mixture per hour, an amazing load capability indeed. Unfortunately there are no other reports on the use of this apparatus, which appears to me to be of a simple, yet very smart design.

2.1.17. Steady-state rheoelectrolysis

Recently, Rilbe (1978) has published a completely new approach, called 'steady-state rheoelectrolysis'. By this technique, it is possible to create stable pH gradients by steady-state electrolysis of ordinary buffer solutions in conjunction with superimposed, external liquid flows

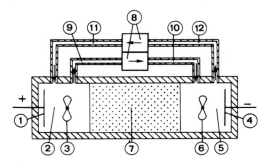

Fig. 2.19. Principle of steady-state rheoelectrolysis with liquid exchange between anolyte (2) and catholyte (5). The separation takes place in the convection-free portion (7), whereas anolyte and catholyte are constantly homogenized by the stirrers (3) and (6). The double pump (8) transfers equal volumes of liquid in both directions, and in the separation zone there should be no liquid flow. (From Rilbe, 1978.)

between the anolyte and catholyte compartments. This principle is illustrated in Fig. 2.19: the separation takes place in a convection-free section of the cell, while the anode and cathode compartments are continuously stirred, and liquid is pumped externally in both directions, at such a rate that the hydraulic flow through the pumps will exactly balance the added mass transport due to electric migration and diffusion of each ion constituent. If constant electric and hydrodynamic flows are allowed to continue until a steady state is reached, a smooth and stable pH gradient will develop in the separation chamber between the electrodes. Theoretical pH courses have been derived for rheoelectrolysis of a buffer composed of a weak acid and its salt with a strong base, and vice versa. If a buffer containing a salt of a weak acid and a weak base, having a ΔpK

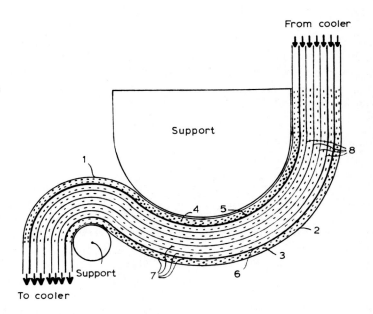

Fig. 2.20. Transverse section of the apparatus: 1 and 2 = impermeable plastic sheets stretched over formers to hold the sandwich of membranes watertight; 3 = cathode, 4 = anode; 5 = cation-exchange membrane; 6 = anion-exchange membrane; 7 = gauze separators; 8 = isoelectric membranes. (From Martin and Hampson, 1978.)

= 1, is subjected to rheoelectrolysis, a smooth pH gradient of 2 to 2.5 pH units is generated, centered on the two pK values. Experimentally, pH 4–5 and pH 8.5–9.5 gradients have been obtained by steady-state rheoelectrolysis of an acetate and a borate buffer, respectively. A similar approach has been described by Martin and Hampson (1978). In this last case, the electrolyser is of the multimembrane type, but the diaphragms are coated with amphoteric compounds, and are used under conditions in which they are isoelectric. A drawing of this apparatus can be seen in Fig. 2.20. Even though this chamber is built on a different principle, the basic idea for creating and maintaining a stable pH gradient in this system is still the concept of 'steady-state rheoelectrolysis'. In fact, in both cases, similar equations are derived for steady-state conditions and similar pH ranges can be obtained. Quoting from Martin and Hampson: 'the use of monovalent buffering ions limits the range of pH to about 1.5 units in the case where one ion only is buffering and to about 3.5 units if both ions are buffers and are chosen with a suitable interval between their pK values'. Both systems, if commercially available, could have virtually unlimited capability. A variant of both approaches utilizing two component buffer solutions in a multicompartment apparatus, has been recently reported by Jonsson and Fredriksson (1981).

2.2. Preparative IEF in gels

2.2.1. Fawcett's continuous-flow apparatus

We have already seen in §2.1.15 the horizontally flowing density gradients of Fawcett (1973, 1976). This author has also described vertical, continuous-flow chambers packed with a Sephadex bed as an anticonvective medium. While the former method requires rather complicated equipment, the latter makes use of a rather simple experimental design. In one version Gianazza et al. (1975) have fractionated synthetic ampholytes in a chamber with only twelve outlets without recirculation of anolyte and catholyte. This system can be run for weeks with very little attention by arranging a continuous sample input fed by constant hydrostatic pressure via a Mariotte flask. Fawcett's technique is based on the continuous-flow electrophoresis method first described by Svensson and Brattsten (1949) and by Grassman (1950). The difference between the two principles is illustrated in Fig. 2.21. The separation chamber con-

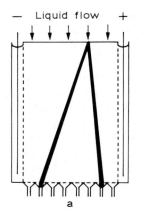 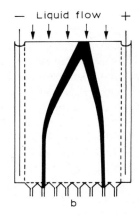

Fig. 2.21. Diagram illustrating the principle of: (a) continuous-flow electrophoresis and (b) continuous-flow IEF. (From Fawcett, 1973.)

sists of two cooling plates 23 cm wide, 30 cm high and 0.3 cm apart. Semipermeable membranes fixed to the sides separate the trough from the electrode vessels. The membranes are porous polyethylene sheetings, impregnated with polyacrylamide gel. The cell is packed with Sephadex G-100 beads (or polyacrylamide gel particles) supported by a filter membrane. The fifty-four exit tubes on the chamber floor are connected to a multichannel peristaltic pump, built on the delta principle. Continuous-flow IEF offers distinct advantages over density gradient stabilized columns. The zones remain under focus from the applied potential right until removal from the apparatus. They are, therefore, not subject to diffusion effects or to remixing during elution as commonly occurs in density gradient columns. Large quantities of material can be fractionated by continuous-flow IEF, since high concentrations of protein can be tolerated in focused zones. Fawcett reports separations of 500 mg protein/day. Furthermore, the system can be run in a cascade form, that is, first in a wide pH range for an initial screening and then, sequentially, in a narrower pH gradient for better separation of components of interest.

2.2.2. Quarmby's continuous-flow cell

This is one of the latest developments in continuous-flow system (Quarmby, 1981). A three-dimensional view of the separation cell and sections of it are shown in Fig. 2.22. The flow through chamber has a width of 15 cm, a length of 20 cm and a thickness of only 2 mm. This allows cooling only from the bottom surface, as usually performed in analytical systems (see chapter 3). The cell is stabilized by a capillary network made of Sephadex G-75, medium grade, and is usually operated at a 45° angle to the horizontal. Perhaps the most astonishing feature of this system is the collection manifold, made out of a block of Silicone rubber, consisting of 188 outlet tubings, spaced at 0.8 mm centers along the length of the material, each channel being 0.4 mm in diameter. This allows collection by very small liquid increments, resulting in high resolution of the outflowing sample zones. Table 2.3 lists and compares various flatbed systems which have been used for continuous-flow IEF either in free-flow systems or in capillary stabilized cells.

2.2.3. IEF in granulated gels

Radola (1973a,b, 1975) has described an excellent method for large scale preparative IEF in troughs coated with a suspension of granular gels, such as Sephadex G-75 'superfine' (7.5 g/100 ml) or Sephadex G-200 'superfine' (4 g/100 ml) or Bio-Gel P-60, minus 400 mesh

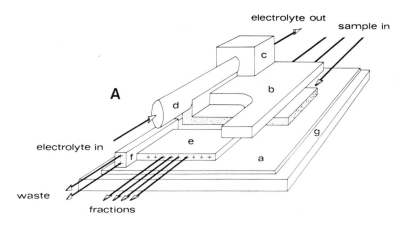

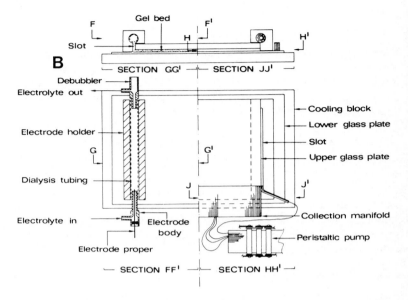

Fig. 2.22. (A) Three-dimensional view of the separation cell: (a) lower side of cell (thin glass), (b) upper side of cell (plate glass), (c) electrode holders, (d) dialysis tubing, (e) collection manifold, (f) waste manifold, (g) cooling block. The stippled area indicates the gel stabilization bed. (B) Sections of the completed separation cell. (From Quarmby, 1981.)

(4 g/100 ml). The trough consists of a glass or quartz plate at the bottom of a lucite frame (Fig. 2.23). Various sizes of trough may be used; typical dimensions are 40 × 20 cm or 20 × 20 cm, with a gel layer thickness of up to 10 mm. The total gel volume in the trough varies from 300 to 800 ml. A slurry of Sephadex (containing 1% Ampholine) is poured into the trough and allowed to dry in air to the correct consistency. Achieving the correct consistency of the slurry before focusing seems to be the key to successful results. Best results seem to be given when 25% of the water in the gel has evaporated and the gel does not move when the plate is inclined at 45°. The plate is run on a cooling block maintained at 2–10°C. The electrical field is applied via flat electrodes or platinum bands which make contact with the gel through absor-

TABLE 2.3

A comparison of the physical features of various flat-bed systems which have been used for continuous-flow isoelectric focusing and electrophoresis (from Quarmby, 1981)

Type of system	Dimension of separation cell (mm)			Details of outlets used to collect separated fractions		Systems used in the purification of:	References
	Thickness	Length	Width	Total number of outlets	Outlets per mm width of cell		
Recycling IEF	20	200	20	10	0.5	Protein	Egen et al. (1979) Bier et al. (1980)
Free-flow (vertical)	0.5	360	110	48	0.44	Proteins	Seiler et al. (1970)
	0.7	510	120	92	0.77	Erythrocytes	Just et al. (1977,1972) Hannig et al. (1975, 1978)
Free-flow (horizontal)	0.5	440	500	48	0.1	Pancreatic amylase	Prusik (1974)
						Amino acids	Prusik (1975)
	0.4	320	210	118	0.56	Inorganic compounds	Wagner et al. (1978)
Capillary-stabilized	3	300	230	54	0.23	Proteins	Fawcett (1973)
	9.5	250	200	36	0.18	Haemoglobin, albumin	Hendeskog (1975)
	6	225	140	12	0.09	Synthetic ampholytes	Gianazza et al. (1975)
	2	200	150	188	1.25		Quarmby (1981)

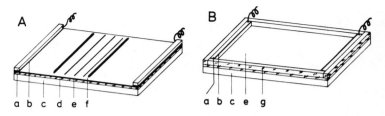

Fig. 2.23. Preparative isoelectric focusing. (A) Small scale separations. (B) Large scale separations: (a) electrode; (b) filter paper pad soaked with electrode solution; (c) cooling block; (d) glass plate; (e) gel layer; (f) focused proteins; (g) trough. (From Radola, 1975.)

bent paper pads soaked in 1 M sulphuric acid at the anode, and 2 M ethylene diamine at the cathode. In most preparative experiments initial voltages of 10–15 V/cm and terminal voltages of 20–40 V/cm were used. As much as 0.05 W/cm were well tolerated in 1 cm thick layers. Samples may be mixed with the gel suspension or added to the surface of preformed gels either as a streak or from the edge of a glass slide. Larger sample volumes can be mixed with dry gel (approx. 60 mg Sephadex G-200 per ml) and poured into a slot in the gel slab. Samples may be applied at any position between the electrodes. Ideally, the gel slab should be covered with a lid to prevent it from drying out. A suitable apparatus, the 'Double Chamber' is now available from Desaga, Heidelberg, G.F.R. After focusing, proteins are located by the paper print technique (Radola, 1973a). Focused proteins in the surface gel layer are absorbed onto a strip of filter paper which is then dried at 110 °C. The proteins may then be stained directly with dyes of low sensitivity such as Light Green SF or Coomassie Violet R-150 after removing ampholytes by washing in acid. Alternatively, proteins can be detected directly in gels cast on a quartz plate by densitometry in transmission at 270–280 nm. The pH gradient in the gel can be measured in situ with a combination microelectrode sliding on a calibrated ruler. Radola's technique offers the advantage of combining high resolution, high sample load and easy recovery of focused components. As much as 5 to 10 mg protein per ml gel suspension may be fractionated in wide pH ranges. Radola (1975) fractionated 10 g pronase E in 800 ml of gel suspension and obtained excellent band resolution at this remarkable load capacity. At these high protein loads, even uncolored samples can be easily detected, since they appear in the gel as translucent zones. Proteins are easily recovered in a small volume at relatively high concentrations by elution. The absence of sucrose is a further advantage in the subsequent separation of proteins from ampholytes. As there is no sieving effect for macromolecules above the exclusion limits of the Sephadex, high molecular mass substances, such as virus particles, can be focused without steric hindrance. The system has a high flexibility, since it allows analytical, small scale and large scale preparative runs in the same trough, merely by varying the gel thickness. When no suitable methods are available for detecting specific biological activities by the paper print technique, it is usually necessary to fractionate the gel and test for activity in eluates. This can be a rather laborious procedure. Fawcett (1975a) has described an alternative approach to IEF in granulated gels with semiautomatic sample collection. Instead of a flat bed, he uses a vertical glass column such as described by Brakke et al. (1968) to support the granulated gel. After IEF, the upper electrolyte is removed and the Sephadex gel is displaced upwards by a dense sucrose solution and into a fraction-collecting device. The fractionating device removes aliquots of the emerging gel and the fractions are washed through a funnel and into a collecting tube. Fawcett was able to obtain fractions equivalent to 3 mm sections of the gel with only moderate disturbance of focused zones. For simplicity, flexibility, high resolution, and high load capacity, Radola's technique has much to offer. Yet, even though this technique was first described as long ago as 1969, there were only a few users at that time. Its major defect was that pH gradients were unstable and distorted zones were often obtained. These problems were thought to be due to carboxyl groups on the Sephadex matrix which Radola (1973a) suggested might be removed by treatment with propylene oxide. This problem has been thoroughly investigated by Winters et al. (1975) and Winters (1977) who have solved most of the drawbacks of this technique. They have stressed three important points in the practical procedure: (1) the Sephadex itself has to be pre-washed; (2) the Sephadex gel should be of the superfine grade, G-75 being the best when the print technique has to be used; (3) the final water content of the gel has to be carefully controlled. If too dry, the gel bed cracks, if too wet the protein zones tend to sediment into the lower part of the gel. Pre-washing of the Sephadex is extremely important. Most Sephadex batches contain charged low-molecular mass contaminants which interfere with the pH gradient formation. Many of these contaminants are removed by washing swollen Sepahdex with distilled water on a Buchner funnel. The water within the gel is then displaced by ethanol and the gel is subsequently dried under vacuum (clean Sephadex is now commercially available from LKB as Ultrodex). Winters et al. (1975) have also developed a trough, for preparative IEF in Sephadex gels, which can be fitted onto the LKB Multiphor 2117 apparatus (see §3.2.14). This trough, equipped with a sample applicator and with a stainless steel frame divided into 30 channels, for gel fractionation, is available from LKB Produkter AB (see Fig. 2.24). This is a useful addition, since it allows analytical and preparative runs with the same equipment. When using this trough, Winters et al. (1975) suggest a gel thickness up to 5 mm, since a greater thickness will result in band distortion due to more efficient cooling at the bottom as compared with the gel layer at the top of the gel slurry. The sample load in this trough

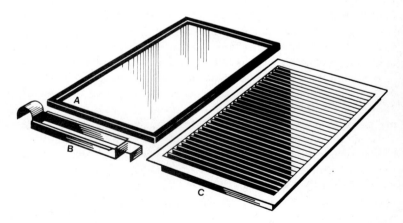

Fig. 2.24. Trough (A), sample applicator (B) and fractionation grid (C) for preparative IEF in granulated gels using the LKB 2117 Multiphor apparatus. After IEF, grid (C) is lowered on trough (A) thus fractionating the gel layer into 30 fractions. Each fraction is scooped up from each channel with a spatula. (Courtesy of LKB Produkter AB.) (From Winters et al., 1975.)

ranges from 200 to 400 mg protein, depending on sample heterogeneity. The gel slurry (approximately 5 g dry gel in 100 ml of 2% Ampholine) is poured into the tray, without waiting for complete gel swelling. Water is evaporated from the slurry with the help of a small, desk-type fan mounted 1 m above the tray, until 20–50% of the water in the gel has evaporated (this requires 2–3 h). The gel is then transferred to the cooling plate of the Multiphor, the sample applied with the sample applicator (see Fig. 2.24) and the experiment run at 10 °C, for about 10 h, at 10 W, with a constant wattage power supply. The electrodes are usually placed on the short side of the trough, so that proteins are separated along the long axis (25 cm). Thirty fractions are collected by compartmentation of the gel with the aid of the fractionation grid (see Fig. 2.24) and by scraping off the gel with a spatula. The protein is eluted from the gel simply by placing the gel fraction into a syringe equipped with glass-wool as bottom filter, adding a volume, and ejecting the eluate with the syringe piston (Winters et al. 1975).

The development of the technique of preparative IEF in granulate beds has spurred Holmquist and Broström (1979) to design a simple yet effective scanning accessory for in situ detection of focused bands. The scanner is built up around a brass box attached to a Beckman DU spectrophotometer. The light beam from the monochromator is deflected by a mirror in the brass box and allowed to pass through a quartz plate carrying the gel. With this arrangement the gel surface will always be available for marking the position of detected protein zones. The advantage is that the location procedure is rapid and detects whether the zones are distorted. There is no loss of proteins as required for staining a replica.

Another important application of Radola's technique is the electrophoretic desorption of substances tightly bound to affinity gels (Haff et al., 1979; Lasky and Manrique, 1980). Once the granulated IEF flatbed is prefocused, a narrow zone of gel from near one end of the plate is removed, and substituted with the affinity gel containing the substance to be desorbed. Fastest desorption is obtained when the charge of the desorbing molecule is maximized, i.e., when the affinity gel is applied distant from the anticipated focused band position. The focused, desorbed bands are then located and separated from gel and ampholytes by gel filtration. The several potential applications are: splitting of antibodies bound to immobilized antigens and vice versa; separation of proteins bound to hydrophobic gels and recovery of proteins from agarose and polyacrylamide gels. As a variant of Radola's technique, O'Brien et al. (1976) and Johnson et al. (1976) have described IEF in Sephadex G-15 columns plugged at the two extremes with polyacrylamide gel (in meshed nylon nets for easy removal). After IEF, the two gel plugs are removed, an elution plunger placed at the column bottom, and the column content recovered by gel filtration with ampholyte–glycerol solution or with distilled water. 96% sample recoveries and loads up to 300 mg per protein component are reported.

As an alternative to granular Sephadex or polyacrylamide, Otavski et al. (1977) and Harper and Kueppers (1980) have explored other charge free anti-convective media for use in granulated gels. They have obtained encouraging results with flat beds made of Pevikon, C-870 (a copolymer of polyvinyl chloride and polyvinyl acetate; see also § 1.9), an inert plastic in the form of granules 100 μm in diameter. This material has little if any electroendoosmosis and forms very stable beds with high liquid retention. It may be packed by gravity or with thin-layer spreaders and forms a cake which may be cut with excellent precision. Focused proteins are readily recovered by centrifugation through a coarse membrane. This material appears promising for both analytical and preparative uses.

2.2.4. IEF in multiphasic columns

Stathakos (1975) has designed a new type of vertical column for multiphasic isoelectric focusing. The idea is somewhat similar to the principle of multicompartment electrolysers described by Rilbe et al. (1975). The column is designed on the principle of building blocks, i.e., it is composed of alternating separable gel segments and liquid interlayers extending between two electrode compartments. The column, constructed in a modular fashion, can be shortened or lengthened according to experimental needs. The scheme of the apparatus, assembled in six units of equal size, is shown in Fig. 2.25. This type of column allows great experimental flexibility and offers several interesting features. Samples can be introduced practically anywhere in the column: at the top, in any of the liquid interlayers (2.5 mm high) between adjacent blocks, or in

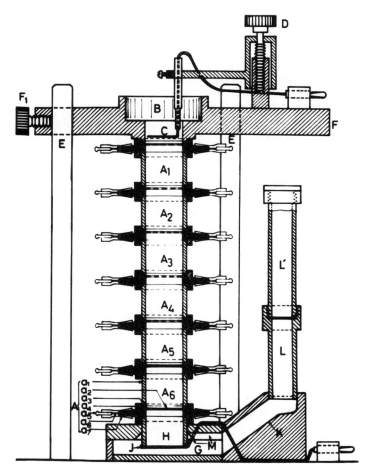

Fig. 2.25. Vertical section through a six-unit column for multiphasic isoelectric focusing. $A_1 - A_6$, separable independent gel units; a_1, tubular part; a_2, supporting membrane; a_3, silicone rubber O-ring; a_4, structural ring; a_5, glass plug; a_6, tygon jacket; a_7, outlet nipple; B, upper electrode chamber; C, upper electrode, movable vertically by means of screw D; E, metal rods; F, upper platform; F_1, screw for securing platform; G, lower electrode chamber; H, base cylinder; J, lower electrode; K, sloped connection to tube L; L, auxiliary counterpressure tube; L', tube extension; M, groove for escape of electrolysis gases. (From Stathakos, 1975.)

one or more of the building blocks. The pH can be monitored at any time through the various liquid interlayers (which form small chambers with inlet and outlet nipples, as shown in Fig. 2.25) by withdrawing a few μl of liquid or by pumping out the whole solution, and returning it after pH determination. Since the porosity of each gel unit can be varied independently, unwanted compounds in the sample can be separated by making the gel 'restrictive' for them, or vice versa. The proteins collected in any gel block can be further purified by rerunning in a shallower pH gradient simply by incorporating this gel block into a second column. The procedure can be repeated in continuous 'blow-up' or 'cascade' experiments, until the state of 'one gel unit-one protein band' is reached (see Fig. 2.26; Stathakos et al., 1980). Finally, the proteins collected in a block can be retrieved electrophoretically, and at the same time separated from Ampholine, by a method similar to the one described by Suzuki et al. (1973) (see §4.6).

2.2.5. IEF in polyacrylamide gel cylinders

Because of the excellent resolution and load capacity afforded by IEF in cylinders of polyacrylamide gel, Righetti and Drysdale (1973) have

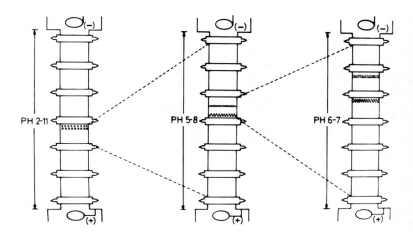

Fig. 2.26. 'Blow-up' experiment for the isolation of a hypothetical protein focusing at pI 6.7 in progressively shallower pH gradients. (From Stathakos et al., 1980.)

scaled up an analytical apparatus for preparative purposes by substituting an interchangeable core with larger tubes. The tubes are held between rubber grommets in a water-tight compartment through which coolant at 1 °C is circulated. The platinum electrodes are circular and are positioned close to the extremities of the tubes to minimize loss of ampholytes from the gel and thus reduce the cathodic drift of the pH gradient. The core accomodates six gels: three of 50 ml capacity (2 cm i.d.), two of 20 ml (1.2 cm i.d.) and an indicator gel of only 2 mm diameter and 2 ml capacity. All gels are cast in 16 cm long glass or plastic tubes. High porosity gels ($T = 4\%$; $C = 4\%$) (nomenclature of Hjertén, 1962, see p. 174) are used to minimize molecular sieving effects. After polymerization gels are stored at 4 °C before use. The maximum sample load in the larger tubes is approximately 200 mg protein. Samples may be applied directly to the top of the gel or may be incorporated into the gel solution before polymerization. Resolution is comparable to that given by the analytical system. As in other gel systems, so much protein can be focused in a zone that uncolored proteins are often clearly visible as opaque discs. The methodology is essentially that described in detail for the analytical system (§3.2.7). However, because of the poorer heat transfer in the thicker gels, power inputs should not be increased proportionally to gel volumes. An adaptation of this apparatus (which can be seen in the analytical version, in the drawing of Fig. 3.21) is commercially available from MRA, Boston, MA. One of the major drawbacks of preparative IEF in polyacrylamide gels is in sample detection and recovery after IEF. Sectioning the gel and eluting with buffer can give incomplete sample recoveries and high dilutions. As an alternative, Suzuki et al. (1973) have described an ingenious method for sample retrieval after IEF gels. They seal one end of a gel tube with dialysis tubing and introduce a narrow layer of a dense buffer. A polymerizing solution of polyacrylamide is floated on top of the buffer to form a short plug of acrylamide gel. The desired section of the focused gel is now minced and packed into the tubes on top of the stacking gel in dense buffer solution. The tube is then filled with buffer and the sample recovered by electrophoresis in the same apparatus used for IEF. This allows quantitative sample recovery, in a highly concentrated form, while also eliminating ampholytes which pass through the dialysis membrane. A similar apparatus has also been described by Chrambach et al. (1973).

2.2.6. *IEF in polyacrylamide and agarose gel slabs*

In addition to Sephadex beds, it is also possible to conduct preparative IEF in slabs of polyacrylamide and agarose gel. Because of the superior heat dissipation in slabs, this approach may be preferable to preparative IEF in large rods of polyacrylamide which are difficult to cool efficiently. Moreover, focusing in an acrylamide slab may give slightly superior resolution to Sephadex beds (Graesslin and Weise, 1974). The apparatus required is similar to that for analytical IEF in thin slabs of polyacrylamide, the thickness of the gel being altered by use of appropriately larger spacer frames or gaskets. Samples may be applied directly to the gel surface as a streak or on an absorbent paper rectangle. Alternatively, samples may be applied in small slits or troughs, either in paper or

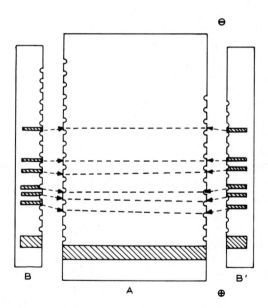

Fig. 2.27. Technique for localization of separated proteins in preparative thin-layer IEF. A, unstained middle part of the gel slab. B and B', Coomassie Blue stained gel strips. The places of corresponding protein bands are indicated by marker holes. (From Graesslin and Weise, 1974.)

in free solution. The resolution given by this method is comparable to that given in thinner analytical gels. Focused proteins on quartz plates may be detected by direct UV scanning with a Zeiss spectrophotometer. It is also possible to assess the position of proteins from rapidly stained longitudinal strips taken from both edges of the sample track (see Fig. 2.27). The proteins can be recovered from the gel by elution or by electrophoretic procedures (p. 295). Although many investigators have reservations about the efficiency of eluting proteins from polyacrylamide, Graesslin and Weise (1974) obtained 74% recovery of total proteins by elution with gentle agitation. With the advent of almost charge-free agarose (Isogel from Marine Colloids, Rockland, Maine) preparative IEF in agarose beds has become a reality (Chapuis-Cellier and Arnaud, 1981). A 0.8% agarose matrix is gelled containing 2% carrier ampholytes. Weak electrolytes have to be used, e.g., 0.1 M NaOH and 0.1 M acetic acid. In a 84 × 30 × 2 mm tray, focusing is over in 8 h at 15 W (initial conditions, 500 V and 30 mA; final conditions 1500 V and 10 mA). The gel is divided into 30 fractions, with the grid of Fig. 2.24, ground and extracted at least 3 times with appropriate buffer. These authors have reported fractionations of up to 1 g serum proteins, with recoveries between 68 and 82%.

2.2.7. Chromatofocusing and ampholyte displacement chromatography

Chromatofocusing was first described by Sluyterman and Wijdenes (1977, 1978, 1981a,b) and by Sluyterman and Elgerson (1978). They demonstrated that proteins travelling down an ion-exchange column in a pH gradient are subjected to focusing and emerge from the column at pH values approximately equal to their pI values. Thus chromatofocusing offers the high resolution obtained by separations based on differences in pI values, together with the high capacity of ion-exchange techniques. Peak widths can be in the range of 0.04–0.05 pH units, and samples containing several hundred milligrams of proteins can be processed in one step. Sluyterman and Wijdenes (1981a) have in fact calculated that the width of a protein band in a moving pH gradient in chromatofocusing is proportional to:

$$\Delta pH \simeq \pm \sqrt{\frac{dpH}{dV} \bigg/ \phi \frac{dZ}{dpH}} \qquad (50)$$

where ϕ is defined as:

$$\phi = F\psi/RT \qquad (51)$$

and ψ is the Donnan potential, dpH/dV the slope of the pH gradient per unit cross-sectional area and dZ/dpH the slope of the titration curve of a protein near its pI. The similarity between this equation and Rilbe's (1973a) Eq. (23) on the resolving power in IEF should be appreciated; the voltage gradient E of Eq. (23) being substituted in Eq. (50) by ψ, the Donnan potential and the protein mobility in the former being substituted in the latter by the protein charge, Z (the diffusion coefficient D, being apparently neglectable in chromatofocusing) (Sluyterman and Wijdenes, 1978). How does the technique operate in practice? First of all, a proper anion exchange column is chosen and is adjusted at the highest pH to be utilized. Next, the protein sample is dissolved in a buffer at the lowest pH to be used, dialysed against the same buffer and applied to the column. The pH interval is chosen so that the pI values of the proteins of interest fall roughly in the middle of the pH gradient. For best results, the separation should take place over a maximum interval not exceeding 3 pH units at a time. At this point, the column is eluted with the lowest pH buffer: the acid-base exchange between buffer

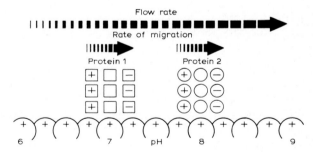

Fig. 2.28. Behavior of protein molecules in chromatofocusing. The pH increases down the length of the column. Protein molecules above their isoelctric point in the gradient become negatively charged and bind to the gel. Molecules below their isoelectric point are repelled from the gel. Molecules at their isoelectric point are neither bound nor repelled. Protein 1 has a hypothetical pI = 7, protein 2 a pI = 8. (From Pharmacia Fine Chemicals.)

and ion exchanger produces a pH gradient within the limits of the two initial pH values, which moves slowly down the column. In this pH gradient, proteins elute in order of their isoelectric points.

The behavior of proteins in chromatofocusing is illustrated in Fig. 2.28. As the protein migrates down the length of the column from the starting zone (lowest pH buffer) the surrounding pH will be, most likely, lower than its pI, so that the protein will carry a positive charge, will be repelled by the positively charged matrix and travel down with little or no retention. However, when it has travelled sufficiently far down so that the pH is greater than the pI, the protein reverses its charge and binds to the ion exchanger. The molecule remains bound to the ion-exchanger until the developing pH gradient causes the pH to drop below the pI of the protein. The protein is then carried along in the eluent buffer again until the pH rises above the pI and it rebinds. This process is repeated until the protein emerges from the column at its pI.

The focusing effect in chromatofocusing is shown in Fig. 2.29. When a protein is applied to the column, it will migrate down in the eluent as far as its pI, from where it migrates more slowly until it elutes. If, during this process, a second aliquot is applied, it will migrate down the column with the eluent front until it meets the slower moving first sample. The two sample aliquots will thus merge in a single zone and proceed down to the bottom of the column where they co-elute. In the laboratory,

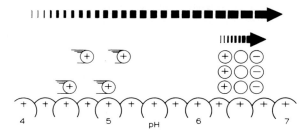

Fig. 2.29. Focusing effect in chromatofocusing. Protein molecules at the end of the zone are repelled from the gel and migrate more rapidly than proteins at the front. If a second sample of the same protein is applied to the column after the first, it will catch up and co-elute with the first. If a second sample of a protein with a higher pI is applied, it will pass through the first zone and elute before it. (Courtesy of Pharmacia Fine Chemicals.)

chromatofocusing can be performed with buffers and ion-exchangers developed by Pharmacia. Two types of buffers, called Polybuffer 96 (buffering in the pH range 9–6) and Polybuffer 74 (buffering in the pH range 7–4) are used in connection with an ion exchange resin called Polybuffer exchanger PBE 94. It is also possible to perform chromatofocusing at more alkaline pH, in which case the matrix used is the Polybuffer exchanger PBE 118 and the column is eluted with Pharmalyte pH 8–10.5. What are these chemicals? Polybuffers 96 and 74 can only be carrier ampholytes buffering, respectively, in the pH 6–9 and pH 4–7 ranges and meant for separations of proteins whose pI values encompass these pH intervals. The two ion-exchangers contain groups which buffer in the pH range 4–9 (PBE 94) or in the pH range 8–11 (PBE 118). These charged groups are coupled via ether linkages, to the monosaccharide units of a Sepharose 6 B matrix. The difference between PBE's and conventional ion exchangers is that, while the latter contain only one buffering group with a single pK (e.g. DEAE Sephadex, pK 9.5 or CM-Sephadex, pK 4.5) the former contains several types of buffering groups with several pK values. In other words, in order to form a reproducible pH gradient along the column length, in a given pH interval, the buffering power of the anionic groups fixed to the resin has to match, in a given pH interval, the buffering power of the carrier ampholytes used as eluants. In fact, PBE 94 has an even buffering capacity in the pH range 4–9 (on the average, 3–3.5 mequiv./100 ml gel per pH unit) while PBE 118 has an even higher buffering power in the pH range 8–11 (average of 4–4.5 mequiv./100 ml gel per pH unit). Probably, PBE 118 is obtained by the synthesis of Sluyterman and Wijdenes (1981b), by linking a polyethylene-imine (PEI) to a Sepharose 6B via a bisoxirane bridge. Since PEI would be largely deprotonated at alkaline pH (see also Righetti and Hjertén, 1981) the weakly basic imino groups are partly converted into strongly basic guanidino groups by reaction with O-methylisourea. Not much is known on PBE 94, but conceivably it could be made by coupling oligoamines (e.g. TETA, TEPA and PEHA, see §1.7.1) to Sepharose 6B. Guidelines for selecting the type of gel, the starting buffer and elution buffer in different pH intervals are given in Table 2.4. As a recent extension on chromatofocusing Hearn and Lyttle (1981) have described a technique for generating internal pH gradients using a mixture of common amphoteric and non-amphoteric buffers.

TABLE 2.4
Buffers and gels for chromatofocusing in different pH ranges[a]

pH range	Gel	Start buffer	Eluent
9–10.5	PBE 118	25 mM triethylamine–HCl, pH 11	Pharmalyte, pH 8–10.5–HCl, pH 8.0
8–9	PBE 94	25 mM ethanolamine–HCl, pH 9.4	Pharmalyte, pH 8–10.5–HCl, pH 8.0
7–8	PBE 94	25 mM Tris–HCl, pH 8.3	Polybuffer 94–HCl, pH 7.0
6–7	PBE 94	25 mM imidazole–CH$_3$COOH, pH 7.4	Polybuffer 94–CH$_3$COOH, pH 6.0
5–6	PBE 94	25 mM histidine–HCl, pH 6.2	Polybuffer 74–HCl, pH 5.0
4–5	PBE 94	25 mM piperazine–HCl, pH 5.5	Polybuffer 74–HCl, pH 4.0

[a] From: chromatofocusing with polybuffer and PBE, Pharmacia Fine Chemicals leaflet, November 1980.

Things are not so clear regarding ampholyte displacement chromatography (ADC), because theory is also lacking. ADC was first reported by Leaback and Robinson (1975) and applied in a number of investigations (Young and Webb, 1978a,b; Young et al., 1978; Young and Webb, 1980; Pagé and Belles-Isles, 1978; Chapuis-Cellier et al., 1980; Emond and Pagé, 1980; Francina et al., 1981). In ADC a column of conventional ion exchanger is eluted using carrier ampholytes; however, gel and eluent are at the same pH, so no pH gradient is formed. The elution mechanism in ADC is not clear, but is thought to involve competition between bound proteins and ions in the eluent for charged sites. Due presumably to the lack of a pH gradient, proteins are not always eluted in order of their isoelectric points in ADC and there is no focusing effect as opposed to chromatofocusing. Moreover in ADC high concentrations of ampholytes are used for elution (as high as 60 g/l) while in chromatofocusing, since the protein is eluted primarily by losing its charge, and not by a displacement process, much lower concentrations are used (of the order of 10 g/l). As an alternative to chromatofocusing and ADC, isoelectric chromatography (Petrilli et al., 1977) could be used. If the ion-exchange chromatography is performed consecutively on anion and cation exchangers at the isoelectric pH of the protein to be purified, the chromatographic process results in the absorption of more acidic and more basic mixture components, respectively, on the anion and on the cation exchangers. On the other hand, the required protein remains unabsorbed. The resolving power of this method, however, is low and is of the order of 0.2– 0.3 pH units in pI differences between the proteins to be separated (in chromatofocusing it is 0.05 pH units).

2.2.8. Concluding remarks – load capacity

The limited survey presented here clearly attests to the ingenuity of investigators in designing, building or modifying apparatus to capatilize on the excellent resolution offered by IEF in adapting the technique for both analytical and preparative purposes. An article by Fawcett (1975a) makes excellent reading for other suggestions for constructing simple but effective systems for preparative IEF.

We have seen that with some methods it is possible to fractionate large amounts of proteins, in the gram scale. In the field of IEF, it has generally been believed that sucrose density gradients could only carry a limited amount of protein. As a representative figure, in the 440 ml LKB column, a maximum load of only a few hundred mg protein is recommended. However, in view of the results and theoretical considerations by Rilbe and Pettersson (1975) (see also § 1.3.8) I think that this point of view should be changed. According to these authors, the theoretical mass content of a protein zone (m) in a linear sucrose density gradient ranging from 0 to 0.5 g/cm, cannot exceed the following inequality:

$$m < 0.625 \, Vr^2 \, \text{g/cm}^3 \qquad (52)$$

where V is the total column volume and r is the zone breadth. For a 100 cm^3 column, one has:

$$m < 62.5 \, r^2 \, \text{g} \qquad (53)$$

Thus, the load capacity rises with the square of the zone breadth. Rilbe and Pettersson (1975) have verified this experimentally. In a 110 ml column, for an *r* value of 0.125, according to the above inequality, one should be able to load a maximum of 1074 mg protein in a single zone. In fact, working with myoglobin, these authors have loaded 1,050 mg in the column, and approximately 800 mg (that is, 76% of the applied sample and 74% of the theoretical maximum) were confined within the main peak (MbI). Janson (1972) has reported the fractionation of 7.3 g of cytophaga Johnsonii cytoplasmic extracts in a 440 ml sucrose density gradient. Fawcett (1975a) using a narrow-range (pH 6.3–7.3) Ampholine, in the 110 ml LKB column, has applied up to 1 g of protein from red cell haemolysates. Goode and Balwin (1973), in the purification of *Staphylococcal* α-toxin, have loaded 1.8 g protein while Smyth and Arbuthnott (1974), working with *α-toxin* from *Clostridium perfringens*, have loaded as much as 3.87 g protein to density gradient columns. In their gigantic multicompartment electrolyser, Jonsson and Rilbe (1980) have fractionated 15 g crude pepsin and 13.8 g of bovine whey proteins; within a single isoelectric zone (hemoglobin) up to 7 g protein could be loaded. These high loads in liquid systems compare well with gel systems. However, at these high sample inputs, protein precipitates may form and sediment along the density gradient. Therefore, high loads in liquid systems are of limited applicability, that is when the protein of interest focuses away from the zone of heavy precipitates. Thus, with regard to protein carrying capacity, gel systems appear to be superior to liquid systems. Although precipitates may form, they are confined to the pI and usually do not interfere with the behavior of other proteins.

On the basis of these considerations, and of the equipment presently available, I think IEF has the prerequisite to become one of the leading techniques for large scale separations of macromolecules. Before ending this chapter, I must confess that although this survey of preparative techniques seems exhaustive, it does not cover all the possible approaches to preparative IEF. I believe there is still at least one new, quite revolutionary preparative approach. However, as this still belongs to the secret projects of my laboratory, I am afraid you will have to hold your breath till the evening edition of this book.

CHAPTER 3

Analytical IEF

A good analytical system should conveniently provide high resolution separations at a low cost in time and materials. Ideally, the system should not be too demanding in experimental technique and should allow simultaneous processing of multiple samples and a convenient means for assessing and recording data. Although the original systems for IEF in vertical density gradients were used for both analytical and preparative purposes, they were not very well suited for routine screening and analyses of multiple samples. The smallest column then commercially available had a volume of 110 ml and required substantial amounts of material for analysis. Moreover, a typical experiment usually took several days to complete. Electrolysis periods were of 2–3 days duration and the subsequent analyses of fractions eluted from the gradients were tedious and laborious. Several attempts were made to develop smaller columns for more rapid and convenient analyses. Many ingenious methods were devised but few were of general applicability. In addition to the inconvenience in analysing gradients after fractionation or with flow cells, most systems were not suitable because of problems associated with the use of sucrose density gradients. These problems included convective mixing and isoelectric precipitation during electrolysis and loss of resolution by diffusion or by mixing on elution. Resolution was also lost by collecting fractions with a greater volume than that of focused zones. Although many of these difficulties may be overcome for example by scanning the column in situ or by using flow cells to monitor eluates, few systems in sucrose density gradients met the requirements for a convenient rapid, small-scale method for simultaneous fractionation of multiple samples.

This section will deal first with analytical IEF in liquid media and will then cover more thoroughly the field of IEF in gels, since this last system, on the analytical scale, has at present attained wide popularity. It is somewhat difficult to draw a line between analytical and preparative techniques. In theory, any analytical technique which allows sample recovery could also be called preparative. The term 'preparative' also depends on the experimental needs, either the laboratory scale or at the industrial level. Some of the systems for IEF in liquid media, described in the next section, which I have grouped under 'analytical' techniques, could also be used for laboratory scale, preparative applications. I have therefore drawn an arbitrary line between analytical (in the μg to mg scale) and preparative (in the mg to g scale) techniques.

3.1. IEF in small density gradient columns

Parallel to the development of IEF in polyacrylamide gel, several research groups explored other designs and apparatus for IEF in small-scale density gradients. Weller et al. (1968) built two U columns for small scale IEF, one with a diameter of 1.0 cm and a working volume of 11 ml, the other with a diameter of 1.9 cm and a working volume of 33 ml. In these U-tubes, one of the arms (the dummy arm) contains only the electrolyte (usually catholyte) and the other (separational arm) the sucrose density gradient to perform the separation and the anolyte. The U-tube is usually cooled in a cold room or immersed in a tank with cold water. When the system is at equilibrium, the anode is removed and a siphon inserted in the ground joint of the separation arm. When solution is added to the dummy arm, a siphon is formed and the column emptied dropwise in a series of test tubes. A similar technique for rapid IEF of multiple samples on a micro-scale has been described by Godson (1970). He performs IEF in J-tubes, made of glass of 1.1 cm outer diameter and 0.9 cm inner diameter, having a total volume of 10 ml. The J-tubes fit standard acrylamide gel electrophoresis apparatus, so that up to eight electrofocusing columns can be run at once (see Fig. 3.1). The long arm (25 cm) contains the sucrose density gradient and acts as the separation chamber. The short arm (16 cm) contains a cushion of heavy sucrose solution and usually, terminates with the cathode solution. The entire J-tube is almost completely immersed in the lower electrolyte chamber, which is stirred with

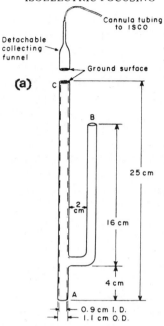

(a)

Cannula tubing to ISCO

Detachable collecting funnel

Ground surface

C

B

25 cm

2 cm

16 cm

4 cm

A

0.9 cm I.D.
1.1 cm O.D.

Scale: 5 mm = 1 cm

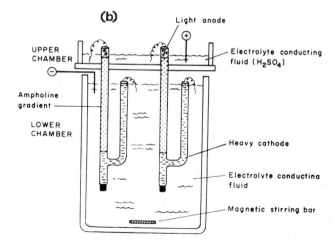

(b)

Light anode

UPPER CHAMBER

Electrolyte conducting fluid (H_2SO_4)

Ampholine gradient

LOWER CHAMBER

Heavy cathode

Electrolyte conducting fluid

Magnetic stirring bar

a magnetic bar and cooled by an outer cooling jacket. The two extremities of the J-tube are connected to the anolyte and catholyte by short platinum wire loops. At the end of the experiment, the column is fractionated by fitting a conical glass funnel to the top of the long arm of the J-tube, and pumping heavy sucrose into the short arm. Similar U-tube systems have also been described by Koch and Backx (1969). In the last few years, Rilbe's group has been developing increasingly short and strong density gradients in conjunction with stable natural pH gradients. Rilbe (1970) has described a parallelepipedic quartz column 14 cm long with a volume of 11.2 ml. Focusing in this column could be completed in 6–7 h, but the price of the quartz column and its accessories was very high. In a search for commercially available apparatus, Rilbe's group found that a spectrophotometric cell designed for flowing solutions and equipped with mantles for thermostating medium satisfied practically all demands to be put on a column for IEF. Rilbe (1973b) and Jonsson et al. (1973) have described for use in IEF the flow cuvette type code 167-QI from Hellma GmbH (Mullheim, Baden, West Germany). The central compartment, useful as a column for IEF, has the dimensions $4 \times 10 \times 35$ mm = 1.4 ml. Thus focusing in this cell requires a very small volume and can be completed in a very short time, usually less than 2.5 h, inclusive of preparations of solutions for the density gradient and of absorption scanning. Since IEF is performed in a spectrophotometer, scans can be made during or after focusing simply by equipping the spectrophotometer with a cell elevator and a fixed horizontal slit. To this purpose, Rilbe has used a Vitatron Universal Photometer. In addition to scans, spectra can also be taken of individual protein bands: this allows the study of chemical reactions of proteins when reactive ionic species are migrated through already focused proteins. Thus, by injecting a small amount of sodium dithionite at the cathode, Rilbe (1973b) was able to monitor the conversion of ferric sperm whale myoglobin into the ferrous form. At this point, the addition of a small amount of potassium ferricyanide at the cathode reconverts the ferrous into the

Fig. 3.1.   J-tubes for isoelectric focusing in an acrylamide gel apparatus: (a) dimension of the J-tube and arrangement for collecting; (b) J-tubes in their running position. (From Godson, 1970.)

ferric form, as seen by the specific spectra of the two forms. The evaluation of the pH course in these small volume cells would appear to be a very difficult proposition. Yet Fredriksson (1972) has described a microfractionation method which allows the recovery of 60 μl fractions from the flow-cell used for IEF. For pH measurements, the Radiometer E 5021 microelectrode unit (Radiometer, Copenhagen, Denmark) was used. The pH sensitive glass membrane is shaped as a horizontal capillary tube that, via a vertical polyethylene tube, can be filled simply by suction. In this electrode, as little as 20 μl of sample volume are required for pH measurements.

A host of other apparatus for small-scale IEF in density gradients has been described. Some of them use conventional polyacrylamide gel electrophoresis equipment, in which the tube bottom is plugged with a semiporous plug of agarose (Massey and Deal, 1973) or of polyacrylamide (Behnke et al., 1975) or with a dialysis membrane (Holtlund and Kristensen, 1978). Others utilize as separation chambers chromatographic columns of 10–15 ml volume (Kint, 1975) or again variants of U-tubes of 10 and 30 ml size (Osterman, 1970; Doi and Ohtsuru, 1974) or variations of the J-tubes (Korant and Lonberg-Holm, 1974; Azuma et al., 1977).

3.2. Analytical IEF in gel media

As long as we insist on living at 1 g gravity, there is no way to avoid gravitational instability of protein zones surrounded by a free liquid medium. In fact, each time a protein or other macromolecule is not uniformly distributed in the conducting medium, the density in this zone will be higher than in the surrounding solution. Unless the resulting pressure gradients at the zone boundaries are compensated by some means, convection currents will arise, tending to eliminate the concentration differences built up by the current. Thus, pure protein zones can only be obtained by supporting the conducting liquid by a capillary system, which eliminates disturbances due to convection. In conventional electrophoresis, all sorts of zonal stabilizing media were described. Perhaps the earliest was electrophoresis in filter paper strips, as reported in 1939, in Brazil, by Von Klobusitzky and Köning for separation of snake venom from Bothrops Jararaca. A host of stabilizing media were then employed: potato starch (Kunkel and Slater, 1952), cellulose powder (Haglund and

Tiselius, 1950), ethanolyzed cellulose (Flodin and Kupke, 1956), glass powder (Hocevar and Northcote, 1957), minerals (Strain, 1939), asbestos (Butler and Stephen, 1947), silica gel (Consden et al., 1946), rubber sponges (Mitchell and Herrenberg, 1957), gelatin (Van Orden, 1971) and gels of polyvinyl alcohol (Reich and Sieber, 1966). These supports cannot be indiscriminately used in IEF, since many of them bear fixed charges which, due to the extremely low ionic strength typical of IEF (Righetti, 1980), would severely interfere with formation of pH gradients. I will limit my discussion to three anticonvective media, which are today most commonly used in IEF: agarose and polyacrylamide gels and cellulose acetate foils.

3.2.1. Structure of agarose matrices

Agarose is a purified linear galactan hydrocolloid isolated from agar or recovered directly from agar-bearing marine algae, such as the class Rhodophyta (red algae). It was first introduced as a support for zonal electrophoresis by Hjertén (1961, 1962b) as a replacement of agar gels. Indeed, agar was suggested (Araki, 1956) to be a mixture of a neutral 'agarose' and an ionic 'agaropectin' polymer. The formula for the repeating unit, agarobiose, was shown to be an alternating 1,3-linked β-D-galactopyranose and 1,4-linked 3,6-anhydro α-L-galactopyranose (I).

(I)

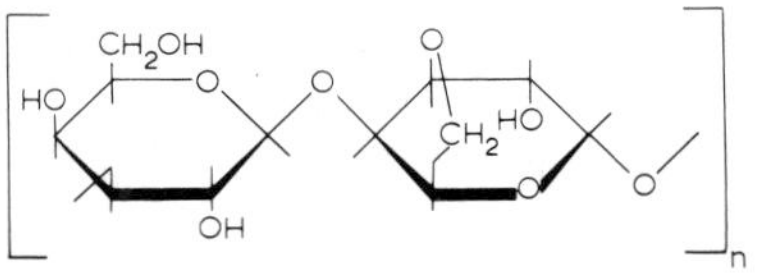

Indeed, even the 'neutral' fraction, agarose, contains some charged groups, consisting of L-galactose-6-sulfate moieties (II):

(II)

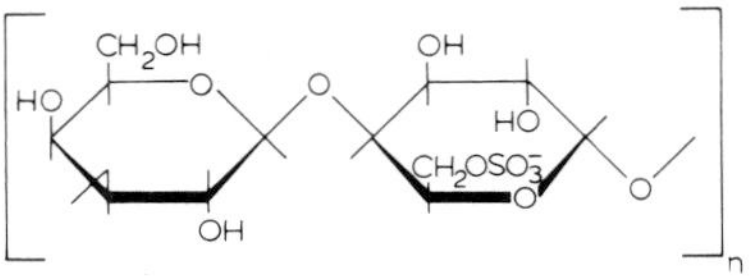

and of pyruvic acid derivatives in the form of ketal 4,6-(1-carboxy-ethylidene)-D-galactose (III):

(III)

Low-charge agarose (with a sulphur content from 0.09 to 0.18%) was found to be suitable for most electrophoretic separations, but was still disastrous for any IEF separation. Better preparations were obtained by purification through a QAE-Sephadex (Johansson and Hjertén, 1974), by DEAE-Sephadex chromatography (Duckworth and Yaphe, 1971), by alkaline treatment (Porath et al., 1971) or by alkaline desulphation in the presence of sodium borohydride followed by reduction with lithium aluminium hydride in dioxan (Låås, 1972).

Agarose is a particularly attractive matrix for IEF, because it can be gelled with an extremely large pore size, unmatched by any other gel. (Righetti et al. (1981b) have titrated the porosity of agarose gels and found that the most diluted gel which can be cast with agarose for IEF (0.16%) has an average pore diameter of 500 nm. Serwer (1980), using Miles HGT (P) agarose, has described gels containing as little as 0.075% solids, which should have an average porosity (diam.) of ca. 800 nm. How can agarose present such an open pore structure? It has been demonstrated (Arnott et al., 1974) that this polysaccharide in solution exists as a double helix (Fig. 3.2, left) and therefore it is considerably more rigid than a polyacrylamide strand. Moreover, 7 to 11 such helices form bundles which extend as long rods (Fig. 3.2, right) thus further strength-ening the architectural framework of the gel. It does not seem possible to cast hydrophilic gels with porosities greater than the 500–800 nm range reported, since their structure will collapse under 1 g gravity. In fact agarose gels are non-equilibrium structures, and undergo a slow collapse or tightening of the gel matrix with time. This causes syneresis or exudation of water from the gel upon standing. The only way to obtain porosities greater than 1 μm is to use siliceous skeletons and the capillary system should be a ceramic body (Haller, 1965).

3.2.2. Structure and properties of polyacrylamide gels

Polyacrylamide gels, as first described by Raymond and Weintraub (1959) and subsequently characterized by Ornstein (1964) and Davis

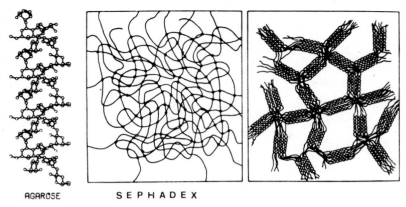

AGAROSE   S E P H A D E X

Fig. 3.2. *Left:* the agarose double helix viewed perpendicular to the helix axis. The hydroxymethyl groups are located along the helix perimeter. *Center and right:* a schematic representation of the agarose gel network (*right*), in comparison with a network such as Sephadex (*center*), formed from free chains at similar polymer concentration. Note the lateral aggregation of double helices in agarose gels, which strengthens the supporting gel structures. (From Arnott et al., 1974.)

(1964) are among the most versatile and most popular matrices for electrophoretic separations. Their popularity stems from several fundamental properties:

(a) optical clarity, including ultraviolet (280 nm) transparency;

(b) electrical neutrality, due to absence of charged groups (as opposed to carboxyls in starch and carboxyls and sulphates in agaroses);

(c) availability in a wide range of pore sizes.

Their chemical formula, as commonly polymerized from acrylamide and N,N'-methylene bisacrylamide (Bis), is shown in Fig. 3.3, together with the three, most widely employed catalysts, persulphate (ammonium or potassium) riboflavin (or its 5'-phosphate derivative) and N,N,N',N'-tetramethylethylenediamine (TEMED). Their polymerization kinetics, as a function of different crosslinkers (Gelfi and Righetti, 1981a), of temperature (Gelfi and Righetti, 1981b) or of catalysts (Righetti et al., 1981c) have been recently described. The chemical structures of the various crosslinkers used are given in Table 3.1. The general properties of polyacrylamide gels have been reviewed by Chrambach and Rodbard (1971), Chrambach et al. (1976) and Tanaka (1981). Hydrophilic gels

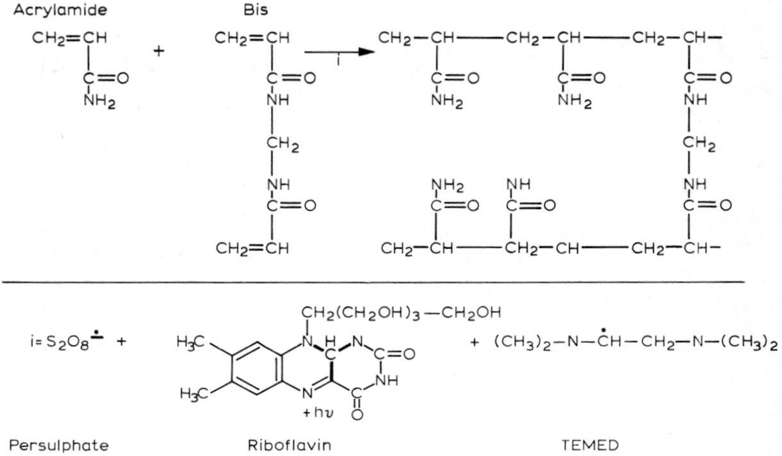

Fig. 3.3. The polymerization reaction of acrylamide. The structure of acrylamide, N,N'-methylenebisacrylamide and of a representative segment of cross-linked polyacrylamide are shown. Initiators, designated by i, shown are persulfate, riboflavin, and $N,N,N'N'$-tetramethylenediamine. Light is designated as $h\nu$. (From Crambach and Rodbard, 1971.)

are considered to be a network of flexible polymer chains, into whose interstices macromolecules are forced to migrate by an applied potential difference, according to a partition governed by steric factors. Large molecules can only penetrate into regions where the meshes in the net are large, while small molecules find their way into tightly knit regions of the network, closer to the cross-links. The partition coefficient, thus, corresponds to the part of the whole space that is 'permitted' to a given macromolecule (Flodin, 1961). Different models of the gel structure have been described: geometric, statistical and thermodynamic. A typical assembly of geometrical models is given in Fig. 3.4. One of the oldest is that of Porath (1963) who has considered them to be conical (Fig. 3.4B). Squire (1964) has used a broader collection of constraining shapes, by assuming that the pore size distribution corresponds to that produced by equal amounts of cones, hollow cylinders (Fig. 3.4D) and crevices. This last pore model could be seen as 'cracks on the wall' and could be

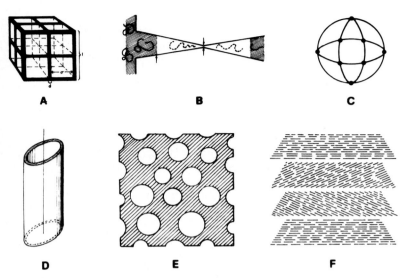

Fig. 3.4. Geometric pore shapes in gels as suggested by Ornstein (1964) (A), Porath (1963) (B), Casassa (1967) (C,D,F), Ackers (1967) (D,E) and Squire (1964) (B,D, F). (From Righetti, 1981.)

exemplified as parallel planes (Fig. 3.4F). Cylindrical pores have also been assumed by Ackers (1964), who has also regarded them as circular and lying in a plane perpendicular to the direction of movement of the molecules (Fig. 3.4E). The distribution probability of flexible macromolecules within voids represented as spherical cavities (Fig. 3.4C), cylindrical pores and slab-shaped cavities has been calculated by Casassa (1967) using random-flight statistics. Ornstein (1964) has assumed that the chains in a polyacrylamide gel follow the edges of cubes in a cubical matrix (Fig. 3.4A). Actually, one of the oldest models is the 'random meshwork of fibers' of Ogston (1958), who has calculated the fraction of space available to a sphere with radius r_s if straight cylindrical rods with radius r_r are distributed in the space in a random fashion with an average density of L length units of rods per volume unit of space. Laurent and Killander (1964) have extended this model by assuming that the molecule is a sphere of a given radius, while the network is described as infinitely long straight rods that are randomly located in space. Among the 'random' models, Giddings et al. (1968) have also hypothesized a gel structure consisting of an isotropic network of random planes in which all plane orientations are equally represented.

In order to circumvent the necessity of postulating specific geometric shapes for the pores within gel partitioning systems, statistical calibrating functions, which relate the molecular size of the macromolecule to the volume available to it, have been proposed. Hohn and Pollmann (1963) assumed that K_{av} is related to molecular mass by a Boltzman distribution, and found this relationship to fit well for oligonucleotides. Ackers (1967) assumed that the volume of the domains in the gel, where molecules of a certain radius could just reach, could be described by a normal (Gaussian) distribution. Rodbard (1976) has suggested that the pore sizes follow a 'log-normal' distribution, i.e., it is the logarithm of the pore size that obeys a Gaussian curve; this would provide a more satisfactory fit for the data of Fawcett and Morris (1966) on pore radius distribution in polyacrylamide gels. As a further refinement of this model, Rodbard (1976) has proposed a 'logistic' distribution, since it provides simple equations which can be fitted more easily to small desk-top calculators or drawn directly on logit-log paper.

Another way to avoid specific pore geometries, is to apply classical thermodynamic theories to the partitioning of a solute between two phases. Thus, Albertsson (1960) has proposed a thermodynamic relationship of the Brönsted-type to the partitioning of cells between non-miscibile liquid phases. This approach has been extended by Fischer (1969) who has calculated thermodynamic parameters for solute transfer between bulk liquid and gel phase. In another line of thinking, Bode (1980) has explained molecular sieving in polyacrylamide gels as stemming from the properties of a hypothetical 'viscosity-emulsion' composed of two interlacing fluid compartments endowed with different frictional coefficients. In this model, the gel is imagined as a visco-elastic matrix composed of layers of parallel sheets each of which comprises fluctuating polymer chains inserted into an unspecified backbone. Fluctuations due to the thermal agitation are centred symmetrically around the median plane of each sheet. The motions of the polymer chains give rise to elastic forces directed towards any compact object

$$\bar{p} = K d \big/ \sqrt{C} \qquad (\text{RAYMOND and NAKAMICHI, 1962})$$

$$\bar{p} = \frac{K'd}{C} + K'' \qquad (\text{TOMBS, 1965})$$

$$\bar{p} \propto 1 \big/ \sqrt[3]{C} \qquad (\text{RODBARD and CHRAMBACH, 1970})$$

$$\bar{p} = 140.7 \times C^{-0.7} (\text{RIGHETTI, BROST \& SNYDER, 1981})$$

Fig. 3.5. Proposed equations linking the mean pore size $(\bar{p})$ to the gel concentration (C). The first three equations have been derived for polyacrylamide gels, the fourth for agarose matrices only. (From Righetti, 1981.)

which tends to invade the volume otherwise available to the polymers for molecular reorientation. No matter how we try to describe gel porosity, a more interesting question is if there is a way to measure or to correlate a given pore diameter with the concentration of (monomeric or polymeric) material used to cast the gel. Four different equations, which link the mean pore diameter $(\bar{p})$ to the gel concentration (C), have been described (Fig. 3.5). Thus, in turn, the proportionality has been found to be: $\bar{p} \propto C^{-0.5}$ (Raymond and Nakamichi, 1962) or $\bar{p} \propto C^{-1}$ (Tombs, 1965) or $\bar{p} \propto C^{-0.3}$ (Rodbard and Chrambach, 1970) or $\bar{p} \propto C^{-0.7}$ (Righetti et al., 1981b). How is it that there is such a wide disagreement among different groups? In reality, only the first two equations are in conflict since they refer to the same type of regularly cross-linked ($3-5\%$ C) polyacrylamide gel (indeed Rodbard [1976] for these types of gels is heavily in favor of $\bar{p} \propto C^{-0.5}$). The last two equations have been derived for different types of gels and thus different gel structures. The proportionality $\bar{p} \propto C^{-0.3}$ reflects the aggregate network of highly cross-linked polyacrylamides (see ahead) while $\bar{p} \propto C^{-0.7}$ refers to highly diluted agarose gels (0.16% to 1%), which probably exhibit a close-to-ideal behavior. At higher concentrations, such as commonly found in polyacrylamides, the structure regularity could be lost, for instance due to chain entanglements (Ziabicki, 1979) or other non-ideal behavior so that the pore size could diminish more rapidly than as expressed in our equation. As stated ($\S 3.2.3$) the pore limit for polyacrylamides appears to be around 500 nm, while for agaroses 500–800 nm have been reported. This maximum pore size, however, should be considered as a dynamic, rather than a static, value. Since the measurements have been made by moving latex particles electrophoretically through the gel, it is quite possible that they physically force apart the gel strands to make way through the gel maze, in which case the pore diameter 'at rest' should be smaller than the one measured (Righetti et al., 1981b). This is in agreement with an observation by Gordon (1975) that, when diffusion alone is occurring in a gel, the pore size through which a given macromolecule can pass is smaller than under the conditions of electrophoresis. As a practical consequence of this, recovery of macromolecules from gels is best achieved by electrophoretic, rather than diffusional, means (see §4.6).

3.2.3. Highly cross-linked gels. Use of different cross-linkers

There are two ways to increase the pore size in polyacrylamide gels: one is to form very diluted gels (the lowest concentration being 2% T, 2.2% C) (Shaaya, 1976) the other is to increase progressively the percent cross-linker (% C) at fixed amounts of solids (acrylamide + cross-linker, % T) (Fawcett and Morris, 1966). Highly diluted polyacrylamide gels are gluey, extremely difficult to handle and do not appear to reach a pore size greater than 80 nm (Righetti et al., 1981b). The approach to high pore size through high levels of cross-linker appears to be more promising. The first to suggest this was indeed Davis (1964) who recommended 3% T, 20% C_{Bis} gels for the large pore segments used in disc electrophoresis. Rodbard et al. (1972) demonstrated that 50% C gels were practically non-sieving for macromolecules up to a few million molecular mass. What happens in these gels has been elegantly demonstrated in the electron microscope work of Rüchel and Brager (1975) and of Rüchel et al. (1978). The cross of Fig. 3.6 effectively illustrates the change in gel

---

Fig. 3.6. Transmission electron microscope images of polyacrylamide gels with varying concentrations of total acrylamide (T) and varying concentrations of the cross-linker bisacrylamide (C). The values shown are T/C; the bar equals 400 nm. In the horizontal series, C remains constant and T increases left to right. This series reveals a decreasing cell as well as pore size, with increasing T, a relationship of gel structure proportional to the known sieving properties of the gels. In the vertical series, T remained constant and C increased from bottom to top. This series reveals a nearly parabolical function of the gel structure with a minimum size of ca. 5% C and confirms previous findings concerning the relationship between cross-linkage and sieving in the gel. (From Rüchel et al., 1978.)

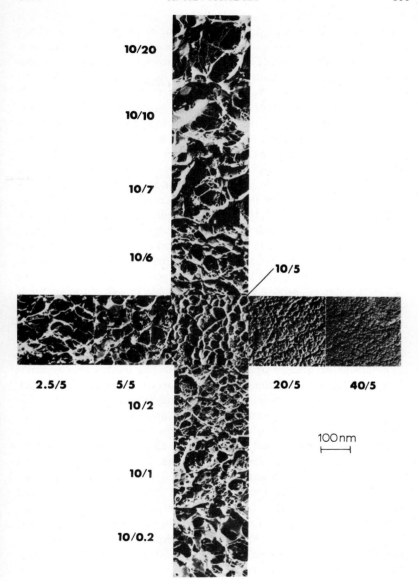

10/20

10/10

10/7

10/6

10/5

2.5/5   5/5   20/5   40/5

10/2

10/1

10/0.2

100 nm

structure that results from alterations in the T/C ratios. The horizontal arm shows the structural changes which occur by progressively increasing % T at fixed values of cross-linker (5% C throughout). From a large pore structure for dilute gels (2.5% T) the pore size progressively and mono-tonically decreases to a tightly knit sieve in high % T gels (40% T). If we now run a series at fixed % T (10%) and varying amounts of cross-link, the behavior of the matrix is quite different. As shown in the vertical arm of Fig. 3.6, highly porous structures can be found at both, very low % C (0.2%) and again at high % C (20% and higher). Thus the porosity changes follow a nearly parabolic function with structural units of a minimum size at about 5% C. Incidentally, this is the morphological confirmation of the empirical fact that gels of the 5–7% C range pro-duce maximum retardation for migrating macroions (Fawcett and Morris, 1966; Hjertén et al., 1969; Margolis and Wrigley, 1975). The explanation to the high porosity in high % gels has been given by Gelfi and Righetti (1981a,b) and Righetti (1981). A low % C gel (2–4%) is a homogeneous gel, i.e., the average chain density in the matrix is quite constant in any direction of the space, and the most abundant topological elements are singlets (chains connected with their two ends to two different cross-links). Faulty topologies (or gel irregularities), namely loops (chains connected with both ends to the same pair of junctions) are extremely rare. However, as % C increases, the equilibrium during gelation is shifted toward the formation of faulty topologies (loops and doublets) with the simultaneous occurrence of two phenomena: the linear polyacrylamide chains grow shorter and thicker and the knots (or junctions) grow larger. As % C is further increased, the chains tend to disappear and the pre-dominant topological elements left are concatenated or annealed loops, which eventually grow into beads or spheres. In fact, the size of these 'clusters' or 'coagula' must approach the size of visible light (300–700 nm) since these gels are strongly opaque. Thus, opaque (high % C) gels have the following general properties: (a) they are 'faulty gels', i.e., the distribution of polyacrylamide chains within the matrix is non-homogeneous (domains of high and low strand density); (b) they are highly porous; (c) they have lower elasticity than transparent gels; (d) they have lower capacity for swelling in the solvent than transparent gels. The explanation to the enormous porosity of very high % C gels could be the following: in regularly cross-linked gels, the macromolecules have to move through the gel network, through pores within the chain framework; in highly or pure cross-link gels, the macromolecule will move around the gel grains (or beads), in the void volume in between the gel 'coagula'. Thus, paradoxically, highly cross-linked gels resemble gel filtration media. As demonstrated by Righetti et al. (1981b) 50–60% C gels can reach a pore size as high as 500–600 nm, which could render them particularly attractive for IEF fractionation, but there are practical limits to their use. First of all, they are mechanically unstable and totally unelastic (above 50% C) so that they can only be used in horizontal, thin-layer sheets and must be glued to the glass support via a silane bridge (Bianchi Bosisio et al., 1980). Secondly, above 30% C_{Bis} they become hydrophobic and the matrix keeps exuding water and collapsing with time. This apparent hydrophobicity is not just due to the use of Bis, because it is apparent also with gels cast with more hydrophilic cross-links, such as DHEBA, but is probably due to the fact that the high density of bridges prevents the matrix from swelling in the solvent so that polymer–polymer contacts are favored over polymer–solvent interactions. A happy compromise are thus 30% C_{Bis} or 40% C_{DHEBA} gels, which combine a high pore size (200–250 nm) with acceptable handling properties and swellability.

The versatility of polyacrylamide gels is also shown by the large number of cross-linkers, besides Bis, which can be used to cast gels with peculiar properties for different fractionation purposes. I have listed them in Table 3.1. O'Connell and Brady (1976) have reported the use of DHEBA for casting reversible gels, since the 1,2-diol structure of DHEBA renders them susceptible to cleavage by oxidation with periodic acid. The same principle should also apply to DATD gels, as described by Anker (1970). Alternatively EDA gels (Paus, 1971) could be used, since this cross-link contains ester bonds which undergo base-catalyzed hydrolytic cleavage. The poly (ethylene glycol) diacrylate cross-links, as reported by Righetti et al. (1981b) belong to the same series of ester derivatives. As the latest addition to the series, BAC gels, which contain a disulfide bridge cleavable by thiols, have been described (Hansen, 1976; Hansen et al. 1980). These gels appear to be particularly useful for fractionation of RNA, and are the only ones which can be liquefied under very mild and almost physiological conditions. Practically any cross-link can be used, but definitely DATD and TACT should be

TABLE 3.1
Structural data on cross-linkers (from Righetti et al., 1981)

Name	Abbr.	Chemical formula	$\bar{M}_r$	Chain length
N,N'-methylene bisacrylamide	Bis (MBA)	$CH_2=CH-C-NH-CH_2-NH-C-CH=CH_2$ (both $C=O$)	154	9
Ethylene diacrylate	EDA	$CH_2=CH-C-O-CH_2-CH_2-O-C-CH=CH_2$ (both $C=O$)	170	10
N,N'-(1,2-Dihydroxyethylene) bisacrylamide	DHEBA	$CH_2=CH-C-NH-CH-CH-NH-C-CH=CH_2$ (with OH OH, both $C=O$)	200	10
N,N'-Diallyltartardiamide	DATD	$CH_2=CH-CH_2-NH-C-CH-CH-C-NH-CH_2-CH=CH_2$ (with O OH OH O)	288	12
N,N',N''-Triallylcitric triamide	TACT	$CH_2=CH-CH_2-NH-C-CH_2-C-CH_2-C-NH-CH_2-CH=CH_2$; $HO-C-NH-CH_2-CH=CH_2$	292	12–13
Poly(ethylene glycol) diacrylate 200	PEGDA$_{200}$	$CH_2=CH-C-O-CH_2OCH_2CH_2O-C-CH=CH_2$	214	13
N,N'-Bisacrylylcystamine	BAC	$CH_2=CH-C-NH-CH_2-CH_2-S-S-CH_2CH_2NHCCH=CH_2$	260	14
Poly(ethylene glycol) diacrylate 400	PEGDA$_{400}$	$CH_2=CH-C-O-CH_2CH_2OCH_2CH_2OCH_2CH_2-$ $-O-CH_2CH_2OCH_2CH_2OCH_2CH_2-$	400	25

avoided since, being allyl derivatives, they are inhibitors of gel polymerization. Their use at high % C is simply disastrous (Gelfi and Righetti,
1981a,b; Bianchi Bosisio et al., 1980).

3.2.4. IEF in thin agarose slabs

Agarose suitable for IEF is now available from Marine Colloids (FMC
Corp., Rockland, ME) as Isogel agarose, from LKB Produkter (as agarose
EF) and from Pharmacia (as agarose IEF). While the process for the
production of the first two matrices is still secret, in the last case it is
known that the agarose, which still contains negative charges (mostly
sulphate groups) is rendered suitable for IEF by a balancing process, i.e.,
by covalently fixing to the polysaccharide enough positive charges to
counterbalance the effect of the negative ones. This process had already
been described by Grubb (1973) and by Ragetli and Weintraub (1966)
except that Pharmacia has probably used quaternary amino groups, with
pH values outside the pH ranges used in IEF. Agarose IEF has been independently described by Rosén et al. (1979a,b), Rosén (1980) and by
Saravis et al. (1979, 1980a and b) and Saravis and Zamcheck (1979).
Excellent resolution of very high $\bar{M}_r$ proteins such as α_2-macroglobulin
(780,000 daltons), 19 immunoglobulins M (IgM) (900,000 daltons),
kehole limpet hemocyanin (3×10^6 daltons) and zinc glycinate human
tumor marker (2×10^6 daltons), could be achieved by IEF in agarose
matrices. Besides its much larger pore size, agarose also presents other
significant advantages over other supports, such as polyacrylamide:
(a) it is non-toxic, while acrylamide and Bis are neurotoxins; (b)
it is gelled without the aid of catalysts, which in polyacrylamide
produce long-lived radicals; (c) it is fully compatible with subsequent
immunofixation and crossed electrophoretic techniques; (d) it allows
quick and efficient staining and destaining and it is easily dried for
permanent records (these last properties are now shared by ultrathin
polyacrylamide films, as will be described later). A simple gel casting
procedure has been made possible by the use of clear, flexible, polyester-
based plastic film (Gelbond, from Marine Colloids) rendered hydrophylic on one side for adherence of gels. Gel Bond is available in 0.1
(flexible) and 0.2 (semi-rigid) mm thickness as well as in sheets and
rolls (I suggest you buy sheets, as rolls curl when cut to desired size).
A piece of Gel Bond film (usually 12.5×12.5 cm) is placed, hydro-

phobic side down, (it is the side where a drop of water does not spread, but stays as a bead), on a leveling table and, if needed, rolled flat with the aid of a rubber roller (Fig. 3.7A,B). In order to avoid too rapid a heat dissipation due to contact with the metal table, which could result in uneven gelling, thermal insulation is suggested by placing a sheet of polystyrol foam beneath the Gel Bond film (LKB instruction leaflet). Any concentration of agarose between 0.5% and 1.25% (w/v) can be used for IEF, however, since the higher concentrations retain water better than the lower ones, a good compromise is to cast a 0.8% agarose gel, which still has an excellent porosity (170 nm). The desired volume of agarose solution (or a bigger volume, which can then be dispensed in aliquots in individual glass tubes and stored at $4\,^{\circ}C$ until remelted when needed) is boiled in a water bath for 10 min. The flask should be covered with a cold finger or inverted small beaker to prevent evaporation (Parafilm is inadequate because it melts or tears!). When agarose is fully dissolved, the flask is transferred to a water bath at $56-58\,^{\circ}C$, and allowed to equilibrate with added pre-heated carrier ampholytes to a final concentration of 2.5% and poured directly on the Gel Bond

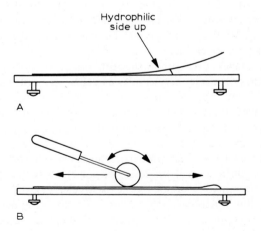

Fig. 3.7. Placing (A) and rolling flat (B) a Gelbond film on a levelling table for casting agarose gels. (From Saravis and Cook, courtesy of Marine Colloids Division; instruction leaflet, 1979.)

horizontal sheet (Fig. 3.8A). The volume of agarose solution is chosen so as to have a thickness of about 1 mm of gelled matrix. If thinner gels are required (e.g., 0.8 mm), it is better to let the agarose gel in a vertical cassette, as regularly used for polyacrylamide gel polymerization (Fig. 3.8B), in order to ensure uniform gel thickness, particularly along the borders. The casting assembly will have to be prewarmed and shall contain, on one face, the Gel Bond film. The gelled plates are usually aged overnight at 4 °C to increase the mechanical strength of the gels. However, it has been suggested (Saravis et al., 1980a) that it is preferable to keep them at room temperature, to eliminate the syneresis (gel collapse and water exudation) observed upon storage at 4 °C. The samples are applied as 3 μl droplets onto the gel surface, or soaked into Paratex filter pieces (LKB 2117-103), or with the aid of a 52 μm thick Mylar sheet containing loading slits or, in the case of biopsies, by direct tissue application onto the matrix (Saravis et al., 1979). The focusing process is usually over within 60 min at a constant wattage of 10 W (1100 V at steady-state) (Vesterberg, 1980). Since agarose is still not completely charge-free, and there is still a considerable water transport to the cathode, it has been suggested that it is desirable to drain the gel continuously with a thin, cellulose paper foil at the cathode (Thymann, 1980), or to run the IEF experiment at 15 °C instead of 4 °C (Vesterberg, 1980). However this seems to apply only to early batches of

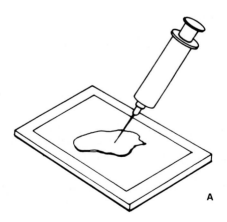

A

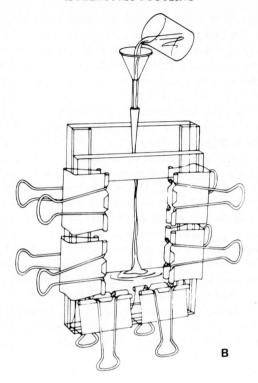

Fig. 3.8. (A) For casting agarose gels 1 mm thick or thicker, the gel solution is poured directly on a horizontal sheet of Gelbond film (do not use a syringe, though, but an Erlenmeyer!). (B) Cassette for casting agarose gels thinner than 1 mm. The casting apparatus should be pre-warmed at ca. 50 °C and should contain the Gelbond film on one face. (Both figures from Saravis and Cook, courtesy of Marine Colloids Division; instruction leaflet, 1979.)

agarose IEF, which had not been properly 'balanced': with present-day brands, this residual cathodic drift should be strongly reduced. At the end of the IEF run, the agarose plate is quickly fixed in 33% methanol, 5% TCA and 3.5% sulfosalicylic acid in water. After the fixation step, the agarose gel is dried onto the Gel Bond film by the sequence illustrated in Fig. 3.9. It is covered with filter paper and blotting towels and left for 30 min under a 500 g to 1 kg weight, then fully dried with a hair dryer and freed from the gel-adhering filter paper by a quick wetting step (steps 3 and 4 in Fig. 3.9). At this point, the dried agarose film can be quickly stained and destained by classical Coomassie Blue methods or, as suggested by Saravis et al. (1980b), by using Crowle and Cline (1977) stain (2.5 g Crocein Scarlet, 150 mg Coomassie Brilliant Blue R-250, 50 ml glacial acetic acid, 30 g TCA in 1 liter final volume).

When running proteins under denaturing and/or reducing conditions (8–9 M urea and 2-mercaptoethanol) it would be ideal to use an open-

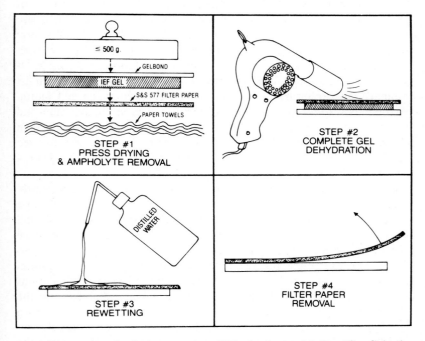

Fig. 3.9. Procedure for drying an agarose IEF gel prior to staining. After fixing in TCA, the gel is pressed for 30 min against filter paper and paper towels with a weight of 0.5 to 1 kg/dm². The dehydration is completed with a hair-dryer (step 2) and then the filter paper removed from the agarose film by a quick re-wetting step (3 and 4). (From Saravis and Cook, courtesy of Marine Colloids Division; instruction leaflet, 1979.)

pore matrix like agarose, since upon unfolding the stokes radius of a protein is increased (Creighton, 1979). Unfortunately agarose gels and urea are not quite compatible, since this matrix is extensively hydrogen bonded and urea is a typical H-bond breaking agent. This problem has been solved by Olsson and Låås (1981) by simply increasing the agarose concentration for IEF runs from 0.8% to 2% (w/v), adding 10% sorbitol to the gelling solution and letting the 8 M urea, 2% agarose and 10% sorbitol solution gel in a cassette for 15–20 h at 21 °C. If there is a danger that cyanate ions, in equilibrium with urea (their concentration is 20 mM in 8 M urea at 20°C and pH > 6), will carbamylate amino groups in proteins, 1% β-mercaptoethanol (an effective scavenger of cyanate ions) can be gelled in the agarose matrix. Agarose IEF is also suitable for a variety of applications. Saravis et al. (1980b) have described an immunoperoxidase labelling by which focused antigens can be detected, with agarose gel, at an antibody dilution of <1:2,500. If this sensitivity is not high enough, a 100-fold amplification can be obtained by exploiting the strong binding between avidin and biotin. A complex is formed among biotinylated secondary antibody, avidin and biotinylated horseradish peroxidase: since this is much more stable than the anti-primary (goat anti-mouse)–peroxidase conjugate, the focused antigen can be detected at antibody dilutions of up to 1:250,000, a remarkable increment in sensitivity.

Another application of agarose IEF is in crossed immunoelectrofocusing. This technique involves the focusing of proteins in the first dimension followed by perpendicular electrophoresis in the second dimension, in an antibody containing agarose matrix. In the past, since the IEF dimension was in polyacrylamide gel, this could only be performed by overlaying the IEF-polyacrylamide strip onto the agarose-antibody gel (Söderholm et al., 1975). Today, since both dimensions can be run in agarose, the IEF-strip can be directly moulded into the second dimension agarose plate, with more reproducible results. As a last approach, agarose IEF can be used in connection with SDS-electrophoresis in the second dimension, to display $pI/\overline{M}_r$ maps, after the classical O'Farrell technique (1975). In this case, by substituting in the first dimension the polyacrylamide gel with an agarose matrix, there are better chances that very large macromolecules will reach a true pI position, thus allowing more reproducible data from run to run.

3.2.5. IEF in cellulose acetate foils

Cellulose acetate, first introduced for zone electrophoresis by Kohn (1957) (it became so popular, that it used to be called Kohn's electrophoresis tout court) would also be an ideal support for IEF, since it is practically a non-sieving matrix. However this membrane, as prepared by manufacturers, contains enough carboxyl groups to fully disrupt the IEF process. The strong electroendoosmotic flow is further enhanced by the extremely low ionic strength of focused carrier ampholytes. Harada (1975) and Harada et al. (1980) were the first to describe the use of an acetate membrane, treated with a surface active agent (available from Fuji Photo, Tokyo, Japan, as Separax-EF) for use in IEF, but the method has found little application, possibly due to a still substantial electroosmosis. More recently, Ambler (1978a,b) has described an extensive methylation process of cellulose acetate strips leading to a support suitable for IEF. Here, again, since the electroosmotic flow is not completely abolished, the run has to be performed in 8% Ampholine (against the customary levels of 2% in polyacrylamide) and 10% glycerol. In narrow ranges, weaker catholytes and anolytes, such as 0.2 M lysine and 0.2 M acetic acid, respectively, should be used. Ambler's methylation process with boron trifluoride in methanol could be applied only to cellulose acetate gel membranes (e.g., Cellogel) but was unsuitable with dry, non-gel membranes, such as Sepraphore III (Gelman, Ann Arbor, MI, USA). Ambler and Walker (1979), have now described a modified methylation process which can also be applied to this last group of membranes, by more gradually processing them through a series of aqueous solutions of increasing methanol concentrations. In my experience, however, the process of IEF in cellulose acetate matrices is still quite erratic and often difficult to control. More work needs to be done in this field, both to improve the quality of the support and to standardize the method. It is also highly desirable, for future use of this technique, that properly treated and carefully controlled cellulose acetate foils are made available to the scientific community by a commercial firm. In fact, just now, Iso Sepraphore membranes (Gelman) have been described for use in IEF (Janik and Dane, 1981). Among the few applications reported, Boussios and Bertles (1978) have described the IEF separation of human globin chains in acetate plates impregnated with 5 M urea and 5% β-mercaptoethanol. The good separations obtained might be due to the fact that urea stabilizes the pH gradients by strongly reducing the cathodic drift (Gianazza et al., 1979).

3.2.6. IEF in thin layers of granulated gels

This is the analytical version of the Radola technique, already discussed on p. 129. Granulated gels offer some advantages over continuously polymerized polyacrylamide gels. With granular gels, there are no risks due to polymerization catalysts that are known to produce artefacts in polyacrylamide gels (see p. 178). Gel porosity is not a limiting factor. There is no steric hindrance on focusing very large molecules.

The methodology is essentially the same described previously for preparative application (see p. 129). Glass plates of several sizes (20 × 10 cm; 20 × 20 cm or 40 × 20 cm) are used and coated with a layer of Sephadex G-75 (superfine) of 0.6 to 1 mm thickness for analytical purpose or 2 mm for small-scale preparative purposes. The thick gel slurry, containing 1% Ampholine, is dried in air until the gel coat does not move when the plate is inclined at an angle of 45°. This corresponds to a loss of approximately 25% water from the gel slurry. The sample is applied usually at the middle or at the anodic side, soaked in pieces of filter paper or as a drop or a streak on the gel layer. The plate is then placed on the cooling block of the Desaga 'Double Chamber'. The electrical field is applied through the flat carbon electrodes sitting on pads of a thick paper (such as type MN866 with a surface weight of 650 g/m², Machery, Nagel & Co., Duren, West Germany) soaked with 1 M sulphuric acid at the anode and 2 M ethylene diamine at the cathode (Radola, 1969, 1973a,b, 1975). Proteins are located after focusing by the print technique with a filter paper (type MN 827, surface weight 270 g/m², Machery, Nagel & Co.) gently rolled on the gel layer. The prints are stained with one of the following stains: Light Green SF, Coomassie Brilliant Blue R-250, Coomassie Violet R-150 and Coomassie Brilliant Blue G-250.

A variant of this technique has been described by Ziegler and Köhler (1976). The Sephadex slurry is made to also contain acrylamide monomer and a cross-linking agent. After IEF, the whole layer is polymerized by spraying a solution containing polymerization initiators in a concentrated buffer onto the gel. The advantages are the possibility of focusing very large molecules without steric hindrance combined with the ease of handling a polymerized gel. There is a risk, however, that during the IEF process proteins could react with unpolymerized monomers, especially at alkaline pH. Other granulated gels could be used, such as polyacrylamide beads (Bio Gel P) and even use of plastic grains (Pevikon C-870) has been reported (Harpel and Kueppers, 1980). In this last case, however, the bed should be mixed with a hydrophilic polymer, since plastic does not absorb water and the ohmic resistance in the gel bed could be too high.

3.2.7. IEF in polyacrylamide gels

The development of analytical techniques for IEF in polyacrylamide gel and the advantages of this method over IEF in liquid media were described in a flurry of papers published independently and almost simultaneously (Awdeh et al., 1968; Dale and Latner, 1968; Fawcett, 1968; Leaback and Rutter, 1968; Riley and Coleman, 1968; Wrigley, 1968a,b; Catsimpoolas, 1969a). These papers clearly demonstrated the considerable potential of this new technique. However, during the transition from IEF in all liquid media to gels, few systematic studies were made to optimize factors such as gel composition and electrolysis conditions. Many of these early systems, particularly those using tubes, were plagued by a marked instability in the pH gradients which somewhat restricted their general applicability (the 'cathodic drift' of Righetti and Drysdale, 1971 or 'plateau phenomenon' or Finlayson and Chrambach, 1971). Thus, it is doubtful that at the beginning of gel IEF steady-state conditions and sufficiently stable pH gradients allowing macromolecules to reach their pI positions have ever been achieved. Fortunately, today's systems have been developed enough to allow, in most cases, stable banding patterns to be obtained and pI positions to be reached long before pH gradient decay becomes significant. Moreover, with the revolutionary technique of immobilized (or grafted) pH gradients (§ 1.11.7) the problem of 'cathodic drift' has been completely eliminated: the pH gradients are stable indefinitely and the protein zones never move away from their pI position in the polyacrylamide gel. For all these reasons, and due to the high versatility of polyacrylamide gels (the possibility of being cast in narrow to wide pore size; in tubes and slabs, in ultrathin layers and as easily liquefiable matrices) I feel that this gel has still much to offer and will continue to be widely used for most IEF applications.

3.2.8. Gel composition

Selection of an appropriate gel composition is a key consideration for polyacrylamide gel IEF (PAGIEF). Two factors are particularly important: gel porosity and the concentration of ampholyte in the gel. Ideally, one should use the highest gel porosity consistent with satisfactory mechanical strength. Both properties are largely determined by the relative amounts of acrylamide and cross-linker which are usually expressed by the relationship (Hjertén, 1962):

$$T = \frac{\text{g acrylamide} + \text{cross-linker}}{100 \text{ ml solution}}$$

$$C = \frac{\text{g cross-linker}}{\% \ T}$$

The lowest acrylamide concentration that will form a satisfactory gel for IEF is about $T = 3\%$, $C = 4\%$ (Bis), but these gels are very soft and gluey Gels of composition $T = 5\%$, $C = 3\%$ (Bis) (Vesterberg, 1971b) or $T = 4\%$, $C = 4\%$ (Righetti and Drysdale, 1971) have been found suitable for many purposes. The upper size of proteins that can be focused in these gels is about 1.5×10^6 daltons, but such large proteins often require very long focusing times. On the other hand, smaller proteins of 100,000 daltons or less focus quite rapidly. These gels are rather soft, and require careful handling. However, since most proteins are smaller than 200,000 daltons, more robust gels, e.g., $C = 4\%$, $T = 5\%$ may be preferred and are, in fact, to be recommended. These gels are easily handled for staining and densitometry.

Another important requirement for PAGIEF is the development of sufficiently stable pH gradients to ensure that proteins overcome restrictive effects of the gel to reach their equilibrium points. Also, since the position of a focused protein will depend on the actual pH gradient, it is desirable to develop reproducible pH gradients. Failure to develop stable and reproducible pH gradients can lead to considerable difficulties in correlating banding patterns from different experiments or with different batches of carrier ampholytes of the same nominal pH.

We have found that a certain minimum concentration of 2% (w/v) carrier ampholytes is required for the development of stable pH gradients in polyacrylamide gels. Below this level and particularly below 1% (w/v) pH gradients in gels are often unstable (Finlayson and Chrambach, 1971, Righetti and Drysdale, 1971). Considerable local heating may occur in these gels. A likely explanation is that a certain level of ampholyte is required to provide sufficient overlap in adjacent ampholyte zones for good conductivity throughout the gel. A minimum concentration of ampholytes is also required to buffer large amounts of protein in a pH zone. Ideally, the carrier ampholytes should provide a buffering capacity of at least 0.3 μequiv/mg at pH values close to their pI (Vesterberg, 1973c). In addition to ensuring adequate levels of ampholytes in different pH ranges, additional precautions are also required in pH gradients that do not include ampholytes buffering near pH 7 but which, nevertheless, must include pH 7 between the electrodes. In such cases, much of the electrical potential drop will occur at this point, at the expense of other parts of the gradient which may be underfocused. This situation may be rectified by adding a small amount of pH range 3–10 or 6–8 to the gel ampholytes (Haglund, 1971).

3.2.9. Gel polymerization

I will describe in this section a general methodology valid for both gel cylinders and slabs. For detailed information on polymerization kinetics as a function of different cross-linkers, of temperature and of catalyst, the reader is referred to Gelfi and Righetti (1981a,b) and Righetti et al. (1981c). Reagents required: recrystallized acrylamide and Bis (or any other appropriate cross-link except DATD); N,N,N',N'-tetramethyl ethylendiamine (TEMED); ammonium (or potassium) persulphate, riboflavin or riboflavin-5'-phosphate (FMN); carrier ampholytes; optional gel additives, e.g., sucrose, sorbitol, glycerol, urea, neutral or zwitterionic detergents. The following stock solutions should be prepared:

Gel solution	28.8 g acrylamide and 1.2 g Bis in 100 ml distilled water (30% T, 4% C)
Carrier ampholytes	40% (w/v) solutions [except for pH 2.5–4 and pH 9–11 Ampholine, which are 20%

(w/v) solutions]

TEMED	1 in 10 dilution in distilled water
Ammonium persulphate	40% (w/v) solution prepared fresh weekly
Riboflavin (or FMN)	4 mg/100 distilled water

Care should be taken in handling acrylamide since it is a neurotoxin in its monomeric form. Contact with skin should be prevented. There is usually no need to maintain separate solution of acrylamide and Bis except when using high cross-linked gels. Solutions of riboflavin and acrylamide are sensitive to light and should be stored in a bottle wrapped in aluminium foil (a brown bottle does not give much protection from light) at 4 °C. It is best to prepare stocks which will last no longer than a month, as acrylamide in solution tends to hydrolyse very slowly producing free acrylic acid.

Procedure (1): Add the desired volume of gel cocktail (a 5 to 6% T, 4% C_{Bis} or 4% C_{DHEBA} or 4% C_{BAC} and 2% carrier ampholyte is a universal recipe) to a conical flask with a side arm.

(2) Stopper the flask and degas thoroughly on a water or vacuum line for 5 min. For reproducible operation, I prefer a mechanical pump giving a vacuum of ca. 0.1 mmHg and connect the Erlenmayer to a nitrogen tank via a three-way tap, so that, once the vacuum is interrupted, the flask is filled back with nitrogen instead of air. Degassing, in order to be efficient, should be done while warming the gel mix and not while keeping it in ice.

(3) Add 5 μl/ml of TEMED and 1 μl/ml of ammonium persulphate (in this order) and mix rapidly.

(4) Immediately pour the gel cocktail in the slab cassette or in the gel tubes. In the former case, the mould is filled to the brim, while in the latter case the cylinder is filled to about 0.5 to 1 cm from the rim, so that the gel solution should be overlayed with a dilute catalyst mix to ensure uniform gel polymerization and to avoid a gel meniscus.

(5) Set the gels aside to polymerize at room temperature. This should take no more than 30 min: by this time, regularly cross-linked gels (2–5% C) should have achieved 95% conversion into polymer. Highly cross-linked gels ($> 10\% C$) might require overnight polymerization. If needed, gels can be thermostated at 45–50°C: this will greatly accelerate the polymerization process. Precast gels may be kept for several weeks before use if properly sealed.

Notes: 1. All gel reagents should be of the highest quality. In particular, acrylamide should be recrystallized to remove free acrylic acid, a common contaminant in reagent grade acrylamide. Acrylic acid will generate fixed charges in the gel causing electroendoosmosis and instability of the pH gradient. In addition, unreacted monomeric acrylamide can react with amino, sulphydryl and phenolic hydroxyl groups at elevated pH. This reaction proceeds slowly at pH 9, but is quite fast (30 min) at pH 11 (Dirksen and Chrambach, 1972). The acrylamide may be recrystallized from chloroform and Bis from acetone, according to Loening (1967). Several satisfactory preparations of purified acrylamide are commercially available (e.g., from Bio Rad, BDH, LKB and Polysciences).

2. Gels may also be photopolymerized in glass or quartz tubes by using riboflavin or FMN at a final concentration of 656 μM. The solutions should be placed at 10 cm distance from fluorescent tubes, e.g., Philips TL 20 W/SS at room temperature for at least 8 h. 1 h photopolymerization, as commonly accepted in the literature (Brackenridge and Bachelard, 1969) barely affects 60% conversion of monomers into polymer. This is why gels photopolymerized for 1 h usually have a soft and gluey consistency, just like gels cross linked with DATD.

3. The porosity and mechanical strength of the gels may be altered by varying the amount of acrylamide solution and/or compensating volume changes of water.

4. Other gel additives, e.g., sucrose, urea, non-ionic detergents may also be added at step 1.

5. The addition of TEMED is not essential since the carrier ampholytes serve a similar function in gel polymerization (Riley and Coleman, 1968).

6. Several factors can affect the polymerization and physical properties of acrylamide gels (see Rodbard and Chrambach, 1971; Chrambach and Rodbard, 1972). Experimental procedures should, therefore, be standardized as much as possible. Particular attention should be paid to the following variables: (1) purity of reagents, (2) degassing conditions, (3) amount and type of catalyst added, (4) temperature of polymerization, (5) duration of polymerization. Perhaps the most troublesome factor is the amount of persulphate required as catalyst. Different preparations may vary in their potency and an appropriate level should be determined by trial and error. Only the minimum amount required for satisfactory gelling should be used since persulphate may oxidize carrier

ampholytes and slightly alter the resulting pH gradient. Excess persulphate also generates potentially troublesome free-radicals (Brewer, 1967, Fantes and Furminger, 1967; Mitchell, 1967) which, if not removed prior to sample application, may create artifacts, e.g., by oxidizing cysteine residues to cysteic acid. However, according to Dirksen and Chrambach (1972), even riboflavin as a catalyst has oxidizing properties on thiol groups. Free radicals seem to persist in polyacylamide gels, even after prolonged electrolysis (Peterson, 1971). Only pre-electrophoresis of thioglycolate into the gel, in amounts equivalent to the catalyst added (1 to 5 mM), results in reducing conditions in the gel (Dirksen and Chrambach, 1972).

7. pH range of Ampholine. It should be noted that the pH gradients given by carrier ampholytes may not always be linear or have a uniform field strength over the indicated pH range (Vesterberg, 1975). Consequently, it may be advisable to fortify deficient regions with ampholytes of the appropriate pH range. For example, Vesterberg (1973c) recommends the addition of pH 9–11 Ampholine to the wide range pH 3.5–10 Ampholine to extend the linearity of the gradient in the alkaline region. He also recommends the addition of pH 4–6 and 5–7 Ampholine to improve the distribution of the field strength. It is also possible to produce specially tailored pH gradients by expanding one section of the gradient by the addition of the appropriate pH range Ampholine. This often allows a better separation of components of particular interest while displaying all components in the wider pH range.

3.2.10. Choice of electrode solutions

Non-volatile free acids and bases are in general used as anolyte and catholyte, respectively. For gel slabs, most authors recommend stronger electrolyte solutions than are commonly used for gel tube apparatus, since in horizontal slabs the elctrolyte chambers are reduced to thick filter paper strips impregnated with anolyte and catholyte. Solutions of 1 M phosphoric acid and 1 M sodium hydroxide have been found satisfactory for most purposes. Alternatively, as suggested in the Bio Rad technical bulletin 1030 (1975), a solution containing a mixture of NaOH and $Ca(OH)_2$ could be used as catholyte. Calcium hydroxide will form insoluble calcium carbonate and prevent the build up of carbonate in the basic electrolyte, thus eliminating the migration of these ions into the gel. Although strong acid and base are useful for most pH ranges, better results are often obtained by using other electrolytes such as ampholytes, whose pH ranges encompass that of the gel ampholyte. Thus Vesterberg (1975) used 1% Ampholine pH range 4–6 and 6–8 as anolyte and catholyte, respectively, to maximize the spread of a pH 5–7 Ampholine in a gel. Suggested formulations for anode and cathode solutions are given in Table 3.2 (as suggested in LKB, Pharmacia and Bio Rad technical bulletins). As seen from the table, there is a general agreement to use strong, concentrated acids and bases in the wide pH range (pH 3–10) while, in narrow ranges, progressively weaker anolytes and catholytes, or mixtures of carrier ampholytes, are recommended. When

TABLE 3.2
Suitable electrode solutions for IEF in gel slabs

pH range	Cathode (−)	Anode (+)
LKB		
3.5–9.5	1 M NaOH	1 M H_3PO_4
2.5–6	0.5% Ampholine, pH 5–7	1 M H_3PO_4
5–8.5	0.1–1 M NaOH	0.1–1.0 M H_3PO_4
	or 1% Ampholine, pH 8–10	or 1% Ampholine, pH 5–7
7.5–10.5	1 M NaOH	0.1% Ampholine, pH 7–9
Pharmacia		
3–10	1 M NaOH	40 mM aspartic acid
2.5–5	0.2 M histidine	0.1 M H_2SO_4
	or 0.1 M NaOH	
4–6.5	0.2 M histidine	40 mM glutamic acid
5–8	1 M NaOH	40 mM glutamic acid
6.5–9	1 M NaOH	0.25 M HEPES[a]
8–10.5	1 M NaOH	0.25 M HEPES
Bio Rad		
3–10	1 M NaOH	1 N H_3PO_4
2.9–5.0	2% Biolyte, pH 6–8	1 N H_3PO_4
3.7–6.2	1 M NaOH	1 N H_3PO_4
4.4–7.2	1 M NaOH	1 N H_3PO_4
5.6–8.2	1 M NaOH	2% Biolyte, pH 4–6
7.3–9.5	1 M NaOH	1 N HEPES
8.2–10.0	1 M NaOH	1 N HEPES

[a]   HEPES: *N*-2-hydroxyethylpiperazine-*N′*-2-ethane-sulfonic acid.

focusing in gel tubes, the electrolytes should be diluted to less than 0.1 M concentration, so as to ensure high conductivity in the liquid phase and a minimum of voltage drop between the platinum wire and the gel extremities. On one occasion (Swanson and Sanders, 1975) an ammonia buffer, pH 10, has been used as catholyte and an acetate buffer (pH 4) as anolyte, but this seems to me contrary to the principle of IEF fractionation. When focusing in gel slabs, the wicks should be wetted with appropriate solutions before application to the gel. Care should be taken to ensure that the wicks are uniformly wetted. A convenient and satisfactory method is to place the wicks on a glass plate and pipette a standard amount of electrolyte between the interface of the wick and the plate. This ensures uniform wetting and the removal of any air bubbles that might be trapped in the wick. Excess liquid should be removed by blotting (note: the wicks should have a rather dry appearance; if excess anolyte and catholyte form pools on the gel surface, this results in a strong distortion of the focused ampholyte pattern).

3.2.11. IEF in polyacrylamide slabs

I will describe here the general methodology and instrumentation developed for IEF in thin layers. (The hottest topic is, of course, IEF in ultrathin matrices so, if you just cannot hold your breath, skip to §3.2.12.) IEF in thin slabs of polyacrylamide offers several advantages. Perhaps the most important is the excellent comparison and convenience offered by analyzing multiple samples in parallel tracks of a single gel rather than in several individual tubes. Possible modification of sample by contact with electrolytes, an ever present hazard with gel tubes, is minimized in gel slabs since samples may be applied at any point onto the gel surface. In addition, the thin-layer technique simplifies many of the methods used to analyze focused patterns such as conventional staining, autoradiography, immunofixation, and zymogram techniques involving overlayers (see later). The rectangular cross-section of slabs allows densitometry and photography with less risk of optical artifacts, as compared to gel tubes. Developed gels may also be dried onto plates for convenient assessment or storage. Finally the thin-layer technique usually allows better control of heat dissipation and is less subject to pH gradient instability. Perhaps the only advantage of gel rods over gel slabs run horizontally is the possibility, in the former system, of running oxygen-sensitive enzymes under strictly anaerobic conditions. Suitable apparatus for isoelectric focusing in thin slabs of polyacrylamide gel is now available from several sources. The gel may be held either vertically or horizontally between cooling plates. I prefer horizontal systems since the gel is subject to less mechanical stress. Simpler electrode arrangements are possible and there is more flexibility in methods for sample application. Gels are usually only 1−2 mm thick and separations may be conducted either along the long or the short axis. Gels are in general cooled through their bottom surface and the electrodes are placed directly on the gel surface. Evaporation is minimized with a close fitting

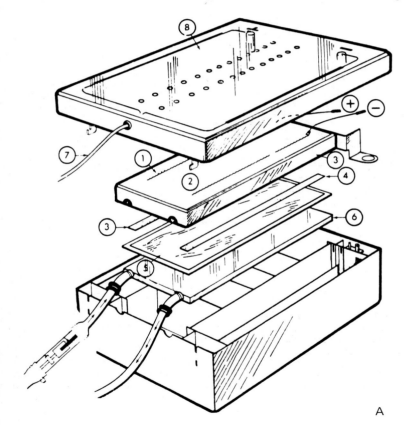

A

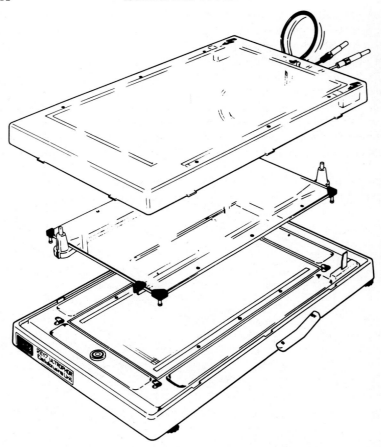

Fig. 3.10. (A): drawing of the LKB 2117 Multiphor. Arrangement for IEF in the short side (10 cm). (1) and (2) Anodic and cathodic platinum wires, respectively, for runs in the short side. (3) Platinum electrode holder and gel cover lid. (3) and (4) Anodic and cathodic filter paper strips, respectively. (5) Thin-layer polyacrylamide gel cast on a 1 mm thick glass slab. (6) Cooling block. (7) Cable connection to power supply. (8) Safety cover lid. (B): drawing of the Ultrophor 2217 chamber. Notice the movable electrodes and tightly fitting lid for atmospheric control. The chamber also has adjustable feet for levelling. (Courtesy of LKB Produkter AB.)

lid over the gel. A typical apparatus design for thin layer is shown in Fig. 3.10A (LKB Multiphor 2117). This has been the prototype on which most other chambers have been modeled. Other apparatus are: the 'Double Chamber' (Desaga), the Pharmacia Flat Bed (FBE 3000) and the Bio Rad Chambers models 1045 and 1415. The Pharmacia system has the following features: (a) adjustable feet for levelling; (b) a large (25 × 25 cm) aluminium cooling plate covered by a Teflon cloth; (c) movable electrodes, which permit choice of different gel size. The Bio Rad Chambers are available in two sizes, the model 1405 with a cooling platform of 12.5 × 22 cm and the model 1415 with a cooling block of 12.5 × 43 cm. An interesting feature is that the cover lid is fitted with a condensation coil, just larger than the gel surface, which prevents condensation on the gel. All these instruments are supplied with lateral electrolyte chambers for use with several other electrophoretic techniques, such as immuno- and SDS-electrophoresis (this explains Multiphor: multiple electrophoretic analysis) (see also Righetti et al., 1980). As a latest addition, LKB has introduced the Ultrophor 2217, a highly compact, shallow chamber only for IEF analysis, provided with movable electrodes and atmospheric control via a tightly fitting cover lid (Fig. 3.20B). Practical advice for the use of a typical thin-layer system is described below. As with tube systems, minor variations are required for different apparatus, but specific procedures are usually adequately covered in instruction manuals. I will not give here any practical recipes for gel polymerization, since they vary according to experimental needs. General guide lines are given in §3.2.8 and §3.2.9. In most horizontal systems, the gel is cast between two glass plates. One of these plates may be plastic if a template for sample application is required. The plates are usually separated by 1 or 2 mm by a rubber gasket or by a spacer frame lightly coated with water repellent grease. The plates are clamped together to form a watertight chamber (Fig. 3.11). One of the two plates may also be used as a template to imprint indentations into the gel for convenient sample application. These indentations should not extend into more than 30–50% of the gel thickness. Deeper troughs create conductivity problems and may cause local skewing of the pH gradient. Should this happen, the trough should be filled with a slurry of Sephadex and gel ampholyte to improve the conductivity. When only a few analyses are required or when different analyses are planned for several

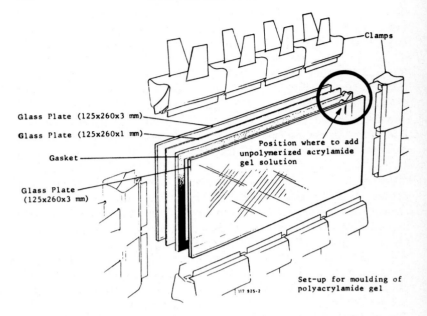

Fig. 3.11. Polyacrylamide moulding cell for the LKB 2117 Multiphor apparatus.
(From Davis, 1975.)

samples, the gel chamber may be subdivided by suitable frames. Gels are usually cast in a vertical position (see Fig. 3.12). Both plates must be thoroughly clean. However, one may be coated with Kodak Photo-Flow (1:1,000 dilution) before use to facilitate its separation from the gel surface. The gel solution should be applied to the chamber with a fine tipped pipette or with a syringe. It is advisable to apply the first portion by tilting the assembly on a lower corner to avoid air pockets. As the chamber fills, it should be gradually lowered to the horizontal. The height of the gel should correspond to the distance between electrodes. After filling the chamber to the appropriate level, the rubber gasket is closed, to prevent air from entering the chamber. Since the gel chamber is completely sealed by the rubber gasket, there is no need to add water to the top of the gel solution. In chambers with an open top, the gel

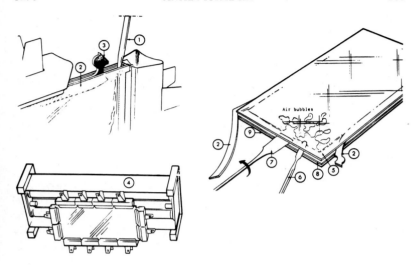

Fig. 3.12. Sequential procedure for filling a moulding cell (*upper left*), photo-polymerizing the polyacrylamide gel (*lower left*) and removing the glass cover slab from the thin layer gel (*right side*). This last operation is best done with two spatulas; while slightly pressing down the upper part of the polyacrylamide gel with a thin spatula (6), a broader spatula (7) is slowly twisted as indicated by the arrow. As air bubbles begin to penetrate the space between the upper gel surface and glass plate (5), the broader spatula is gently twisted to a vertical position and the upper glass plate (5) removed. During this operation care should be taken not to loosen the gel from the lower glass plate (8) as this would cause severe band distortion during subsequent IEF. (Courtesy of LKB Produkter AB.)

solution should be overlaid with water containing traces of catalyst, to give a flat surface and uniform polymerization. When the gel has completely polymerized, the gel chamber should be placed in a horizontal position in the refrigerator or on a cooling block. The resulting contraction usually causes a partial separation of the gel from the cover plate which simplifies its subsequent removal. Shortly before use, the cover lid should be carefully separated. To do this, the gel chamber should be placed on a flat surface and the top plate pried apart from the gel surface by gentle leverage with a spatula against the bottom plate (see Fig. 3.12). It is extremely important not to disturb the adhesion of the gel to the bottom plate, since this may alter the electrical conductivity at that point and cause a local distortion in the pH gradient. Once the cassette has been opened, the thick filter paper strips soaked in anolyte and catholyte, as described in §3.2.10, are layered on the gel surface. It is important that these strips are positioned straight and parallel to each other and at such a distance that the electrode platinum wire (in the case of chambers with fixed electrode lids) can sit centered on top of them when the electric circuit is closed. For this purpose, most apparatus are provided with graph grids which serve as templates for electrolyte strip positioning. The gel plate should now be placed on top of the cooling block. To ensure good contact with a fast heat transfer it is essential to have a water layer between the gel plate and the cooling face. This may be achieved by wetting both the cooling surface and the gel plate with a solution of non-ionic detergent. Alternatively, kerosene or light paraffin oil can be used. The gel plate should be lowered slowly over this solution so that a straight water front, without trapped air, forms between the two faces (see Fig. 3.13). The electrode is now placed in position in preparation for the IEF run. Horizontal polymerization is also possible as illustrated in the principle of the casting tray adopted by Bio Rad. Fig. 3.14A shows the gel casting process and Fig. 3.14B a vertical section of the cast gel. A plastic tray is provided with two lateral ridges of different thickness, from 0.5 to 2 mm. By lowering a glass slab onto these ridges (Fig. 3.14A), a chamber is automatically formed of the desired size and thickness. The gel solution to be polymerized is gently expelled from the tip of a pipette moved along the glass edge, so that the chamber is filled by gravity and capillary suction. LKB has adopted a similar principle, called the 'capillary gel casting kit', which consists of two 0.5 mm thick polyvinyl chloride spacers glued to a 3 mm thick glass slab. The top plate which has to form the chamber is 20 mm shorter than the bottom glass, so that the gel solution can be easily injected in the 0.5 mm gap and fill the mould by capillary attraction. A detailed description of this procedure is given by Esen (1981). Horizontal gel casting is particularly attractive when preparing ultrathin gels. For additional details for focusing in thin-layer slabs the reader is referred to articles by Davies (1975), Karlsson et al. (1973) and Vesterberg (1975). The following general aspects should also be considered:

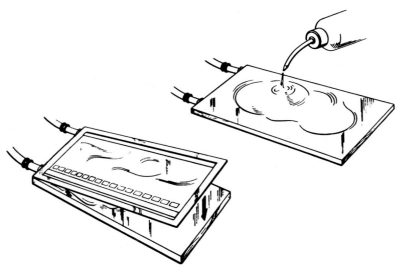

Efficient cooling of the gel plate

Fig. 3.13. Setting of the polyacrylamide gel plate on the cooling block. The cooling block is first wetted with a solution of non-ionic detergent, or with kerosene or light paraffin oil, and then the gel plate is gently lowered on the cooling block, so that air bubbles are completely excluded. This ensures uniform cooling. (Courtesy of LKB Produkter AB.)

(1) The efficiency of gel cooling. The efficiency of heat transfer per cm^2 surface area will depend on the materials used to construct the cooling block and the thin slab plate (beryllium oxide blocks, as used in heat shields by NASA, have been suggested).

(2) More efficient cooling is achieved in slabs of high surface area/gel volume.

(3) The thicker the gel, the more heat will be generated for any given applied potential difference. Ideally, one would like to focus in a thin film of gel, that is why ultrathin gels have been developed (§3.2.13).

(4) Temperature of coolant. The lower the temperature of the coolant, the higher the electrical load that can be applied.

(5) Concentration of ampholytes. The current carried in the gel is

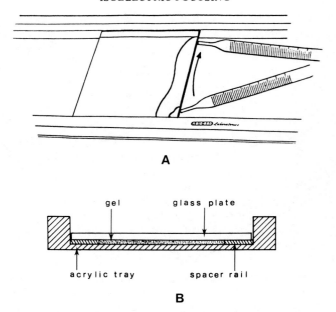

A

gel   glass plate

acrylic tray   spacer rail

B

Fig. 3.14. (A) Gel casting procedure with the BioRad apparatus. The acrylamide solution is slowly released while moving the tip of the pipette along the edge of the glass plate. (B) Vertical section of the moulding chamber for gel polymerization. (From BioRad Labs., instruction manual.)

directly related to the level of ampholytes. As mentioned before, a minimum level of 2% ampholine is, however, required for most purposes to allow the development of smooth and stable pH gradients. Gel additives such as sucrose may increase the resistance in a gel by increasing the viscosity.

(6) The gel path. The important value in determining electrical load is the potential drop per cm. Thus higher potential differences may be tolerated when running the samples lengthwise rather than across the width of the gel.

(7) Variation in conductivity. As mentioned previously, the electrical field need not be uniformly distributed throughout the gel. In the wide pH range Ampholine, pH 3.5–10, the areas close to the electrodes heat more than the remainder of the gel in the initial stages of IEF (Davies,

1975). As the pH gradient develops these warm zones broaden and move towards the middle of the gel. At very high electrical loads, these 'hot spots' correspond to areas of condensation on the inner surface of the electrode lid.

3.2.12. Precast plates

LKB has introduced ready-made polyacrylamide matrices, called 'Ampholine PAG plates', which are 1 mm thick and are cast on a thin plastic film which allows easy handling of the gel slabs. They are available in different pH intervals: a broad (pH 3.5–9.5) and narrow pH ranges: 4–5, 4.0–6.5 and 5.5–8.5. The pH 4–5 plate is particularly made for the phenotyping of α_1-antitrypsin (α_1-AT) since all the α_1-AT family focuses in this region. In addition, there exists a PAG-plate for HbA_{1c} (the glycosylated derivative of normal human adult hemoglobin, whose quantitation is important for diabetic conditions) covering a nominal pH range 6–8. This plate, however, contains probably a 'separator' which flattens the pH gradient around pH 7, thus allowing resolution between the adjacent bands of HbA and HbA_{1c}. Recently, Serva has introduced a whole series of precast plates, called 'Servalyt Precotes'. They have two basic features: they are cast on a strong polyester sheet (180 μm thick) called 'Gel-fix' to which they adhere by covalent bond; they are ultrathin (100 μm gel thickness) made of 5% T, 3% C monomer mixture, containing 3% Servalyt and 10% sorbitol (for ultrathin gels, see §3.2.13). They are available in several pH intervals: one wide pH 3–10 range; four encompassing 3 pH units: pH 3–6; pH 4–7; pH 5–8 and pH 6–9; two Precotes spanning 2 pH units: pH 4–6 and pH 5–7 and lastly 2 pre-cast gels covering 1 pH unit spans: pH 4–5 and pH 7–8. The availability of ready-made gel plates appears to be particular useful in clinical analysis, where a large number of samples has to be screened daily. I must mention here that, in addition to the carrier ampholyte ranges reported in §1.6.2, Serva now has available a series of Servalyt covering a 3 pH unit span: pH 3–6; pH 4–7; pH 5–8 and pH 6–9; a series of 1 pH unit ranges: pH 2–3; pH 3–4; pH 4–5; pH 5–6; pH 6–7 and pH 7–8 and even some intervals spanning only 1/2 pH unit: pH 4–4.5; pH 4.5–5; pH 5.0–5.5 and pH 5.5–6. In addition to that a new broad range (pH 3–10) called 'iso-dalt grade' is now available, which is claimed to give reproducible IEF patterns for two-dimensional separations. Also LKB has recently marketed a new series of Ampholine (1818) in the following ranges: pH 3.5–9.5; pH 2.5–4.5; pH 4.0–6.5 and pH 5.0–8.0 which are carefully balanced mixtures yielding linear pH gradients in both polyacrylamide gels and agarose matrices.

3.2.13. IEF in ultrathin polyacrylamide layers

This technique represents one of the most interesting recent breakthroughs in polyacrylamide gel slab IEF. The method has been developed by Görg et al. (1977, 1978, 1979a,b, 1980a). Since ultrathin polyacrylamide gels would be impossible to handle and would break apart during the process of staining and destaining, they have to be supported by a suitable, tear-resistant backing. At the beginning, cellophane sheets were used, but today we tend to prefer polyester foils (available from Serva as 'Gel fix' sheets or from LKB as 'PAG-foils'). These polyester supports should not be confused with the 'Gel-bond' film developed by Marine Colloids. The latter has been processed as a support for agarose IEF, contains a film of agarose dried onto one of the sides and thus has one hydrophilic and one hydrophobic side; the former is a polyester which has been partially hydrolyzed in 6 N NaOH and then coated on both surfaces with a silane ester which is supposed to fix covalently the gel to the plastic surface (Radola, 1980). (I doubt that the polyacrylamide layer is really chemically bound to the polyester; probably the chemical treatment generates a micro-etched surface to which the gel sticks.) When the gel is polymerized in detergent, it slips away from the plastic foil, indicating the absence of covalent bonds (see Valkonen et al., 1980). Gel layers of 120, 240 or 360 μm thickness can be cast by using as gaskets in the gel cassette either one, two or three U-shaped frames cut out from Parafilm sheets. The chamber is assembled as shown in Fig. 3.15: a thick glass plate (3 mm) is sprinkled with some water, the polyester foil lowered onto it and pressed flat with a rubber roller (see Fig. 3.7), the U-gasket placed next and finally the cover lid lowered onto it. The cover, if made of glass, should be coated with Photo Flow (Kodak) or with 0.01% Nonidet P-40 solution for ease of removal. I prefer a thick (0.8–1 cm) Perspex slab cut, in the upper edge, 2 cm higher than the bottom glass plate: this greatly facilitates filling of the chamber and ensures proper adhesion of the gel to the polyester foil (see also Fig. 3.8B). Since filling of the chamber is not an easy operation,

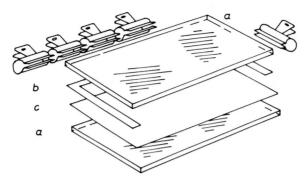

Fig. 3.15. Procedure for casting ultrathin polyacrylamide gels onto polyester foils. Assembly of the gel cassette: (a) 3 mm thick glass (b) U-shaped Parafilm gasket (1 layer = 120 μm); (c) polyester foil. (From Görg et al., 1978.)

especially with 120 and 240 μm thin gels, the two upper clamps are removed and two paper clips introduced between the glass plates, thus allowing the insertion of the needle of the syringe containing the polymerization mixture (Fig. 3.16). When the clips are removed and the clamps fastened in place, the liquid level in the chamber rises to the desired height. Several advantages are inherent in this technique:

(a) resolution is markedly improved in ultrathin gels as compared with conventional gels;

(b) heat transfer is much more efficient, thus allowing higher field strengths and sharper zones than in 1–2 mm thick slabs;

(c) by adhering to the polyester foil during all operation steps, the gels can be handled very conveniently and are protected from fracture;

(d) staining, destaining and drying are completed in a fraction of the time needed for thicker gels;

(e) the demand for carrier ampholytes and other reagents is drastically lowered,

(f) zymograms can be developed within a few minutes, thus retaining the high band sharpness of the IEF dimension. For even thinner gels, in the 50–100 μm range, a modification of the above method, called the 'flap technique', has been described by Radola (1980a,b). Since it would

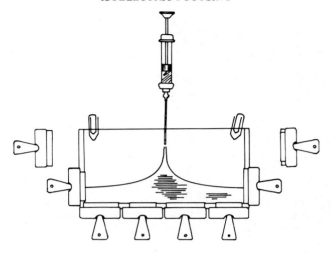

Fig. 3.16. Moulding chamber for ultrathin gels. The two upper clamps are removed and two paper clips inserted between the glass walls, for easier pouring of the polymerization solution. After the gel mixture has been completely poured in the cassette, the paper clips are removed and the two clamps placed back in the frame, so that the liquid rises to fill up the chamber. (From Görg et al., 1978.)

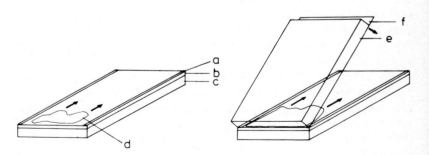

Fig. 3.17. Gel preparation with the flap technique: (a) 50 or 100 μm spacer strips; (b) silanized glass plate or polyester film; (c) glass base plate; (d) polymerization mixture; (e) glass cover plate; (f) cover film, 100 μm hydrophylic polyester film protruding at the upper end for easier removal of the polymerized gel. (From Radola, 1980a.)

be impossible, by the conventional casssette technique, to cast such ultrathin gels, the problem has been solved by using a horizontal glass plate with appropriate spacer strips, onto which the total volume of polymerization mixture is poured. The chamber is sealed by slowly lowering on it the cover plate, so that the gel mixture is spread evenly between the two plates (Fig. 3.17). The thin gel veil can be made to adhere to polyester films, or can be covalently bound to the glass surface by using a silane coupling agent (methacryloxypropyltrimethoxysilane) having the following formula:

$$CH_3-O-\underset{\underset{OCH_3}{|}}{\overset{\overset{OCH_3}{|}}{Si}}-CH_2-CH_2-CH_2-O-\underset{\underset{O}{\|}}{C}-\underset{\overset{CH_3}{|}}{C}=CH_2$$

In this last case a true covalent bond is formed since the silane bridge forms a:

$$-\overset{|}{Si}-O-\overset{|}{Si}-$$

link with the glass surface, while at the other extremity of the molecule, methacryl groups are available for copolymerization into the polyacrylamide gel. This reagent is available as Silane A-174 from Pharmacia or as Polyfix 1000 from Desaga or, for a few dollars per gallon, directly from the producing company (Union Carbide, Sisterville Plant, West Virginia, USA). Covalent bonding of polyacrylamide to glass, by the use of Silane A-174, has been independently described by Bianchi Bosisio et al. (1980) when working with highly brittle, highly Bis-cross-linked polyacrylamide gels. I recommend the following: make up a 0.2% Silane A-174 in anhydrous acetone. Just before use, dip a clean, thin glass in this solution for 30 s and let it drip to dryness. The silane solution, kept in the dark and at 4 °C, is stable for several months. Keep the set of silanized glasses separated from untreated glass: the silane coat is still active after use for few gel polymerization cycles and after washing so that, if the gel is inadvertently cast between two silanized plates, upon opening the cassette it will tear into two halves bound to each glass surface.

A third variant of the Görg technique has been described by Ansorge and De Mayer (1980). They are able to cast gels, by a sliding technique, in the thickness range 100 to 300 μm and up to considerable lengths (60 cm). A sketch of their set-up is shown in Fig. 3.18. The chamber consists of a flat, horizontal supporting bed provided with guiding studs on each side, between which a flat plate, to be coated with the gel, is laid down. Next to this plate, leaving a gap, a dummy plate having the same thickness is placed. PTFE spacer strips are laid over the plates next to their outer edges, and a third plate is placed on the spacers, so that it covers the dummy plate and a few millimeters of the plate that is to be coated. The cover plate can slide over the spacers; the sliding motion is constrained to one direction by the guiding studs. A slow horizontal sliding motion permits the gel solution to be spread evenly between the bottom and top plates. Between the guiding studs, the top and bottom plates are accessible and can be clamped firmly together after the space between them has been filled with gel solution. Ultrathin-layer gels represent one of the major recent innovations in IEF techniques and will probably wipe away previously used IEF methodology.

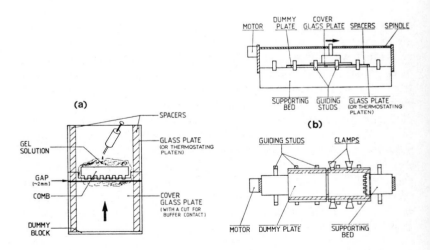

Fig. 3.18. Preparation of ultrathin gels by the sliding technique (schematic): (a) principle; (b) gel casting device. (From Ansorge and De Mayer, 1980.)

They are also particularly attractive in two dimensional techniques utilizing ultrathin matrices not only in the IEF, but also in the SDS step (Görg et al. 1980b, 1981).

3.2.14. Sample application to gel slabs and rods

Samples may be applied in a variety of ways to thin layer slabs. Some form of template, either with preformed indentations or basins that restrict lateral spread is advisable. As many as 20 tracks, each 1 cm wide, may be handled in the LKB gel plate. Small indentations in the gel are convenient for sample application. However, samples may be applied either directly to the gel surface, or on some suitable absorbent. Fig. 3.19 indicates some of the ways for applying samples to thin-layer plates. The choice of absorbent is important. Wadström and Smyth (1975a) have

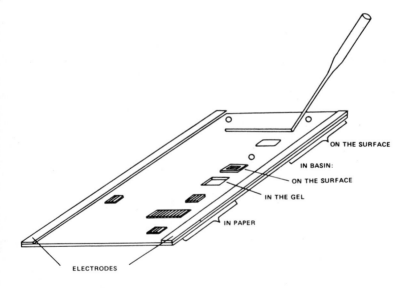

Fig. 3.19. Different ways of applying a protein solution to the gel for isoelectric focusing. From left to right: soaked into a square of chromatography paper; into a rectangle of the same material, and into two squares, one placed close to each electrode; in a basin in the gel; in a basin on the gel surface; as a droplet left on the gel; as droplet spread over a rectangular area; as a streak; and finally as two droplets, one close to each electrode. (From Vesterberg, 1973c.)

observed that many proteins are not readily eluted from the commonly used cellulose filter papers, in which case cellulose acetate or dessicated polyacrylamide may be more suitable. Samples of unknown concentration may be applied from a strip of paper cut in the form of an acute-angled triangle. Focused zones appear as long thin lines of linearly increasing concentrations, thereby facilitating the estimation of appropriate sample loads for subsequent experiments. Dilute samples may be loaded from a rectangular paper or from a suitably sized trough in the gel. As much as 150 μl may be applied in a trough, provided the sample is equilibrated with the same ampholyte concentration as the gel to prevent local discontinuities in the pH gradient. However, since strips of different materials often tend to adsorb uncontrolled amounts of sample, today the method of choice is to use application strips made of silicone rubber and containing slots or circular holes. They were first described by Altland (1977). Use a third common central cathodic or anodic electrode in the gel middle and two silicone rubber bands each with 48 circular holes *et voila, les jeux sont fait*: 96 samples can be focused in a single gel (see Fig. 3.20). A plastic frame with rectangular holes has also been reported by Chuat and Pilon (1977). Usually the strips should be positioned in a gel zone where they do not interfere with the expected focusing pattern. The samples should not be applied too near to the electrodes, where they might be exposed to extreme pH values. The distance between sample and electrodes should be about 10–15% of the total gel length; best separations are obtained when the samples are

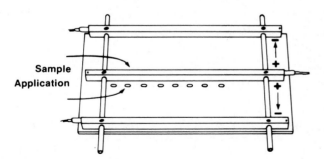

Fig. 3.20. Triple electrode set-up for screening a double number of samples. (From BioRad Catalog G, 1981.)

applied on a prefocused pH-gradient. The sample application strip must lie flat on the gel surface, free from air bubbles. In gel rods, samples may be applied directly to the top of the gel or incorporated into the gel solution before polymerization. Top loaded samples may usually be added at any period during the IEF run. When top loading, the sample zone should be rendered dense by addition of 5–10% glycerol or sucrose. The sample should be equilibrated in 2–4% carrier ampholytes so that it is protected, during the period required for migrating into the gel, from the acidic or basic environment of anolyte or catholyte bathing the gel extremities. The addition of a trace of bromophenol blue or other dye that does not bind to sample proteins greatly facilitates visualization of application of colorless samples. For larger volumes of dilute samples, they can be incorporated directly into the gel before polymerization. Internal loading should only be considered if samples are not adversely affected by gel constituents or by the heat of the polymerization reaction. Gel temperatures may reach as high as 40°C during polymerization (Righetti and Righetti, 1975).

When analyzing body fluids (e.g., urine, cerebrospinal fluid) two major problems are encountered: the protein content is very low and the salts rather high. This prevents direct IEF of the sample as such. This has been elegantly circumvented by Vesterberg and Hansén (1978): the sample is adsorbed by hydrophobic interaction on phenyl-Sepharose gel in small columns, thus eliminating salts and concentrating the proteins. Elution is affected by directly applying the hydrophobic resin to the gel surface: the combined effect of very low ionic strength in IEF, of cold gel temperature and of high field strength are very efficient in driving the adsorbed sample out of the Sepharose matrix to its pI position. Incidentally, this is also an excellent method for eluting proteins tenaciously bound to Affi-gels in affinity-chromatography (Haff et al., 1979).

3.2.15. Gel drying

Although gels can be stored wet sealed between transparent plastic sheets using a kitchen-type bag sealer (Yu and Spring, 1978), for filing and detection of radioactivity (autoradiography, fluorography) a dried film is a must. If the polyacrylamide gels are dried down without any support, they shrivel. However, by drying down onto a filter paper backing, the gel dimensions are preserved as the gel attaches itself to the paper support. Today, gel slab dryers, based on a simultaneous vacuum and heating action, are available from several commercial sources (Bio Rad, LKB, Pharmacia). The gel, after fixing, staining and destaining, is soaked for a few hours (only 30 min for ultrathin gels) in 3–5% glycerol containing enough methanol to prevent swelling. In the case of gradient gels, higher amounts of glycerol (e.g., 10–15%) are needed, otherwise the dense region of the gradient will crack upon drying. Two sheets of 3 MM filter paper are placed on the stainless-steel support screen and wetted with water. The gel slab is then aligned on this, care being taken not to trap air bubbles under the gel since this can result in cracking during drying. This is overlaid with a sheet of pre-wetted cellophane, followed by a porous plastic sheet and finally by the silicone foil (attached to the apparatus) forming a leak-proof seal. Vacuum is supplied by a water aspiration or vacuum pump and the heating block turned on. The time required for drying will have to be found by trial in each lab, as it depends on the type of gel used. If air is allowed to enter the assembly before drying is complete, the gel will crack irreparably. The resulting dried gel is sandwiched between its filter paper backing and a protective cellophane sheet. All this is not needed when running ultrathin gels supported by silanized polyester: the gel, after equilibration in 3% glycerol, can simply be left to dry in the air or in an oven at 40–50 °C. Several home-made methods for drying gels have been described (Lin et al., 1969; Hebert and Strobbel, 1974; Maizel, 1971; Mayer, 1976). When focusing peptides in ultrathin matrices after IEF the gel is pasted to a 3 MM filter paper and immediately placed in an oven at 110 °C: in a few minutes the gel is dehydrated to a film, thus fixing the peptides in their pI zone. Now the paper can be sprayed, as though it were a chromatogram, with a series of specific amino acid stains (Gianazza et al., 1979b).

3.2.16. IEF in gel cylinders

Most methods for IEF in gel cylinders use gel dimensions similar to those used in analytical 'disc' gel electrophoresis; i.e., 5–15 cm in length and 0.2–0.5 cm in diameter. Although some authors advocate the use of conventional electrophoretic apparatus, I think that many of these systems would benefit from more efficient heat control and different electrode arrangements. Fawcett (1968) described a simple method for

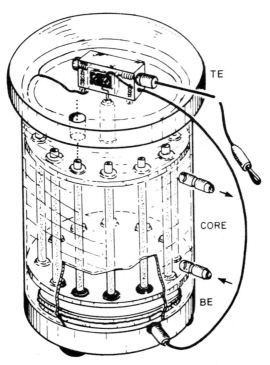

Fig. 3.21. Schematic drawing of the analytical gel-tube apparatus of Righetti and Drysdale (1971, 1973). TE, top electrode compartment fitted with a safety switch. BE, bottom electrode compartment. The central core is interchangeable with the preparative unit. (Courtesy of Medical Research Apparatus.)

overcoming heating problems by thermostating gel tubes in a chilled, circulating electrolyte solution. He also recognized the importance of minimizing the electrolyte volumes and devised the ingenious solution of inserting a platinum electrode directly into a layer of Ampholine on top of each gel tube. Righetti and Drysdale (1971) developed an apparatus whose salient features included efficient gel cooling, small electrolyte volumes and close juxtaposition of electrodes and gels. This apparatus is depicted schematically in Fig. 3.21. The gel tubes are held between the electrodes in a separate compartment through which coolant is circulated. The outer wall of the central core extends about 1 inch above the top end plate to form an electrolyte chamber. The lower electrolyte chamber is a separate unit which forms the base of the apparatus. Platinum wire electrodes are used. The top electrode is on the inner face of the lid while the bottom electrode is set into the inner face of the lower electrolyte chamber. The extremities of the tubes pass through rubber grommets in the end plates to make contact with the electrolytes. The electrodes are positioned so that they are less than 1 cm from the tube extremities when the apparatus is assembled. An appropriately placed aperture in the lid (which rotates freely around a central nylon post) allows samples to be applied directly to the gel surface without raising the lid. This aperture also allows the top compartment to be flushed with a gas if required. Small vents are also cut from the rim of the base for gas escape or flushing. Finally, the lid also contains a safety switch which is wired across both electrodes and only passes current when the lid is properly seated on the central post. This safety switch disconnects current from both electrodes whenever the lid is lifted or the apparatus accidentally knocked over. The apparatus usually holds 12 tubes (10 cm long and 0.3 cm i.d., volume approximately 1 ml) but other sizes may be used by changing the rubber grommets. Routinely, gels 3 mm in diameter are used since they can be cooled efficiently, while plastic is less subject to electroendoosmosis such as may may be of plastic, glass or quartz. Glass tubes dissipate heat more efficiently, while plastic is less subject to electroendosmosis such as may occur in a charged glass surface. However, gel adherence to plastic is less than in glass, which makes glass a preferred gel container. If necessary, the clean pyrex tubes can be coated with 1% Gelamide 250 (Polysciences, Warrington, PA) which also quenches the wall-generated electroendoosmotic flow, a constant problem in vertical tube IEF. General procedures for gel composition and polymerization are given in §3.2.8 and §3.2.9. For choice of electrode solution, see §3.2.10. For gel casting, the usual procedure of 'disc' electrophoresis is adopted: the glass tubes, whose bottoms are sealed with Parafilm, or plasticine or 'Vacutainer tops', stand vertically on a leveling table. After filling up to the desired height (usually 0.5 to 1 cm from the top rim) the gel cocktail is overlaid with catalyst mixture. Gel tubes are a must when samples have to be run under strict anaerobic conditions or at subzero temperatures. Except for these particular uses, the vast majority of scientists today use flat slabs, especially the ultrathin matrices.

3.2.17. Micro-isoelectric focusing

The presently available techniques for IEF in gel tubes or in slabs allow detection of few microgram proteins. However, when lower amounts of sample are available, it could be desirable to have systems which allow detection of protein in the nanogram range. One such system has been described by Grossbach (1972). He performs IEF in quartz capillaries of 50 μm, 100 μm and 300 μm i.d. and 65 mm long (from Heraeus Quarzschmelze, Hanau, Germany). The gels are made to contain 7% acrylamide, 2% Ampholine, 20% sucrose and an amount of sample varying from 20 to 70 ng of each protein component. The sample is mixed with the monomer solution before polymerization and the quartz tubes are filled by capillary action.

After polymerization, the extremities of the capillary are filled with 2% Ampholine solution by means of a micropipette operated by a micromanipulator. The capillaries are then connected to a conventional apparatus for disc electrophoresis by means of a bored silicone rubber stopper and a short length of glass tubing. IEF is performed at room temperature in a voltage gradient of 100 V for 10 to 120 min. Immediately after the run, the gels are pushed out of the capillaries, by means of a tightly fitting steel-wire, into a drop of 20% TCA on a depression slide. The resolution obtained in these gels is comparable to that achieved in standard 3—5 mm gel tubes. However, in the capillary system, a shift of the pattern of proteins along the gel column is regularly observed and, upon prolonged focusing, the zones migrate to one end of the column and finally run out of the gel. This might be due to the high electroendoosmotic flow along the walls of the quartz tube. This flow is probably enhanced by the high ratio of surface/gel volume.

Gainer (1973) has described a micro method capable of detecting proteins at the 10^{-10} to 10^{-9} g levels. He polymerizes 7.5% or 5.5% acrylamide gels, containing 2% Ampholine, in glass capillary tubes (Corning 7740, 0.58 mm i.d., 1.15 mm outer diameter, 7 cm long). After polymerization, the gels are connected to a micro-electrophoresis apparatus (Gainer, 1971) so that the upper parts of the tubes are in the anodic chamber containing 0.3% H_2SO_4, while the lower parts are immersed in the cathodic chamber containing 0.5% ethanoldiamine. The gels are subjected to a pre-run for 30 min, to remove excess persulphate and then the sample (5–50 ng of each protein component) is layered on top of the gel. Equilibrium is reached in approximately 3 h at 120 V at room temperature. The stained protein bands on the micro gel appeared very sharp (less than 0.2 mm in thickness) and it was possible to detect as little as 10^{-10} g protein. Gainer (1973) has reported consistently higher pI values for basic proteins in microgels as opposed to macrogels. It might be that the manipulation of microgels allows for less absorption of atmospheric CO_2, so that this may be more suitable for IEF in basic pH ranges. Additional information on micro-IEF can be found in the book *Micromethods in Molecular Biology* by V. Neuhoff (1973) (pp. 49–56).

Miniaturized slab systems have also been described (Rüchel, 1977; Matsudaira and Burgess, 1978; Ogita and Market, 1979; Poehling and Neuhoff, 1980) mostly to be used for the second dimension in an IEF-SDS system. These chambers are often of stamp size (e.g., 30 × 25 × 0.1 mm) and in general are constructed from microscope slides cut to suitable size. Small strips of plastic sheets of various thickness (e.g., 0.1–0.2 mm) are used as spacers. Miniature clamps, melted dental wax or electric tape are used to seal the chambers laterally. If the spacers are less than 0.25 mm thick sealing of the chamber base is not necessary since capillary forces overcome gravity downflow of the liquid; otherwise the chamber bottom is sealed by pressing into a cushion of plasticine coated with Parafilm. One of the two plates of the gel sandwich can be silanized, so that the gel adheres to it during staining and destaining. Miniaturized, ultrathin chambers, coupled to detection by silver staining (see §3.3.2) can reveal proteins in the sub-nanogram range. IEF in microchambers has been described by Radola et al. (1981): in gels only 3 cm long equilibrium protein patterns could be obtained in only 10 min. When the distance between the electrodes was reduced to 1 or 2 cm, focusing was achieved in 2 and 4 min, respectively. Time for fixation, staining, destaining and drying was a bare 5 min. Microtechniques appear promising albeit special equipment for photographing and densitometry could be necessary.

3.2.18. Isoelectric focusing at sub-zero temperatures

This is a useful extension to electrophoretic techniques of the cryo-biochemistry methods introduced by Douzou (1977). Early attempts at studying mixed hemoglobin tetramers by IEF below zero were reported by Park (1973), who only succeeded in lowering the temperature to about -5 to $-10°$C. Considerable progress has been recently described by Perrella et al. (1978, 1979, 1980). The major problems to be solved were a suitable thermostatting method, a proper polyacrylamide gel which would not exhibit glass transition below $-15°$C and an appropriate aqueous-organic, anti-freezing solvent. These problems were solved by using a modified Righetti and Drysdale (1971, 1973) tube apparatus, fitted with a cooling serpentine also for anolyte and catholyte compartments (Fig. 3.21) and utilizing narrow-bore glass tubes (2 mm i.d.) for rapid Joule heating dissipation. The gel was a copolymer of acrylamide-methacrylate or acrylamide-ethylacrylate which are able to stand temperatures as low as -30 to $-40°$C without matrix modifications. As for the aqueous-organic solvents, best results were obtained by dimethyl-sulphosice (DMSO)–water mixtures, as previously reported by Righetti et al. (1977). In theory, IEF experiments could be carried out down to $-30°$C or $-40°$C; in practice, however, the high viscosity of the aqueous-organic medium and the much lower mobility of proteins at subzero temperature, set a limit to a more comfortable $-20°$C. An additional problem results from precipitation of acidic carrier ampho-lytes (below pH 5), probably due to aggregation, with severe disturbances of the pH gradient in the zone of the precipitate. Best results are thus obtained with pH ranges above pH 5, setting the temperature limit to $-20°$C and choosing a DMSO concentration in water of 37%, which ensures reduced viscosity and is compatible with temperatures down to $-30°$C. By this technique, hybrid tetramers HbA–HbS, starting from the parent molecules HbA and HbS, could be obtained (Perrella et al., 1979), as well as a whole series of intermediates of oxidation of HbCO. Recently, the same authors have also been able to map the course of the pH gradient at $-20°$C by using the dye-indicator method of Douzou (1977). Generally speaking, in the nominal pH ranges 6–8 or 7–9, the combined effect of 37% DMSO and $-20°$C, as compared with focusing in water at $+4°$C, is to increase all pH values of focused Ampholine by about 1 pH unit. In the case of carbonmonoxy hemoglobin A, its pI value shifts toward the alkaline to the same extent. Cryo-isoelectric focusing appears to be a very promising technique for studying subunit exchange in solution, as well as ligand binding and enzyme substrate complexes.

3.3. Detection methods in gels

3.3.1. Staining procedures

Special procedures have to be used for staining proteins by IEF, since the carrier ampholytes form insoluble complexes with many protein stains. This problem can be overcome by first precipitating focused proteins in acid, e.g., 10% TCA, and eluting the acid-soluble ampholytes with extensive washing. After ampholyte removal, proteins can be detected by conventional methods. Today, several direct staining methods have been developed to circumvent this laborious procedure, by solubilizing the dye-ampholyte complexes in alcoholic solvents and/or high temperatures (Righetti et al., 1977). I will give here examples of the staining methods used:

Fast Green FCF (Riley and Coleman, 1968). For detecting bands containing more than 10 μg of protein in 3 mm gels. Immerse gels for 4–8 hr in an aqueous solution of 0.2% Fast Green FCF (Fisher Scientific, NJ, USA) in an aqueous solution containing 45% ethanol–10% acetic acid. Destain in an aqueous solution of 10% acetic acid–25% ethanol. Fast Green FCF stained bands tend to fade away with time.

Light Green SF. This method has been used by Radola (1973a) to stain gel patterns by the paper print technique. After taking the gel print, the paper is washed 15 min in TCA. TCA is subsequently removed by rinsing for a few minutes with a mixture of methanol–water–acetic acid (333:66:10, v/v/v) and then the print is stained for 15 min by immersion in the same mixture containing 0.2% Light Green SF. For destaining the above mixture is used in the absence of dye. This stain is 3 to 5 times less sensitive than either Coomassie Brilliant Blue R-250 or Coomassie Brilliant Blue G-250. However, Radola (1973a) has used it successfully to resolve protein patterns in regions of very high protein loads, where staining with Coomassie Brilliant Blue would have produced blurred and unresolved zones.

Bromophenol Blue. After Awdeh (1969). Immerse gels for 3 h in an aqueous solution of 2% Bromophenol Blue in 50% ethanol—5% acetic acid. Destain in 30% ethanol, 5% acetic acid in water. Handle carefully. The affinity of these gels for fingers, tissue paper or other dry surfaces is high, so beware!

Coomassie Brilliant Blue R-250. Several direct methods have been described for staining with this popular and sensitive dye. Four variations are offered here.

(1) After Hayes and Wellner (1969).

After IEF the gels are fixed in 5% TCA—5% sulphosalicylic acid solution for at least 1 h. They are then washed in 3 liters of water (at least 1 h) to remove most of the excess acid. Staining is accomplished by transferring the gels to a solution of 0.066% Coomassie Brilliant Blue R-250 in 0.2 M Tris—HCl buffer, pH 7.7. Staining is allowed to proceed for 3 to 4 h. Background stain is removed by washing the gels in dilute buffer (Tris—HCl, 0.001 M, pH 7.7) for 24 to 48 h.

(2) After Spencer and King (1971).

This is a valuable method for detecting focused protein bands rapidly. Proteins absorb dye from a weak solution (0.01%) in 5% trichloroacetic acid, 5% sulphosalicylic acid and 20% methanol in water. Although less sensitive than other procedures with Coomassie Brilliant Blue, background staining is very low and only one staining solution is required. Banding patterns may usually be detected after about 1 h. The intensity of the bands may be enhanced by increasing the level of dye to 0.05% but at the expense of higher backgrounds.

(3) After Vesterberg (1972).

This method stains proteins with Coomassie Brilliant Blue without interference from ampholytes by heating the gels at 60 °C for 15 min in a solution of 0.1% dye in 28% methanol, 11% TCA and 3.5% sulphosalicylic acid, in water. Destaining is also effected at 60 °C in 25% ethanol—8% acetic acid in water. This method is more convenient with gel slabs rather than cylinders. Although rapid, stained patterns often tend to fade. In this method, the dye tends to precipitate on the gel surface and to be trapped in gel cracks. A modification of this procedure has been described by Söderholm et al. (1972). After the experiment, the gel is immediately transferred to a bath containing 2% (w/v) sulphosalicylic acid, 11% (w/v) TCA and 27% (v/v) methanol in distilled water, at 65°C. After 20 min, the gels are washed twice in destaining solution (8.5% acetic acid–27% ethanol in distilled water) at 20°C and then stained with 0.1% dye dissolved in destaining solution. Recently, Vesterberg and Hansén (1977) have described four additional variations, using various combinations of TCA, sulphosalicylic acid, perchloric acid (PCA) and urea in the fixing solution. It appears that urea increases the stain sensitivity by unfolding the protein chains. The detection limit appears to be around 0.2 μg protein per band.

(4) After Righetti and Drysdale (1974).
We have developed a method that combines high sensitivity and low background and is equally suited for slabs and cylinders. Gels are immersed for 4–6 h in a solution of 0.05% Coomassie Brilliant Blue and 0.1% cupric sulphate in acetic acid–ethanol–water (19:25:65). Gels are destained for 4 h in the same solution but containing only 0.01% Coomassie Brilliant Blue and finally in acetic acid–ethanol–water (10:10:80). Gels should be handled with gloves since this method stains fingerprints! Malik and Berrie (1972) described a protein staining procedure with Coomassie R-250 prepared using sulphuric acid, KOH and TCA. This is a rapid staining procedure, but the gel background becomes significantly stained.

Coomassie Brilliant Blue G-250. Diezel et al. (1972) have introduced the use of Coomassie G-250 (a dimethyl substitute of the R-250 dye) as a colloidal dispersion in 12.5% TCA. This causes a selective binding to the protein zones and prevents the dye's penetration into the network of the gel, thus avoiding background staining. Upon storage in 5% acetic acid, the bands are intensified as the protein-bound dye becomes soluble, diffuses into the gel and binds again to the interior portions of the protein bands. Reisner et al. (1975) have found that 0.04% CBB G-250 in 3.5% PCA remains in a leucoform, but changes to the blue form when coupled to proteins. Ampholine does not precipitate, but the gel takes a faint orange to brownish background. By introducing an additional wash in 5% acetic acid after staining Holbrook and Leaver (1976) have claimed a three-fold increase in sensitivity (but the background becomes heavier). Blakesley and Boezi (1977) have described a recent modification which appears to be excellent for IEF gels. To 400 ml of 0.2% dye in H_2O an equal volume of 2 N H_2SO_4 is added. After standing for 3 h, the precipitate is removed through a Whatman 3 MM paper. To the clear brown filtrate, 90 ml of 10 N KOH are added, producing a dark purple solution. To this, 120 ml of 100% TCA are added, and the greenish solution is ready for use. This dye can be used several times, provided the pH is maintained below 1.0 (in fact for IEF gels, the used dye works much better than the fresh solution, probably because a trace impurity, which tends to precipitate neutral Ampholine, is eliminated; Righetti, unpublished). In 0.7 mm thick slabs, maximum color development is achieved in ca. 5 h, but the bands are already visible within 15 min. When placed in water, there is a marked color intensification. This dye works wonders for peptide analysis by IEF, allowing fixation and detection of chains as short as 12–15 amino acids (Righetti and Chillemi, 1978).

Fluorescent stains. It appears that the ultimate analysis in prestaining has been reached with the introduction of fluorescing stains such as dansyl chloride (Talbot and Yaphantis, 1971), fluorescamine (Stein et al., 1973; Handschin and Ritschard, 1976) and 2-methoxy-2,4-diphenyl-3(2H)-furanone (MDPF) (Barger et al., 1976). Protein quantities as small as 1 ng can be detected with MDPF. However, unless trace labeling is used, it is difficult to conceive that these stains will not alter the protein pI values and introduce an artefactual heterogeneity. These methods thus, as well as covalently bonding of reactive cationic dyes (such as Drimarene Brilliant Blue K-BL) (Bosshard and Datyner, 1977) prior to electrophoretic analysis, do not appear at present to be suitable in IEF. In fact it has been reported that, even in SDS-electrophoresis, where proteins should migrate solely according to mass, pre-labeling with fluorescamine alters the mobility of some proteins (Ragland et al., 1974). This fluorescent dye, in as much as it converts an amino group to a carboxyl group when it reacts with proteins, will surely alter its pI, even upon trace labeling. Table 3.3 summarizes different protein staining techniques used after thin-layer IEF in granulated gels, by the method of the paper print (Radola, 1973a). Table 3.4 lists some of the most commonly used stains in electrophoresis and IEF.

3.3.2. Silver stain

This represents one of the most astonishing staining developments. Its

TABLE 3.3

Protein staining in thin-layer isoelectric focusing by the paper print technique (Radola, 1973a)

Dye	Washing with 10% (w/v) chloroacetic acid: No. times	Dye concn. % (w/v)	Methanol–water–glacial acetic acid (v/v/v)		Time factor for destaining[b]	Sensitivity μg protein[c]
			Staining solution	Destaining solution		
Amido Black 10 B	3	0.2	50:50:10	66:33:10	10–20	5–10
Bromophenol Blue	0 or 1[a]	0.1	66:33:10	40:60:10	2–5	3–5
Coomassie Brilliant Blue G-250	3	0.2	50:50:10	33:66:10	3–5	0.5–1
Coomassie Brilliant Blue R-250	3	0.2	50:50:10	33:66:10	3–5	0.5–1
Coomassie Violet R-150	0 or 1[a]	0.2	50:50:10	33:66:10	2–3	2
Light Green SF	0 or 1[a]	0.2	33:66:10	33:66:10	1	3

Prior to staining the carrier ampholytes were removed by washing with trichloroacetic acid (each washing 15–30 min). The trichloroacetic acid was washed out of the paper for a few minutes with a mixture of methanol–water–glacial acetic acid (33:66:10, v/v/v). The paper print was stained for 15–30 min with occasional stirring and then kept for 5–15 min in several changes of the destaining solution. The first two washings were rejected, the others were decolorized with activated charcoal and reused.

[a] Removal of the carrier ampholytes was not necessary but destaining was more rapid and uniform when a single washing was applied.

[b] For Light Green SF the time necessary to get a completely destained background was 20–30 min.

[c] μg protein detectable when the sample was applied as an 18 mm zone.

TABLE 3.4

Electrophoresis stains and tracking dyes[a]

Stain	Color index No.	λ Maximum absorbance	Use	Reference
Amido Black 10B (Buffalo Black, Naphthalene Black 10B)	20,470	620	General protein stain	Wilson (1973)
Coomassie Brilliant Blue R-250	42,660	590	General protein stain ca. 10 times as sensitive as Amido Black	Fairbanks et al. (1971)
Coomassie Brilliant Blue G-250	42,655	595	General protein stain	Wardi and Michos (1972)
Alcian Blue	74,240	630	Glycoprotein	Datyner and Finnimore (1973)
Uniblue A	14,552	–	Protein stain	Peacock and Dingman (1967).
Methylene Blue	52,015	665	RNA, RNase	Boyd and Mitchell (1965)
Methyl Green	42,590	635	Native DNA, acidic or neutral tracking dye	
Fast Green FCF	42,036	610	Protein stain	Gorovsky et al.. (1970)
Basic Fuchsin	42,500	550	Glycoprotein	Zacharius and Zell (1969)
			Nucleic acids, Sialic acid rich glycoproteins	Dahlberg et al. (1969)
Pyronin Y	45,005	510	RNA, acidic tracking dye	Chrambach et al. (1976)
Bromophenol Blue	–	595	Neutral and alkaline tracking dye	Awdeh (1969)
Bromocresol Green	–	–	Tracking dye for DNA agarose electrophoresis	McDonnel et al. (1977)
Crocein Scarlet	26,905	505	Immunoelectrophoresis	Crowle and Cline (1977)
Xylene Cyanole FF	43,535	–	Tracking dye for DNA sequencing	Maxam and Gilbert (1977)
Toluidine Blue 0	52,040	620	RNA, RNase, mucopolysaccharides	Righetti and Drysdale (1972)
Ethidium Bromide	–	–	Fluorometric detection of DNA	Sharp et al. (1973)
Stains All	–	–	General protein stain	Green et al. (1973)

[a] From Bio Rad Labs. Catalogue G (1981). Note: not all stains reported are directly applicable to IEF fractionations.

sensitivity is said to be at least 100 times greater than Coomassie Blue, close to autoradiography. It was first reported in 1979 by Switzer et al., as a modification of the de Olmo's (1969) neural cupric-silver stain. I will describe here three basic variants: a simplified version of Oakley et al. (1980); a photochemical modification by Merril et al. (1981) and the colorful variant of Sammons et al. (1981).

(1) Simplified silver stain of Oakley et al. (1980). This procedure has been developed for 140 × 140 × 0.8 mm gel slabs.

Step 1. Soak the gel for 30 min in 10% unbuffered glutaraldehyde (for IEF gels, this should be preceded by a fixation step in 50% ethanol—10% acetic acid, so as to remove the carrier ampholytes, for at least 45 min in a 0.35 mm thick gel; rinse for another 45 min, in 5% methanol—5% acetic acid). The glutaraldehyde should be biological grade.

Step 2. Rinse the gel in 500—1,000 ml of glass distilled water for 10 min, change the water twice, and allow the gel to soak for at least 2 h (or overnight, if possible).

Step 3. Drain the water and add freshly made ammoniacal silver (for 100 ml of it, add 1.4 ml of fresh NH_4OH to 21 ml of 0.36% NaOH; with stirring add slowly 4 ml of 19.4% $AgNO_3$, this last solution obtained by adding 20 g $AgNO_3$ to 100 ml distilled water. A transient brown precipitate might form, which should clear upon agitation). The gel should float freely in this solution (use 100—150 ml). To prevent silver deposition on the gel surface, stain no longer than 15 min.

Step 4. Remove the gel from its container and place it in a new basin of distilled water to wash for 2 min.

Step 5. Transfer the gel to a freshly made solution containing 0.005% citric acid and 0.019% formaldehyde (made by dilution of 38% formaldehyde in 10—15% methanol). The stained proteins become visible at this step. The gel should be removed from this solution when a dark background begins to develop and immediately washed for at least 1 h. If the gel has been overstained, it may be destained in Kodak rapid fixer. In turn, destaining can be stopped by placing the gel in Kodak hypo cleaning agent in its normal (1:4) dilution.

(2) Photochemical silver stain of Merril et al. (1981).

Gels are fixed in 50% methanol—12% acetic acid for at least 20 min, followed by three 200 ml, 10 min rinses containing 10% ethanol—5% acetic acid. Gels are then soaked for 5 min in a 200 ml solution of 3.4

mM potassium dichromate and 3.2 mM nitric acid. Wash four times, for 30 s, in 200 ml distilled water and place the gels in 200 ml of 12 mM silver nitrate for 30 min. Maximum sensitivity is achieved if the gels are exposed to intense, fluorescent light during the first 5 min in silver nitrate. This is followed by rapid rinsing with two 300 ml portions of the image developer solution, which contains 0.28 M sodium carbonate and 0.5 ml of commercial Formalin per liter. The gels are then washed in a third portion of this solution until the image has reached the desired intensity. Development is stopped by transferring to 100 ml of 1% acetic acid. Sensitivity enhancement is only obtained in gels thinner than 1 mm. A detection limit of as little as 0.01 ng protein per mm^2 has been reported.

(3) Variegated silver stain of Sammons et al. (1981).
The flow-sheet for this stain is given in Table 3.5. The gels are left over-night in 50% EtOH and 10% HAC (gel volume: solution volume = 1:5.5) but this applies mostly to SDS-gels. After IEF, 1–2 h in this solution, depending on gel thickness, should suffice. If stain cannot be performed immediately, the gels can be stored in the last washing solution (10% EtOH–0.5% HAC). It is essential that solutions used in the staining procedure be degassed prior to use. An ideal concentration of AgNO$_3$ in 1.5 mm thick gels is 1.9 g/l and equilibrium should be done for 2 h. The optimum ratio of the gel volume to solution volume is 1:3. The reducing

TABLE 3.5
Staining procedure for the silver-based gel electrophoresis color development system (from Sammons et al., 1981)

Steps	Solutions	Duration of agitation
Fix	50% EtOH 10% HAC	2 h or more
Wash	50% EtOH 10% HAC	2 h
	25% EtOH 10% HAC	1 h 2×
	10% EtOH 0.5% HAC	1 h 2×
Equilibrate gel	AgNO$_3$ (1.9 g/l)	2 h or more
Rinse	H$_2$O	10–20 s
Reduce silver	NaBH$_4$ (87.5 mg/l)	10 min
	HCHO (7.5 ml/l)	
	in 0.75 N NaOH	
Enhance color	Na$_2$CO$_3$ (7.5 g/l)	1 h
	Na$_2$CO$_3$ (7.5 g/l)	1 h, store

EtOH = ethanol; HAC = acetic acid

solution consists of 0.75 M NaOH, 7.5 mg/l formaldehyde (USP 37%) and 87.5 mg/l NaBH$_4$ (gel volume: solution volume = 1:5.5). The formaldehyde should be added to the reducing solution immediately prior to submersion of the gels. After 10 min the gel is removed from the reducing solution and placed in the color enhancing solution. The color develops as the sodium carbonate diffuses into the gel over the next several hours, with a maximum at 6 h. By this staining procedure, the stained polypeptides appear in four basic colors: blue, green, yellow and red. Whether these color shades are due to size, or amino acid composition or both remains to be seen. Visual inspection of Fig. 4 in Sammons et al. (1981) suggests that green colors (in a 2-D separation of platelet proteins) are particularly abundant in basic proteins of relatively small size, but this remains to be verified in general. Chemicals for this procedure are available, as 'Gelcode' from Upjohn Diagnostics, Kalamazoo, MI.

I should like to make some comments on the silver stain. The exact mechanism of staining is not fully established. The imino group of the peptide bond in proteins seems to be the main silver binding organic group. During the reaction of imino groups with silver cations in an acidic environment, the imide hydrogen is replaced by an equivalent of the metal. Silver cations also precipitate from alkaline or neutral solutions with various reducing solutions such as formaldehyde, that is why the developer usually contains this aldehyde. For this same reason, the first method reported (Oakley et al., 1980) in which the first step is a glutaraldehyde fixation procedure, has been criticized on the grounds that it gives a very high background and sensitizes the gel surface at the site of the bands, so that developing proceeds at an uncontrollable fast rate (Boulikas and Hancock, 1981). The incredibly high stain sensitivity, well above the detection limit of colorimetric analysis (usually in the μm range) has been explained by Switzer et al. (1979) as a physical growth of the initially deposited silver on the protein zone, an amplification mechanism similar to the events which occur in exposing and developing a photographic emulsion. A recent modification has been described by Marshall and Latner (1981): instead of using ammonia in the silver solution, they have substituted it with methylamine and have adopted a strategy of overstain followed by destain. According to these authors, this further increases the intensity of the silver stain and prevents the appearance of 'hollow' spots upon subsequent destain. In fact, in the Sammons et al. (1981) technique, spots which contain more than 0.5 µg protein/band appear as hollow rings, i.e., the stain is mostly deposited on the spot contour and not in its center, giving rise to false double peaks upon densitometry (Görg and Righetti, unpublished). As a recent extension, silver staining has been applied also to detection of DNA and RNA (Boulikas and Hancock, 1981) with a sensitivity at least 20 times higher than ethidium bromide detection: as little as 0.6 ng DNA per band could be revealed. Applications of the silver stain to IEF and 2-D analysis have reported by Allen (1980), Merril et al. (1979) and Görg et al. (1981).

It is perhaps too early to assess the impact of silver stain, however a note of caution should be adopted according to the analysis of Poehling and Neuhoff (1981). Quoting from them: 'optimal silver staining is a fine balance between silver deposition on the protein and the level of background staining. Reproducible staining requires empirical standardization of many steps in the procedure. Silver staining in the present form does not stoichiometrically stain proteins, unlike Coomassie Blue. Staining intensity does not bear a linear relationship to protein mass: the increment in sensitivity is both different for individual proteins and dependent on protein concentration'. This view, however, has been challenged by Merril et al. (1982). A critical appraisal on six different silver staining methods can also be found in Ochs et al. (1981).

3.3.3. Densitometry of focused bands

The dimensions of most gel cylinders used in IEF are compatible with scanning devices in most spectrophotometers. Equipment normally used for densitometric evaluation of patterns is obtained by disc-gel electrophoresis. The gel used in the apparatus of Righetti and Drysdale (1971), may be scanned in a 10 cm long cuvette in a Gilford Model 240 recording spectrophotometer fitted with a linear transport device. Quartz or glass gel tubes fit directly into the cuvette holder, so that the focused gels can be scanned in the tube without being extruded. Fawcett (1970) adapted An Amicon SP 800 spectrophotometer either for scanning gels at a constant wavelength or for obtaining spectra of individual components. Catsimpoolas (1973c) has devised the technique of in situ analytical scanning IEF. He performs IEF in a quartz cell held vertically in the chamber of a modified Gilford linear transport device. The focusing column can be monitored continuously and the developing patterns observed. This system gives more reliable analysis of banding patterns in sucrose density gradients, as it permits the detection of closely spaced peaks that might diffuse together during elution in the absence of an electric field. In situ scanning also indicates the end-point in IEF, which is important when the pH gradient becomes unstable on prolonged electrolysis (cathodic drift, Righetti and Drysdale, 1971, 1973). Densitometric evaluation in thin layers has been carried out by Radola (1973b) with a Schoeffel SD 3000 spectrodensitometer (Schoeffel Instruments, Westwood, NJ, USA). The Zeiss MPQ II chromatographic spectrophotometer allows densitometry by transmittance, reflection and fluorescence emission and is compatible with gel cylinders, polyacrylamide thin layers, granulated flat beds and paper or cellulose acetate membranes up to dimensions of 20×20 cm. This scanning spectrophotometer can also be used to obtain spectra of individual bands in the electropherogram. An accessory for a Beckman DH Spectrophotometer, for scanning of 2 mm thick Sephadex G-200 layers has been described by Holmquist and Broström (1979). This allows detection and recovery of focused protein bands without resorting to a paper print. One of the limiting factors in densitometric evaluation of a stained protein pattern is the resolution afforded by the scanning instrument. The absolute resolution of a scanner can be defined as the minimal distance between two discrete points, whose tracing is a valley touching the baseline. By this criterion, with the best scanners presently available, the resolution limit is 230 μm. That is, peaks must be spaced 230 μm apart or more, in order for the recording pen to reach the baseline between the two peaks. In practice, with most scanners, the resolution limit is much higher than that (1 mm or more).

Although narrower slits may be used to improve the resolution, Zeineh et al. (1975) have reported that, with slits narrower than 100 μm, the resolution is not improved, but lost, since the beam diverges after passing through the slit. Very narrow slits are also difficult to use, especially in the UV region, because the amount of energy reaching the sample is too low to allow good densitometry. Therefore, presently available scanners do not do justice to the high resolving power of IEF where, especially in thin layer gels, gel scanners reaching a resolution of 50 μm are often needed.

To overcome this difficulty, Zeineh et al. (1975) have built a soft laser scanning densitometer. This instrument has as a light source a polarized, monochromatic, coherent soft laser light, of 633 nm wavelength, with waves vibrating in the same plane as the slit. This red wavelength is useful for quantitation of Coomassie Blue stained protein patterns or for immuno-precipitin lines or turbidity assays (Ponceau S or red dye stained proteins will be transparent to the radiation). An additional UV soft laser beam at 320 nm for fluorescence or UV absorption of protein bands is also available. This laser scanner does not use the adjustable slit system of conventional scanners and the beam width is internally controlled. The laser beam can be reduced to a microspot having a diameter as narrow as $3-10$ μm. This eliminates the interference problems produced by the slit and minimizes spherical aberration produced by the tube containing the stained gel. Also, the recorder pen goes back to the baseline when the spacing between adjacent bands is more than the beam width. The microbeam is also able to resolve a protein pattern in parallel and uniform Gaussians, even if the actual band profile is skewed, zigzagged or arc-shaped. Two adjacent bands 80 μm apart are fully resolved on this system. Zeineh et al. (1975) have anticipated that, if needed, the resolution of the laser can be improved to a few hundred angstroms, by modifying the presently available equipment. A laser scanner, built on this principle, is now available from LKB Produkter AB (2202 UltroScan).

In case of 2-D separations, proper densitometry of a 2-D slab is not an easy job, since three coordinate values have to be given to each spot: a pI (x axis), a $\overline{M}_r$ (y axis) and a relative intensity (%). Thus, this task is only a little bit less complicated than astronomers cartography, where a fourth coordinate value is needed. At the turn of the century, the Harvard astronomer E.C. Pickering (Ferris, 1979) solved this problem by hiring 'computers', ladies hired at 25 cents an hour to fill blank catalogue pages with tiny black ink numbers; no mistakes were permitted. With the Women's Liberation Front on the loose, and the pressure from multinational computer (machines, this time) companies, this approach would be a bit unrealistic today. Several research groups have independently described computer programs for data acquisition and image analysis of 2-D gels (Garrells, 1979; Bossinger et al., 1979; Kronberg et al., 1980; Taylor et al., 1980; Lutin et al., 1977; Alexander et al., 1980; Capel et al., 1979; Klose et al., 1980). In Fig. 3.22, an

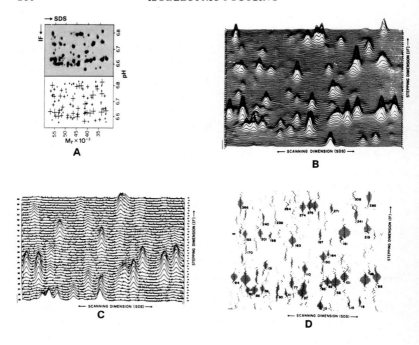

example of such a computer analysis of 2-D gels is given. The protein spots in the fluorogram, or autoradiogram, or Coomassie Blue-stained electropherogram, are converted by the computer into crosses (A), whose area is proportional to the integrated spot density. The crosses are transformed into scan lines (B), whose excursions in the cross region are seen as peaks against the line background. A line analysis is then performed (C) and converted into a chain assembly (D). By integrated density analysis of the chain assembly, a densitogram tracing, resembling the scans of 1-D gels, is finally obtained (not shown).

3.3.4. Autoradiography and fluorography

Proteins can be radiolabeled either during synthesis, using radioactively labeled amino acids, or post-synthetically by iodination (Samols and Williams, 1961), reductive methylation (Kumarasamy and Symons,

1979) or, in the case of antibodies, radioactive hapten (Williamson, 1971), radioiodinated antigen (Keck et al., 1973a,b) and the like. In practice, a host of chemicals which react with amino, thiol, phenolic hydroxyl, imidazole and aliphatic hydroxyl groups, if available radiolabeled, can be used for isotopic tagging of proteins (for a list, see Appendix II in Hames and Rickwood, 1981). In this last case, however, since quite a few of these chemicals alter the protein charge, labeling should be performed after and not before IEF fractionation, so as to avoid artefactual banding patterns due to the multitude of charged species generated by the radiolabeling step. Two basic approaches are possible: the first is to place the gel next to the X-ray film whereupon the radioactive emissions cause the production of silver atoms within silver halide crystals and these are visualized after developing the film. The second method involves slicing the gel after electrophoresis and determining the radioactive protein content of each slice by scintillation counting. This last approach is most commonly used for quantitation, especially in dual isotope experiments. However, its resolution is dependent upon

---

Fig. 3.22. Example of computer analysis of 2-D electropherograms. (A) The film was taken from a fluorogram representing proteins of confluent B103 cells. The gel was run using pH 6—8 ampholytes in the first dimension and 7.5% acrylamide in the second dimension. The small region shown was 36 × 53 mm in area. Normally an area of 150 × 150 mm is scanned, but for explanatory purposes only this small region was processed. The faintest detectable spot represents about 0.5 c.p.m. The computer plot shows the spots detected and the area of the cross is proportional to the integrated spot density. (B) Lines of data generated by successive scans across the film shown in A. Each line contains about 380 readings of film density. These data are received and immediately processed by the computer during the line analysis stage of data reduction. The distance between lines represents a 280 μm step in the stepping (pH) dimension. The vertical bar at the left indicates a density of one absorbance unit on the vertical scale. (C) Line analysis. A special program has plotted each line of data with diagnostic marks to indicate exactly how each line was processed. Each peak is enclosed in brackets and its center is marked by vertical lines. The lines shown are lines 20—59 of B, replotted with greater spacing between them. (D) Chain assembly. The integrated peaks detected during line analysis (C) have been assembled into chains in the stepping (pH) dimension. Each horizontal line depicts the peak intensity and its center is marked by a vertical line. Each descending line points to the previous peak of the chain so that the chains can be followed. The numbered chains are then transformed into integrated density peaks, whose relative areas are calculated by the computer (not shown). (From Garrels, 1979.)

the fraction size, which for gel slicing techniques is not usually less than 0.5 mm. In film-sensitization techniques, four basic variants are possible: direct autoradiography, fluorography, indirect autoradiography and dual isotope detection. I will review them here briefly.

Direct autoradiography. ^{32}P- and ^{125}I-labeled proteins can be easily detected in wet gels simply by sealing the slab gel or rod gel slices with cling bags and placing this against an X-ray film. However, resolution is much improved if the gel is first dried down onto a sheet of Whatman 3 MM filter paper (see §3.2.15). Usually the gel is stained with Coomassie Blue and destained before drying so that the photographic image obtained after autoradiography can be compared with the stained protein banding pattern. To facilitate alignment of the stained, dried gel and the developed X-ray film, it is wise to place a few spots of radioactive ink ($\sim$1 μCi ^{14}C per ml ink) on the filter paper at the gel perimeter prior to autoradiography. The stained dried gel is placed in direct contact with the X-ray film, clamped in a radiographic cassette or between glass or hardboard plates using metal clips, and then wrapped in a black plastic bag or placed in a light-tight box. After exposure, the film is developed according to the manufacturer's instructions.

Direct autoradiography of dried polyacrylamide gel with Kodirex X-ray film will give a film image absorbance of 0.02 A_{540} units (just visible above background) in a 24 h exposure with about 6,000 d.p.m./cm^2 of ^{14}C or ^{35}S, 1,600 d.p.m./cm^2 of ^{125}I, or about 500 d.p.m./cm^2 of ^{32}P (Laskey and Mills, 1975, 1977). The film image absorbance is proportional to sample radioactivity. ^{3}H is not detected since the low energy β-particles fail to penetrate the gel matrix. However, fluorography can detect ^{3}H and also increases the sensitivity of ^{14}C and ^{35}S detection over that possible using direct autoradiography. In addition, Laskey and Mills (1977) have developed a variation of this method, called 'indirect autoradiography' which increases the detection efficiency of ^{125}I and ^{32}P considerably compared to direct autoradiography. The methodologies of fluorography and indirect autoradiography have been recently reviewed by Laskey (1980).

Fluorography (for ^{3}H, ^{14}C, ^{35}S). In fluorography the gel is impregnated with a scintillator (2,5-diphenyloxazole; PPO), dried down and exposed to the X-ray film at $-70°$C. The method is less sensitive when exposure is carried out at higher temperatures. Although low-energy β-particles produced by ^{3}H-labeled proteins are unable to penetrate the gel matrix and expose the X-ray film directly, they are able to interact with the PPO molecules in the gel to convert the energy of the β-particle to visible light which then forms an image on blue-sensitive X-ray film. Thus absorption of β-particles by the sample and gel matrix is overcome by the increased penetration of light. The basic fluorographic procedure is given in Table 3.6. It may be used on either unfixed or stained gels. Exposure to dimethylsulphoxide (DMSO) removes some Coomassie Blue from protein bands such that faintly stained bands may not be

TABLE 3.6
The basic fluorographic technique (from Hames, 1981)

1. Directly after electrophoresis, or after staining/destaining, the gel is soaked in about 20 times its volume of DMSO for 30 min followed by a second 30 min immersion in fresh DMSO[a]. These separate batches of DMSO can be stored for use (in the same sequence) with other gels. Prolonged use is to be avoided since all water must be removed or the PPO will not enter the gel.
2. The gel is immersed in 4 volumes of 20% (w/w) PPO in DMSO for 3 h[b].
3. The gel is then immersed in 20 volumes of water for 1 h. The PPO is soluble in DMSO but insoluble in water. Therefore, PPO precipitates within the gel matrix turning the gel opaque.
4. The gel is dried under vacuum (§3.2.15)[c].
5. X-ray film is placed in contact with the gel and exposed at −70 °C. The most sensitive film is Kodak X-Omat R, XR1, but where this is not available the slightly less sensitive Fuji RX film should be used instead. Kodirex and Kodak No-screen X-ray films, much used in autoradiography, are inefficient at recording the visible light produced by the fluorographic technique and so are not used for fluorography.
6. After exposure, the film is unwrapped before it warms up (to avoid physical fogging) and developed according to the manufacturer's instructions.

[a] Methanol is used instead of DMSO for gels which contain less than 2% acrylamide plus 0.5% agarose. After electrophoresis the gel is soaked in 20 volumes of methanol for 30 min followed by a second 30 min immersion in fresh methanol. Then the gel is immersed in 10% (w/w) PPO in methanol for 3 h and finally dried under vacuum without heat.
[b] Excess PPO may be recovered from the DMSO by adding 1 volume PPO in DMSO to 3 volumes 10% (v/v) ethanol. After 10 min, the suspension is filtered and the PPO precipitate is washed with 20 volumes of water, then air dried.
[c] If gel cracking during drying is a problem, the gel can be successfully dried by soaking in methanolic solutions (up to 30% methanol) containing 3% glycerol after step 3 and prior to gel drying.

visible at the end of the procedure. Furthermore the precipitate of PPO which is formed in the gel obscures the staining pattern. For these reasons a photographic record should be made of stained gels before attempting fluorography. Alignment of stained bands in the gel with the final fluorographic image can be made using radioactive ink spots marked on the dry gel perimeter prior to clamping to the X-ray film. Unfortunately, using untreated X-ray film, the absorbance of the fluorographic image is not porportional to the amount of radioactivity in the sample, small amounts of radioactivity producing disproportionately faint images. This can be overcome by exposing the film to an instantaneous flash of light ($\leqslant$ 1 ms) prior to using film for the fluorographic detection of radioactive bands. The procedure for controlled pre-exposure of of X-ray film is detailed in Table 3.7 (Laskey and Mills, 1975). When the film is pre-exposed to between 0.1 and 0.2 A_{540} units, the absorbance of the fluorographic image becomes proportional to sample radioactivity and the

TABLE 3.7
Pre-exposure of X-ray film for quantitative fluorography (from Hames, 1981)

Pre-exposure is performed using a single flash from an electronic flash unit, e.g., Vivitar 283. It is essential that the duration of the light flash is short ($\leqslant$1 ms) since flashes of longer duration only increase the background 'fog' absorbance without hypersensitizing the film. To overcome slight variations in charging the capacitor, the unit is used only after the charging lamp has been illuminated for at least 30 s. The filters are taped to the window of the flash unit to reduce and diffuse the light output:

(i) An infrared-absorbing filter; placed nearest the flash unit to protect the other filters from the heat generated.
(ii) A colored filter; to reduce light output. A 'Deep Orange' Kodak Wratten No. 22 is suitable for the above flash unit whereas an 'Orange' Kodak Wratten No. 21 is used for weaker flash units.
(iii) Porous paper (e.g., Whatman No. 1 filter paper); to diffuse the image of the bulb so that the film is evenly exposed.

Minor adjustments to illumination intensity are made by varying either the distance between the film and the light source (usually 60–70 cm) or the diameter of an aperture in an opaque mask. The film is backed by yellow paper during pre-exposure, and the surface which has been facing the light source is applied to the gel. The degree of pre-exposure is determined by using any conventional spectrophotometer to measure the increase in background 'fog' absorbance at 540 nm of pre-exposed film compared to unexposed film. Storage of films after pre-exposure is not recommended.

sensitivity of the method increases still further compared to direct auto-radiography such that 400 dmp/cm^2 of ^{14}C or ^{35}S, or 8,000 d.p.m./cm^2 of ^{3}H are detectable. It is essential to note that only under these conditions of film pre-exposure does the absorbance profile of the fluorographic images represent the true distribution of radioactivity in the sample. Therefore, a pre-exposure of 0.15 A_{540} units is used routinely for all quantitative fluorography. Increasing the pre-exposure of X-ray film above 0.2 A_{540} units increases the sensitivity of ^{3}H, ^{14}C, and ^{35}S detection by fluorography still further, but the absorbance of the film image ceases to be proportional to the radioactivity present. Recently, Chamberlain (1979) has introduced the use of sodium salicylate as the fluor in fluorography, instead of PPO. One advantage of salicylate over PPO is that it is considerably less expensive. The other main advantage is that salicylate is water soluble, so that fluorography using this fluor avoids lengthy equilibration of the gel in DMSO followed by washing with water and typically takes 0.5–1 h, instead of about 5 h required with the DMSO–PPO system. Fluorography using salicylate with pre-exposed X-ray film appears to be as sensitive as that using DMSO–PPO, with similar linearity of fluorographic image absorbance to sample radio-activity. As an alternative to both methods, a third variant has been described by Pulleyblank and Booth (1981): the gel is impregnated with PPO dissolved in glacial acetic acid. This method can be applied to both agarose and polyacrylamide matrices (DMSO dissolves agarose gel!). Moreover, unlike DMSO, acetic acid does not cause dye loss from Coomassie Blue-stained acrylamide gels. Residual Coomassie Blue does not affect fluorographic sensitivity since its absorbance at red wavelengths does not overlap with blue emission from the PPO.

Indirect autoradiography (for ^{32}P or ^{125}I). The method involves placing a preexposed X-ray film against the gel (which is either wet or dry, stained or unstained) and then against the other side of the film is placed a calcium tungstate X-ray intensifying screen, the entire sandwich then being placed at $-70\,^\circ$C for film exposure (Laskey and Mills, 1977). Emissions from the sample pass to the film producing the usual direct autoradiographic image, but emissions which pass completely through the film are absorbed more efficiently by the screen where they produce multiple photons of light which return to the film and superimpose a photographic image over the autoradiographic image. The resolution obtained is only slightly less than using direct autoradiography but sensitivity is much increased. Using Kodak X-Omat R film pre-exposed to 0.15 A_{540} units and a Fuji Mach II or Dupont Cronex Lightning Plus intensifying screen, the detection efficiency for ^{32}P is increased 10.5-fold and for ^{125}I 16-fold when compared to direct autoradiography, such that about 100 d.p.m $^{125}I/cm^2$ or 50 d.p.m. $^{32}P/cm^2$ blackens the film detectably. The density of the film image is still proportional to the sample radioactivity allowing quantitative work. The source of the calcium tungstate intensifying screen (which can be used repeatedly) is important, the Fuji Mach II or Dupont Lightning Plus screens being the most efficient for indirect autoradiography. The performance of other intensifying screens can be found in Laskey and Mills (1977).

Dual isotope detection using X-ray film. An interesting technique, especially for comparing 2-D maps generated by two different tissues or cell lines, has been proposed by Choo et al. (1980). The two cell lines are grown in two different isotopes, e.g., ^{14}C and ^3H, then the extracts mixed and analyzed in a single 2-D gel. After fixing and impregnating with scintillator (PPO), the dried slab is fluorographed as described above. The spots developed in the fluorographic process are due to the combined photon emission of both ^{14}C and ^3H isotopes. After this, the gel slab is covered with black carbon paper (which blocks photons emitted from the gel, as well as the weak β-radiation of ^3H) and now autoradiographed, thus revealing only the strong β-spots of ^{14}C emission. Comparison of the fluorograph with the autoradiograph allows an accurate comparison of the gene products of the cell lines.

Similar methods have been described by McConkey (1979) and by Walton et al. (1979). Finally, another method being developed is the use of color negative film (Kronenberg, 1979). This consists of three photographic emulsions each sensitive to one of the three additive primary colors (red, green, and blue). Since the layers are exposed through different thicknesses of gelatin emulsion, emissions of different energies penetrate to different depths and so produce different ratios of exposure in the three layers and so different final colors. When enough radioactivity is present, the radioactive gel can also be conveniently analyzed by a spark chamber, such as the LKB 2105 Radiochromatogram Camera or a similar chamber available from Birchover Instruments Ltd. Recently, it has been demonstrated that the classical strip scanners for thin layers, such as the Packard Radiochromatogram Scanner, are perfectly adequate for scans of thin-layer polyacrylamide gels, provided enough radioactivity (usually at least 0.4 μCi) is present in each band (Righetti and Chrambach, 1978; Magnusson and Jackiw, 1979). For quantitation of labeled proteins after IEF, fractionation of the gel and scintillation counting of the solubilized fractions is still the most widespread method. A particularly convenient method for fractionating cylindrical gels has been described by Bagshaw et al. (1973). Gels are cast in plastic syringe tubes. After focusing, a No. 20 snub nosed needle is placed over the tapered end and a rubber disc over the top of the gel. The gel may now be fractionated into 10 μl aliquots by a plunger driven by a Hamilton Repeating Dispenser. By using plastic tubes where gels do not adhere to the tube walls, band distortion is minimized. Alternatively, cylindrical polyacrylamide gel can be fractionated with an egg slicer device (Heiderman, 1965; Chrambach, 1966) or with an array of steel blades (Matsumura and Noda, 1973) or with a gel slicer equipped with an iris diaphragm (Peterson et al., 1974). The last system allows sectioning in 1 mm thick slices with a precision of 13–34% without freezing or other pretreatment of the gel. Possibly some of these devices (such as the egg slicer or the set of razor blades) could be used for sectioning of slab gels, with only minor modifications. Once the gel has been sliced, for maximum counting efficiency the proteins have to be eluted from the gel. Bis-gels can be solubilized by heating in hydrogen peroxide (Goodman and Matzura, 1971). EDA-gels can be liquefied with alkali (Paus, 1971) and BAC-gels with 2-mercaptoethanol (Hansen et al., 1980) (see also §3.2.3). DATD, of course, should not be used as it inhibits gel polymerization.

3.3.5. Immunotechniques

IEF followed by any type of immunoelectrophoresis has been considered a bidimensional technique and as such will be dealt with further on (see §3.4.1). IEF followed simply by immunodiffusion, possibly within the same matrix used for IEF, will be considered a detection method and will be briefly treated here. Detection by immunoprecipitation of a specific antigen, after IEF, is best done in small gel rods, possibly 1 mm or so in diameter, to reduce to a minimum the amount of antiserum needed for the reaction. After IEF, the gel is extruded in the proper amount of antiserum in a test tube and immunodiffusion allowed to proceed for a few hours (Carrel et al., 1969). Precipitin bands are seen as opalescent rings on the gel surface. The polyacrylamide gel matrix does not seem to hamper the antigen–antibody cross-reaction. This might be due to the fact that, in gel cylinders, proteins tend to focus as hollow rings, with most of the protein packed on the gel surface and little in the gel core. Thus, the protein is readily available for cross-reaction with the antibody.

An alternative to this method consists in cutting the gel rod into discs which are then crushed and transferred to sample wells punched into agar gel. A central trench in the agar slab is then filled with antiserum and diffusion is allowed for 48 h (sectional immunoelectrofocusing; Catsimpoolas, 1969b). Alternatively, the intact gel rod can be embedded in agar. Trenches are then cut parallel to the polyacrylamide gel and filled with appropriate antiserum (Spragg et al., 1973). If focusing has been performed in thin-layer equipment, the gel slab can be overlayered with strips of paper soaked with appropriate antibody (Cotton and Milstein, 1973) or a gel strip can be cut and embedded in agar between parallel trenches filled with antiserum (Dewar and Latner, 1970). A better method for indirect immunofixation consists in overlaying the gel slab, after IEF, with a cellulose acetate strip impregnated with specific antibody; this technique has been used by Arnaud et al. (1977) for phenotyping α_1-antitrypsin and by Johnson (1978) for immunofixation of serum proteins. A direct immunofixation step, for identification of cerebrospinal fluid and serum proteins, has been described by Stibler (1979): after IEF in a gel slab, a proper amount of antiserum is poured over the sample track where the protein to be studied is known to migrate. This is an interesting technique, but it might be quite expensive in antiserum. Other methods have been devised for immunodetection. Thus antigens may be visualized by incubating the washed gel with specific antibody followed by incubation with anti-IgG coupled to horseradish peroxidase (Olden and Yamada, 1977). After further incubation with 3,3'-diaminobenzidine and hydrogen peroxide, the antigen–antibody–peroxidase complexes become visible as stained bands. Alternatively, antigens can be detected by exposing the gel to radioactive antibody followed by autoradiography. A double-antibody method is preferred with the radiolabel on a second antibody directed against the first, or using ^{125}I-labeled *Staphylococcus aureus* protein A (which binds to the Fc portion of the IgG molecule) instead of a second antibody (Burridge, 1978). Careful washing is needed to obtain good ratios of signal to background. A promising modification is to first transfer the proteins to diazobenzyloxymethyl paper where they are coupled covalently (Renart et al., 1979). Specific antigens are then detected by autoradiography after sequential incubation with unlabeled antibody and ^{125}I-labeled *S. aureus* protein A. Antibody and protein A can be removed with urea and 2-mercaptoethanol and the paper challenged again with antibody of different specificity.

3.3.6. Specific stains and zymograms

Most of the specific stains developed for detecting proteins after gel electrophoresis can be adapted for use in IEF. Some of these methods depend on a specific stain for a cofactor such as a metal or prosthetic group, e.g., haem, or other component, e.g., carbohydrate, lipid etc. Others depend on identifying the position of a protein by its biological activity. Some examples of the various possibilities are given below.

Histochemical staining: In many cases, specific stains may be used to detect substances bound to the protein. In the case of metallo-proteins, such as transferrin (Latner, 1973), ceruloplasmin (Latner, 1973) and ferritin (Drysdale, 1970), the protein zone may be located by a specific color reaction for the metal. Alternatively, proteins which bind specific metals can often be detected by instrumental neutron activation analysis. The sample, in sealed ampoules of high purity quartz, is irradiated in a reactor with thermal neutrons. Measurement of gamma ray activity of the long lived radionuclides is performed by using a GeLi detector (Ortec) coupled to a multichannel analyzer (Intertechnique). Schmelzer and Behne (1975) have used this method to detect trace elements bound to serum proteins, such as selenium, chromium, zinc, cadmium, iron, cobalt and antimony. Proteins with cofactors or prosthetic groups such as haem or flavin may also be detected by specific stains. Conjugated proteins offer other possibilities. For example, lipoprotein as may be revealed directly by staining the lipid with Sudan black B (Kostner et al., 1969) and glycoproteins with the periodic acid Schiff stain (PAS) for carbohydrate without removal of ampholytes (Catsimpoolas and Mayer, 1969). Proteins which bind ligands can be revealed after IEF by incubating the gel with an appropriate radioactive ligand. Thus con-canavalin A can be detected with [α-^{14}C]methyl-D-glucoside, the intrinsic factor from gastric juice with [^{14}C] cyanocobalamine and thyroxine-binding globulin with [^{125}I] thyroxine. Keck et al. (1973a) described an interesting approach for the detection of immunoglobulins. After IEF, the focused immunoglobulins were copolymerized with glutaraldehyde in situ, and were thus trapped and immobilized within the gel matrix. The fixed antibodies were then detected by their binding radiolabeled or fluorescent-labeled antigens. Table 3.8 summarizes some common specific stains used in IEF. Many substances may be detected by their biological activity after IEF. By most criteria, IEF is a very mild procedure for enzyme fractionation. Although separations occur in salt-free media, the polyvalent carrier ampholytes have a stabilizing or protective effect on many proteins and help to maintain them in solution. In some cases, carrier ampholytes may afford even greater stabilization than inorganic salts as has been found for α-haemolysin, protease and hexosaminidase (Vesterberg et al., 1967). In some cases, enzyme stabilization may be due to the chelation of inhibitory heavy metals such as Cu^{2+}, Hg^{2+}, and Pb^{2+} by the ampholytes. Many enzymes may be detected by a suitable histochemical stain. Figure 3.23 shows patterns of the isozymes of lactate dehydrogenase separated from crude supernatant fractions from chicken heart by IEF in gel cylinders. Although this enzyme represented only a small fraction of the proteins in the tissue extracts, it was readily detected by its biological activity. In this case, the enzyme activity was revealed by the reduction of pyrimidine nucleotides formed during the oxidation of lactic acid. The formation of the reduced nucleotides was demonstrated by the reduction of tetrazolium salts to form an insoluble colored precipitate. Appropriate steps must, of course, be taken to counteract the buffering effects of the ampholytes in cases where the enzyme focuses at a pH remote from its optimal pH range. Such corrections can usually be effected by carrying out the incubation in strongly buffered solutions or by a brief pretreatment to equilibrate the gel pH. Since the average molarity in gels containing 2% (w/v) ampholyte is in the order of 20 mM (Righetti, 1980), it is usually fairly easy to alter the pH with buffer solutions of 100–200 mM. Wadström and Smyth (1975a) and Gianazza et al. (1980) have discussed various ways to overcome adverse pH conditions in the gel after IEF.

Occasionally, enzymes may lose activity after IEF due to chelation of necessary metal cofactors. For example, Latner et al. (1970) found a

TABLE 3.8
Some specific stains used in situ in gel IEF

Sample	Stain	Authors
Transferrin	2,4-Dinitroso-1,3-napthalenediol (Fe^{3+} reduced with hydroquinone)	Latner (1973)
Apotransferrin	Fe^{3+}-Nitrilotriacetate	Hovanessian and Awdeh (1975)
Ceruloplasmin	p-Phenylenediamine (oxidase activity)	Latner (1973)
Haptoglobins	Haemoglobin/benzidine/H_2O_2 (peroxidase activity)	Latner (1973)
Haemoglobin	Benzidine/H_2O_2 (peroxidase activity)	Latner (1973)
Lipoproteins	Oil Red in 61% EtOH	Latner (1973)
	Sudan Black B	Kostner et al. (1969) Godolphin and Stinson (1974)
Ferritin	Prussian Blue reaction	Drysdale (1970)
Glycoproteins	PAS stain	Hebert and Strobbel (1974)
Immunoglobulins	Glutaraldehyde fixation and then detection with (a) 125-I-labeled antigen; (b) fluorescent-labeled antigen; (c) radiolabeled polysaccharide antigen	Keck et al. (1973a)
Chromatin non-histone proteins	Labeling with ^{32}P	McGillivray and Rickwood (1974)
Concanavalin A	α-Methyl-D-glucoside	Akedo et al. (1972)
RNA	Toluidine blue, methylene blue, or acridine orange	Drysdale and Righetti (1972) Drysdale and Shafritz (1975)
Peptides	Pre-labeling with dinitrofluorobenzene	Kopwillem et al. (1973)
Haemopexin	Binding of haem	Latner and Emes (1975)
Heparin	Toluidine blue	Nader et al. (1974)
Thyroxine-binding-globulin	[^{131}I] Thyroxine	Latner and Emes (1975)
Intrinsic factor (gastric juice)	[^{14}C] Cyanocobalamine	Latner (1973)

LDH ISOZYMES

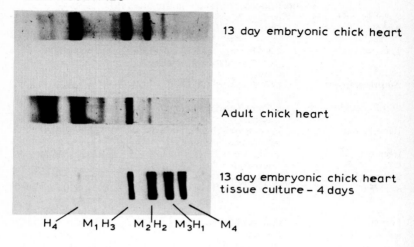

13 day embryonic chick heart

Adult chick heart

13 day embryonic chick heart
tissue culture - 4 days

H₄ M₁ H₃ M₂ H₂ M₃ H₁ M₄

Fig. 3.23. Histochemical staining for lactate dehydrogenase (LDH) isozymes separated from crude supernatant fractions from chicken heart by IEF in gel cylinders. IEF in pH range 3.5–10. (From Righetti and Drysdale, 1973.)

90% loss of activity of alkaline phosphatase on prolonged electrofocusing. However, this enzyme activity could be largely restored by incubation with the zinc cofactor. Additional problems may also arise from anodic oxidation or perhaps electrolytic reduction but these may usually be prevented by the addition of appropriate additives to the electrolyte solution (§4.2). Although some antioxidants may be added directly to sucrose density gradient solutions, many inhibit the polymerization of polyacrylamide gels and should therefore be added to the tops of pre-formed gels or to the electrolytes. In addition to direct histochemical stains with low molecular mass reactants, several 'zymogram' techniques have been developed in which larger reactants are brought into contact with focused enzymes by overlaying the focused gel with another gel impregnated with the substrate or other reactants. Zymogram techniques are best suited for detecting substances separated in thin slabs. Several proteolytic enzymes have been identified by this method. Vesterberg and Eriksson (1972) demonstrated staphylokinase activity by layering fibrin-

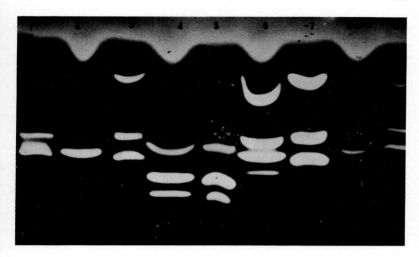

Fig. 3.24. Protease zymogram on casein agar. After IEF in a polyacrylamide gel slab the gel is overlayered with a layer of casein agar. Extracellular proteases of *Serratia liquefaciens (S.l.)* and *Serratia marcescens (S.m.)*. From left to right: *S.I.* strain 2; *S.I.* strain 3; *S.I.* strain 26; *S.m.* strain 6; *S.m.* strain 7; *S.m.* strain 8; *S.m.* strain 10 and *S.m.* strain 11. (From Wädstrom and Smyth, 1975b.)

oclot containing plasminogen over a thin gel slab with the focused enzyme. The staphylokinase converted the plasminogen into plasmin so that the fibrinolytic activity corresponding to the staphylokinase activity appeared as a clear spot on the opaque fibrin plate. Arvidson and Wadström (1973) incorporated casein into an overlaying agar gel to detect staphylococcal proteases after gel electrofocusing while Vesterberg (1973b) detected pepsin activity by the hydrolysis of albumin in an agar overlay. In such cases, the enzyme activity is seen as a clear zone against a blue background when the agar plate is subsequently stained for protein (Fig. 3.24) (Wadström and Smyth, 1975b). Tables 3.9 and 3.10 list some zymogram techniques employing low molecular mass and high molecular mass substrates to detect focused enzymes after IEF. The use of buffering agarose overlayers containing appropriate substrates is particularly attractive for thin-layer gel IEF. Table 3.11 summarizes some overlay techniques in common use (Wadström and Smyth, 1975a). Alternatively, instead of an agarose overlayer, one could use the paper print technique

TABLE 3.9

Zymogram methods used with isoelectric focusing and employing the low molecular
mass substrates

Enzyme	Substrate and chromophore	Authors
β-N-Acetyl gluco-saminidase	Methyl umbelliferyl glycosides	Leaback and Robinson (1974)
N-Acetyl-β-D-hexosaminidase	Napthol-AS-BI-N-acetyl-β-D-glucosamide/fast garnet GBC salt	Hayase et al. (1973)
Acetaldehyde dehydrogenase	Acetaldehyde/NAD/PMS/MTT[a]	Harada et al. (1978a)
Acid phosphatase	4-Methyl-umbelliferyl phosphate	Leaback and Rutter (1968)
Alcohol dehydrogenase	Alcohol/NAD/PMS/MTT	Harada et al. (1978b)
Alkaline phosphatase	α-Naphtyl phosphatase/4-amino-diphenylamine diazonium sulphate	Righetti and Drysdale (1971)
Alkaline phosphatase	β-Naphthyl phosphate/fast blue BB salt, or 4-methyl umbelliferone phosphate	Smith et al. (1971)
L- and D-amino acid oxidase	L-Leu or D-Phe/PMS/triphenyl-tetrazolium	Hayes and Wellner (1969)
Carbonic anhydrase	4-Methyl umbelliferyl acetate	Haglund (1971)
Catechol oxidase	4-Methyl catechol/p-phenylene-diamine	Dubernet and Riberau Gayon (1974)
Creatine phospho-kinase	Creatine phosphate/PMS/nitroblue tetrazolium	Thorstensson et al. (1975)
NAD/NADP dehydrogenase	Substrate/NAD-NADP/PMS/tetrazolium salt	Humphryes (1970)
NADH-Diaphorase	NADH/DCIP/MTT	Tariverdian et al. (1970) Edwards et al. (1979) Fisher et al. (1977)
Esterase	α-Naphthyl acetate/fast red TR salt	Bianchi and Stefanelli (1970)
Esterases	Naphthyl acetate/fast garnet GBC salt	Young and Bittar (1973) Narayanan and Ray (1977)
α-L-Fucosidase	4-Methyl umbelliferyl α-L-fucoside	Omoto and Miyake (1978)
Galactose-1-p-uridyl-transferase	Uridine diphosphoglucose/galactose-1-P/phosphogluco-mutase/glucose 6-P dehydrogenase/NADP	Schapira et al. (1979)

Table 3.9 continued

Enzyme	Substrate and chromophore	Authors
β-Glucuronidase	8-Oxy-chinolin-D-glucuronide/ Fast black K salt	Coutelle (1971)
Glyoxalase I	Methyl glyoxal/glutathione/ DCIP/MTT	Kömpf et al. (1975) Kühnl et al. (1977a)
Lactate dehydrogenase	Lactate/PMS/tetrazolium MTT	Dale and Latner (1968) Chamoles and Karcher (1970a,b) Righetti and Drysdale (1973)
Lipoxidase	Linoleic acid/KI	Catsimpoolas (1969b)
Neuraminidase	2-(3-methoxyphenyl)-*N*-acetyl neuraminic acid	Groome and Belyavin (1975)
Peptidase A	Peptide/L-amino acid oxidase/ amino ethyl carbazole/peroxidase	Kühln et al. (1979b)
Peroxidase[b]	Urea peroxide/guaiacol or *o*-toluidine and mesidine	Delincée and Radola (1970, 1971) Rücker and Radola (1971)
Phosphoglucomutase	Glucose 1-P/glucose 6-P dehydrogenase/NADP/PMS/ MTT	Bark et al. (1976) Brinkmann et al. (1972)
6-Phosphogluconate dehydrogenase	6-Phosphogluconate/NADP/ PMS/MTT	
Procarboxypeptidase A	*n*-Carbo-β-napthoxy-L-Phe/ Diazo Blue B/trypsin	Kim and White (1971)
Superoxide-dismutase A (indophenol oxidase)	MTT/PMS day light	Carter et al. (1976) Harris et al. (1974)
Trypsin and α-chymotrypsin inhibitor	*N*-Acetyl-D,L-phenyl alanine-β-naphtyl ester/trypsin/ α-chymotrypsin	Kaiser et al. (1974)

[a] *Abbreviations:* PMS, phenazine methosulphate; Leu, leucine; Phe, phenylalanine; MTT, 3-(4,5)-dimethyl-thiazolyn-2,5-diphenyl tetrazolium bromide.
[b] Print technique with Whatman No. 1 chromatography paper impregnated with buffered substrate.

TABLE 3.10

Zymogram methods used with isoelectric focusing and employing high molecular mass substrates (Wadström and Smyth, 1975a)

Enzyme	Substrate	Zymogram technique	Authors
Thin-layer gel technique, Staphylokinase[a]	Fibrin clot containing plasminogen	Substrate gel	Vesterberg and Eriksson (1972)
Cellulase	Carboxymethyl cellulose	Paper print	Eriksson and Pettersson (1973)
α-Amylase	Starch-iodine	Starch-film print	Beeley et al. (1972)
Pepsin	Albumin/Coomassie Blue	Immersion	Vesterberg (1973b)
Subtilopeptidase, trypsin, δ-chymotrypsin, crude bacterial proteases	Vitamin-free casein	Agarose overlayer	Arvidson and Wadström (1973)

[a] Plasminogen activator produced by *Staphylococcus aureus*.

TABLE 3.11

Enzymes studied by the use of buffered agarose overlayers containing appropriate substrates (Wadström and Smyth, 1975a)

Enzyme	Substrate[a]	Buffer/ions[a]
α-Amylase	Phadebas amylase test (Pharmacia) 1 tablet/10 ml	0.05 M Tris–HCl, pH 7.4
Deoxyribonuclease	Salmon sperm (Koch-Light) or Calf thymus DNA (Koch-Light) 1% (w/v)	12.5 mM $CaCl_2$ and 12.5 mM $MgCl_2$
Penicillinase	0.2% Soluble starch (Merck) Benzylpenicillin (AB Kabi) 20 mM Lugol's solution: dist. H_2O = 1:1 (v/v)	0.1 M KH_2PO_4, pH 5.9
Phospholipase C	Soya bean lecithin (Sigma) 1.5% (w/v) emulsified	0.15 M NaCl in 0.02 M Tris–HCl, pH 7.4, 1 mM $CaCl_2$ and 1 mM $ZnCl_2$
Protease	Vitamin-free casein (NBC) 1% (w/v), pH 7.4, containing 2 mM KH_2PO_4	1 mM $CaCl_2$
Staphylococcal haemolysin	Rabbit, human or sheep erythrocytes (3%, v/v)	0.15 M NaCl in 0.02 M Tris–HCl, pH 7.4
Elastase	Elastin (Worthington) 0.5% (w/v)	0.01 M Tris–HCl, pH 8.8

[a] Final concentration of substrate and buffer ions in 1.0 or 1.5% (w/v) agarose overlayers.

described by Radola (1973a) or a contact print with Celloclear (Chemetron, Milano). Celloclear is a transparent cellulose acetate membrane (supplied in various sizes and thicknesses) with very large pore size, which allows quick equilibration of high and low molecular mass substrates. The contact print and, if needed, staining and destaining, are performed in a very short time. The membrane can be easily handled and is very resistant to tear (Righetti et al., 1978).

When no suitable histochemical stain is available, biological activities may be detected in gel eluates. For this purpose, the gel rods or slabs are fractionated into thin slices by any of the methods described in the section 'Autoradiography and fluorography' (§3.3.4). The reader should appreciate that it would be impossible, in this book, to cover extensively the vast field of zymogram techniques and that what has been given here is just a limited example of extensive applications. For a deeper insight into

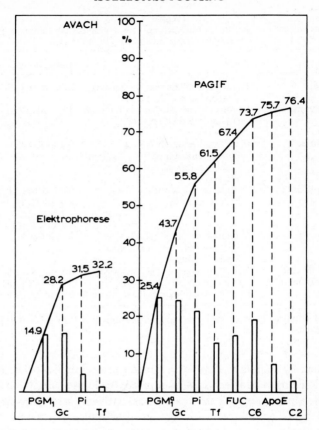

Fig. 3.25. Combined frequency of polymorphism in serum proteins as revealed by conventional electrophoresis (*left*) and isoelectric focusing (*right*). PGM: phosphoglucomutase; Pi: α_1-antitrypsin; Tf: transferrin; Gc: group complement; Fuc: α-fucosidase; ApoE: apolipoprotein E; C_2 and C_6: components of complement Nos. 2 and 6. (From Kühnl, 1977.)

zymogram methods, as applied to electrophoretic separations, readers are referred to the reviews of Gabriel (1971), Shaw and Prasad (1970) and Siciliano and Shaw (1976). An excellent treatise *Handbook of Enzyme*

Electrophoresis in Human Genetics by Harris and Hopkinson (1976 and supplements in 1977 and 1978) deals extensively with zymogram techniques and gives detailed recipes and formulations. Due to its much higher resolving power, IEF has detected a considerably higher polymorphism of serum and tissue proteins as compared with conventional zone or disc electrophoresis. Thus, IEF has become a basic tool for population genetics and in forensic serology (e.g., Kühnl et al., 1977a,b,c, 1978, 1979; Kühnl, 1979, 1981). Figure 3.25 gives an example of the combined frequency of polymorphism in serum proteins as revealed by the two techniques: IEF has discovered many more new alleles in proteins already known to be microheterogeneous and has detected a new polymorphism in proteins considered to be single species by current criteria.

3.4. Two-dimensional procedures

Notwithstanding the high resolving power of electrophoretic techniques Giddings (1969) has calculated that a number of theoretical plates of approximately 2,000 can be achieved in electrophoretic fractionations using reasonable values of path length, voltage gradient and diffusion coefficient) no single electrophoretic method is able fully to resolve all components of a complex protein mixture. This is particularly critical in the case of body fluids, which contain molecules over a thousandfold range of $\bar{M}_r$ values and relative concentrations, with a relatively small range of net charge or free mobility. Often, increasing sample load does not increase the ability to detect minor components, which remain as barely detectable shoulders or irregularities on the tail of the Gaussian distributions of major components (overshadowing effect). The next logical step was thus the introduction of two-dimensional techniques (2-D) with the aim of spreading all the protein spots as far apart as possible and uniformly over the 2-D separation space. A 'macromolecular map' (MM) would thus be generated, in which each spot will be uniquely defined and characterized by a set of coordinates, representing macromolecular parameters of each protein species (e.g., a set of pI-native $\bar{M}_r$ or pI-subunit $\bar{M}_r$ or native $\bar{M}_r$-subunit $\bar{M}_r$ values, etc.). Ideally, if the various species of a multicomponent mixture have to be spread randomly over the 2-D plane, and not aligned along a diagonal in the second run, the fractionation method utilized for the second dimension of 2-D mapping should operate on a principle independent of that used for separation in the first dimension. This concept is very nearly achieved in paper electrophoresis/chromatography (finger-printing) of peptides, where molecules are fractionated on the basis of charge in one dimension and on the basis of their partition coefficients in hydrophobic solvents in the second dimension. The strategy, then, in a 2-D electrophoretic separation of proteins, would be to discriminate molecules on the basis of net-charge (or free mobility) in the first dimension, followed by fractionation according to size, in the second dimension, or vice versa. In this way, one might expect that every region of the 2-D gel slab would be covered with spots.

Smithies and Poulik (1956) were among the first to apply 2-D techniques, by performing electrophoresis on filter paper followed by a run at right angles in a starch gel. The use of two acrylamide gels of different concentrations (Raymond and Nakamichi, 1964; Holmes, 1967; Kaltschmidt and Wittman, 1970a) and even of different pH was then proposed (Kaltschmidt and Wittman, 1970b; Fahnestock et al., 1973; Held and Nomura, 1973). An improved resolution and a (rough) possibility of relating the position of a spot in a 2-D map to the macromolecular parameters of proteins was offered by gel electrophoresis followed by pore gradient electrophoresis as proposed by Margolis and

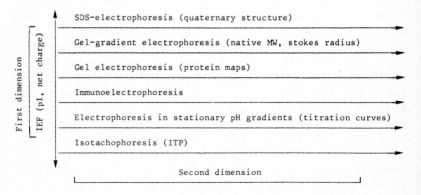

Fig. 3.26. Diagram illustrating various two-dimensional techniques in which one of the dimensions is isoelectric focusing (IEF). (From Gianazza and Righetti, 1979).

Kenrick (1967, 1968). The true innovation, however, came when IEF (which discriminates solely on the basis of charge) in one dimension, was combined with different electrophoretic techniques discriminating either solely on size, or on size and charge, or on biological properties in the other dimension. I will therefore limit my discussion to the combined use of these techniques, as summarized in Fig. 3.26. In this diagram, IEF has been depicted as the first dimension run, even though, in some cases, this combination can be reversed. However, it is customary to use IEF as the first step, because it also permits the use of dilute samples, not suitable for gel electrophoresis, and provides well-sharpened zones for the second stage of fractionation. Furthermore, use of electrophoresis as the second step offers the possibility of removing carrier ampholytes at that stage, thus simplifying the staining procedure.

3.4.1. IEF-immunoelectrophoresis (or immunofixation)

After IEF, the gel may be incubated directly in an appropriately buffered solution containing a specific antiserum (Carrell et al., 1969; Powell et al., 1975). In the phenotyping of α_1-antitrypsin an immunoprint method has been described by Arnaud et al. (1977) by which an antibody-containing cellulose acetate strip is applied onto the polyacrylamide gel. Direct and indirect immunofixation has also been used by Johnson (1978) for detecting α_1-AT after IEF. When focusing in gel slabs, the gel area where the protein to be studied is known to migrate can be spread with undiluted monospecific antiserum (25–50 μl/cm). This allows saving of antisera and detection of different antigens by using different specific antisera over different sample tracks and for different gel areas (Stibler, 1979).

Other variations of immunoelectrofocusing have also been described. Some of the more common are illustrated in Fig. 3.27. For example, after IEF in rods or strips of polyacrylamide gel, the focused gel can be embedded in an antibody-impregnated agarose gel, through which the focused proteins are subsequently caused to migrate by electrophoresis (crossed immunoelectrofocusing). Antigen-antibody complexes form conical precipitin zones whose dimensions give a good measure of antigen concentration (Laurell's rocket technique, 1966). In addition, it is often possible to investigate possible immunological relationships of closely spaced antigens (Alpert et al., 1972). However, problems arise

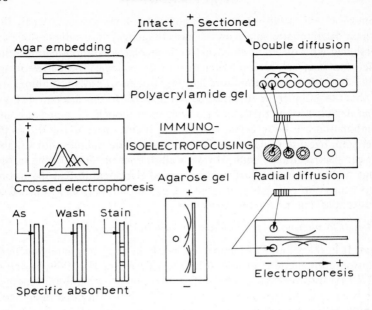

Fig. 3.27. Various forms of immunoisoelectric focusing analysis in gels. (From Catsimpoolas, 1973c.)

when a junction is made between the polyacrylamide gel used in the first dimension and the agarose gel used for the crossed immunoelectrophoresis of the focused antigens. The difference in electroosmotic flow properties between these two gel media results in welling up of water at the anodic interface of the junction and drying-out of the agarose behind the cathodic interface during immunoelectrophoresis. These adverse effects often cause blurring and distortion of immunoprecipitate patterns. These problems have been largely solved by the laying-on procedure devised by Smyth et al. (1977). Figures 3.28a–h illustrate the sequence of steps required for proper operation. The equipment required consists of a scalpel with disposable blades, a long razor blade (20 × 150 mm, e.g., Triumph Glass Cleaner) and a second blade of the same type mounted on a strip of Plexiglass. After IEF, the electrode gel strips are sliced away and removed (Figs. 3.28a,b). The gel is then transferred on graph paper and sliced transversally, with the help of the long rigid blade and scalpel (Fig. 3.28c), along the electrophoretic path of each sample. Now, by using the flexible razor blade, each slice can be sequentially removed without being distorted or stretched, and can be easily lowered on the surface of the agarose gel for the second dimension. Care is taken that the polyacrylamide strip is applied onto the surface of the agarose with the protein-bearing surface forming the contact zone. By this method, highly reproducible results and undistorted immuno-precipitates are obtained (Söderholm and Smyth, 1975). This technique is today greatly facilitated by the availability of polyester foils for gel casting

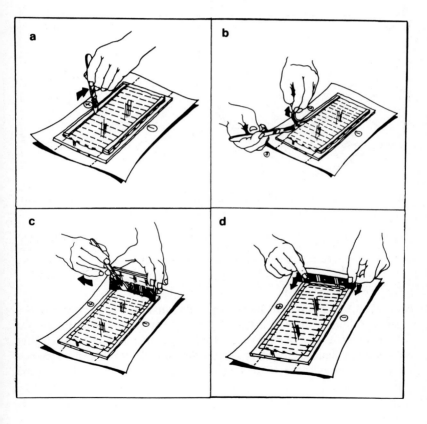

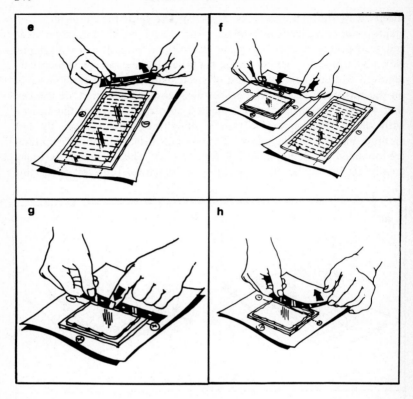

Fig. 3.28. Laying on technique for crossed electrofocusing–immunoelectro-phoresis. The strip containing the (separated) sample is cut away from the poly-acrylamide plate, inverted and layered on the surface of the agarose plate. For further details, see text. (From Smyth et al., 1977.)

(see §3.2.12): cutting of the gel strips, their handling and transfer to the agarose gel are an easy task (see also §3.4.6). Moreover, with the advent of agarose IEF (see §3.2.4) the problem of the junction between first and second dimension gel in crossed immunoelectrophoresis does not exist any longer, since both dimensions can be performed in the same matrix.

A multiple immunoreplica technique, particularly suitable after a 2-D run, has been recently described by Legocki and Verna (1981): the proteins separated in the 2-D gel slab can be subjected to a series of electrophoretic transfers to nitrocellulose paper (partial 'western-blots'), providing several replicas of the gel (for the general population of proteins three replicas can be obtained, but four or five replicas can be given by the most abundant protein spots). Each replica can be reacted with a series of different antisera (at least three), where the preceding antibody is removed by treatment with pH 2.2. The antigen–antibody complexes are visualized using ^{125}I-labeled protein A. By this technique, one gel can be reacted with at least nine different antibodies.

3.4.2. IEF-gel electrophoresis

This is a hybrid method of charge versus charge-size fractionation in the second dimension. Dale and Latner (1969) were the first to demonstrate the potential of this procedure in detecting quantitative and/or qualitative changes in protein maps from normal and pathological sera. After the separation of proteins by IEF in rods of polyacrylamide, the focused gels are embedded into the top of a slab of polyacrylamide and subjected to electrophoresis through the width of the gel. The focused proteins and ampholytes migrate into the gel slab where their electrophoretic mobility is determined by both their charge and their molecular size. The ampholytes are thus effectively separated from the proteins and a characteristic map obtained. The gel slabs can be dried down onto supporting plates, after washing in 7% gelatin and 3% glycerol, to provide a permanent record. Dale and Latner (1969) were able to identify many of the known components of serum in these plates by running standards or by specific histochemical or immunological procedures. This study was followed by the application of this technique to several pathological cases, including nephrotic syndromes, mono- and polyclonal gammopathies, multiple sclerosis, IgG myelomatosis, proteinuria (Latner, 1973) and typing of haptoglobin genotypes (Latner and Ames, 1975). Wrigley (1968b) and Wrigley and Shepherd (1973) used a somewhat similar procedure to analyze wheat grain proteins. On gel electrophoresis, about 20 components may be detected in wheat gliadin preparations, while a similar number are revealed by IEF. However, by combining these procedures, nearly twice as many components were revealed. A similar bidimensional procedure has been used by Macko and Stegeman (1969) and by Stegeman et al. (1973) to map potato proteins. These authors too have found that the combined technique gives a more complex pattern and is more specific for a given potato variety than either electrophoresis or electrofocusing alone.

3.4.3. IEF-electrophoresis in gel gradients

The combined procedures of IEF and gel electrophoresis clearly demonstrated the value of this two dimensional procedure for genetic and taxonomic studies. However, electrophoresis in gels usually separates proteins on the basis of their size and charge and the two may occasionally work against one another to obscure or nullify a separation (Rodbard et al., 1974). Kenrick and Margolis (1970) adopted an earlier two-dimensional electrophoretic procedure to eliminate this complication and to separate focused proteins in the second dimension almost solely on the basis of their molecular size. They used a concave $4.5-26\%$ gradient polyacrylamide gel slab in which the focused proteins migrated electrophoretically to a quasi-equilibrium position according to their molecular size or Stoke's radius. Thus, in addition to increased resolution, this method gives estimates of the pI and the approximate molecular mass of the native protein. Is it actually possible, by pore gradient electrophoresis, in absence of SDS, to measure accurately $\overline{M}_r$ values of native proteins? This highly controversial issue has been examined by Lambin and Fine (1979). They have found that in gradient gels there is a linear relationship between the migration distance of proteins and the square root of the time of electrophoresis. When the slopes of the regression lines thus obtained are plotted against the molecular mass of protein standards, good estimates of M_r values of native proteins are obtained.

A variation of this technique has been described by Felgenhauer and Pak (1975). They perform IEF in the first dimension in a granulated gel layer (Sephadex G-75 superfine) and electrophoresis in the second dimension in a linear gradient of $3-30\%$ acrylamide. The use of granulated gels in the first dimension allows rapid equilibration of large proteins, such as ferritin, α_2-macroglobulin and β-lipoprotein, and quick detection of the focused bands by the paper print technique (Radola, 1973a; see also §2.2.3). During IEF, the Sephadex gel is spread on a thin plastic sheet (0.15 mm). This allows easy cutting of the Sephadex strip and transfer to the polyacrylamide gel slab for the second dimension run. During electrophoresis in the gel gradient, the gel chamber is run horizontally and paper wicks used to provide electrical contact with the electrode reservoirs. As in the previous case, provided that each component is allowed to reach its exclusion limit (Felgenhauer, 1974), a quasi-equilibrium is achieved which allows a good estimation of molecular size (Andersson et al., 1972; Rüchel et al., 1973).

3.4.4. Isoelectric focusing–isotachophoresis (IEF–ITP)

This 2-D technique has been described so far in the reverse combination ITP–IEF. Crossed ITP–IEF should allow an investigation of the differences between the separation principles of these two methods. As presently described, ITP is first run in a thin-layer polyacrylamide gel, then a strip is cut and transferred to the cathodic side of a thin layer (0.5 mm) IEF slab, to be used for IEF separation. Crossed ITP–IEF combines the high resolution of both techniques while overcoming their respective limitations. For instance, proteins with nearly the same isoelectric points can be separated by isotachophoresis at a pH at which their relative pH/mobility curves diverge. On the other hand, proteins with very similar pH/mobility curves at the prevailing pH in ITP fractionation, but with different pI values, will co-migrate in the ITP step, but be separated in the IEF step (Brogren, 1977).

3.4.5. IEF across a urea gradient

An interesting extension to the analysis of quaternary protein structure, is the use of urea concentration gradients in gels (Hobart, 1975). Urea gradient thin-layer gels are poured between glass plates arranged so that the urea gradient is oriented at 90° to the final direction of current flow. The sample is applied uniformly across the gel, in a long paper strip parallel to the electrodes. In this way the sample will focus in regions of progressively increasing urea concentrations. The information gained by this type of experiment is summarized in Fig. 3.29. In the case of covalently-linked structures (which cannot be dissociated by urea), if denaturation is not accompanied by a change in pI, the protein band will be straight (Fig. 3.29a). If there is a pI change, it may occur sharply at a critical urea concentration (Fig. 3.29b) or over a wide urea range (Fig.

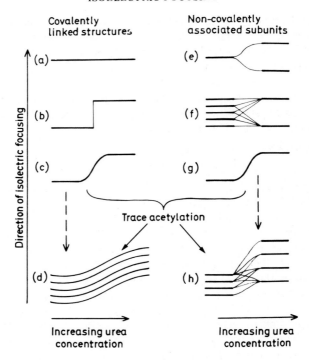

Fig. 3.29. Diagrammatic representation of possible patterns of polypeptide bands obtained by thin-layer gel IEF of native or trace-acetylated proteins in the presence of discontinuous urea concentration gradients. (From Hobart, 1975.)

3.29c). Trace acetylation of covalent structures will give rise to a set of bands which will remain parallel (Fig. 3.29d). In the case of non-covalently associated subunits, the native band may give rise to a number of free subunit bands in high urea regions (Fig. 3.29e). In the case of isozymes made up by random combination of few dissimilar subunits (e.g., LDH), the native insozyme pattern will first give rise to a complex zone with interdigitating bands then at higher urea levels emerge into a zone where only the dissociated subunits are seen (Fig. 3.29f). If the native protein is an oligomer of identical subunits, the denatured pattern will be indistinguishable from that of a covalent structure (Fig. 3.29g). However, identical subunits can be made dissimilar by random charge changes (such as in trace-acetylation). In this case the bands of such artificial isozymes will give rise to a complex pattern (Fig. 3.29h) similar to that described in Fig. 3.29f. Unfortunately this article by Hobart, which I feel is one of the finest in the field of the use of IEF as a structural probe for proteins, has gone largely unnoticed by the scientific community. Most of his predictions, however, have been fully verified by Creighton (1979) in a fine article describing the electrophoretic analysis of the unfolding of proteins across urea-gradient gels.

3.4.6. IEF–SDS electrophoresis

This is by far the most popular technique, as it very nearly achieves the aim of spreading maximally the protein spots on the 2-D gel plane. Barrett and Gould (1973) and McGillivray and Rickwood (1974) were among the first to introduce this method of charge/mass fractionation. However, it was only in 1975, when O'Farrell described a simple method for casting SDS-pore gradient gels and assembling an electrophoretic cell for 2-D runs, and coupled this technique to autoradiography, thus revealing more than a thousand spots on the 2-D gel, that 2-D techniques became popular. At the beginning, most 2-D runs were performed in the sequence IEF–SDS (O'Farrell, 1975; Yeoman et al., 1974; Ames and Nikaido, 1976) since it was believed that the anionic detergent SDS would destroy the pH gradient if present in the IEF dimension. It soon became apparent, however, that proteins exposed to SDS could be run in the IEF gel without severely disturbing the IEF separation (Miller and Elgin, 1974; Danno, 1977). Thus, recently, two methods have been proposed in which the sequence is reversed into SDS–IEF runs. This is not just a mere curiosity, but is in fact necessary when extracting whole cells, since boiling SDS favors practically complete solubilization of membrane-bound proteins while decreasing the likelihood of alteration of cellular proteins by kinases, proteases or other active enzymes in the extract. In one method (Tuszynski et al., 1979), the SDS-gel strip is placed on an IEF gel containing 8.5 M urea and 2.5% NP-40: the combined action of urea and non-ionic detergent in the gel results in the splitting of the SDS-protein complex. Best results were obtained with 2.5–3.5% detergent levels in the gel: when NP-40 was lowered below 2% or increased above 4% resolution was impaired. In the other method (Siemankowski et al., 1978), the SDS gel is briefly exposed to an interfacing solution comprising 9 M urea, 1% DEAE-cellulose, 5 mM ascorbic acid, 0.2 mM EDTA, 0.12% 2-mercaptoethanol, 10% glycerol in 30 mM Tris, 0.225 M glycine buffer, pH 8.6. Each components; 2-mercaptoethanol prevents formation of disulfide bridges and regions of proteins; DEAE-cellulose facilitates removal of SDS; ascorbic acid and EDTA significantly increase the solubility of protein components, 2-mercaptoethanol prevents formation of disulfide bridges and glycerol reduces diffusion of low $\overline{M}_r$ components. A few practical aspects of 2-D runs will be considered here in more detail. There are two basic variants of the technique: (a) vertical systems (O'Farrell, 1975) and (b) horizontal, ultrathin gels (Görg et al., 1980, 1981).

(a) *Vertical systems.* The basic design remains that of O'Farrell (1975) (Fig. 3.30): the gel slab for the second dimension is 16.4 cm wide by 14.6 cm high and 0.8 mm thick and utilizes one exponential gradient from 10 to 16% T and a 4% T stacking gel. Instructions for preparing the first dimension IEF gel cylinder and the second dimension SDS-gel slab are given in Table 3.12A,B. The lysed cells, grown in ^{14}C-labeled amino acids, after treatment with DNase and RNase are extracted with 9.5 M urea, 5% 2-mercaptoethanol, 2% NP-40, 1.6% Ampholine, pH 5–7, and 0.4% Ampholine, pH 3.50–10. If the sample has been radio-

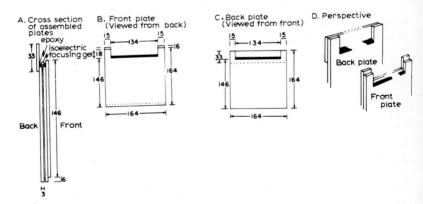

Fig. 3.30. Construction scheme of a slab gel plate. After IEF in the first dimension, the gel rod is embedded on top of a polyacrylamide gel slab for SDS-electrophoresis in the second dimension. All measurements are given in mm. (From O'Farrell, 1975.)

TABLE 3.12
IEF—SDS according to O'Farrell (1975)

A. Preparation of the first-dimensional electrofocusing gel

To make 10 ml of the first-dimensional gel, mix:
    5.5 g of urea (ultrapure)
    1.33 ml 28.38% acrylamide, 1.62% bisacrylamide
    2 ml 10% Nonidet P-40 (NP 40)
    0.4 ml 40% Ampholines (pH 5—7)
    0.1 ml 40% Ampholines (pH 3.5—10)
    1.95 ml water
    5 μl TEMED
    10 μl 10% ammonium persulphate

B. Preparation of the second-dimensional SDS-Gel

Resolving gel

Light solution:
    4 ml 0.4% SDS, 1.5 M Tris—HCl, pH 8.8
    5.3 ml 20.2% acrylamide, 0.8% bisacrylamide
    6.7 ml water
    8 μl TEMED
    25 μl 10% ammonium persulphate
Dense solution:
    2 ml 0.4% SDS, 1.5 M Tris—HCl, pH 8.8
    4.3 ml 29.2% acrylamide, 0.8% bisacrylamide
    1.7 ml 75% glycerol
    4 μl TEMED
    10 μl 10% ammonium persulphate
Stacking gel
    1.25 ml 0.4% SDS, 0.5 M Tris—HCl, pH 6.8
    0.75 ml 29.2% acrylamide, 0.8% bisacrylamide
    3.0 ml water
    5 μl TEMED
    15 μl 10% ammonium persulphate

labeled, a total amount of ca. 20 μg protein will suffice but if the 2-D gel has to be stained by Coomassie Blue, 100—250 μg sample should be loaded to the IEF gel. After IEF (run as in Table 3.12A) the gel rod is equilibrated for two periods of 60 min each in 5 ml of SDS sample buffer: 2.3% SDS, 5% 2-mercaptoethanol, 10% glycerol and 62.5 mM Tris—HCl, pH 6.8. Contact between 1st-D IEF gel rod and 2nd-D SDS-slab is ensured by sealing with stacking gel (4% T) or with 1% melted

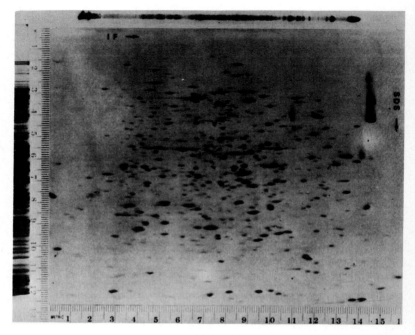

Fig. 3.31. Separation of *E. coli* proteins. *E. coli* was labeled with ^{14}C amino acids *in vivo*. The cells were lysed by sonication, treated with DNase and RNase and dissolved in 9.5 M urea, 2% Nonidet P-40, 2% Ampholine, pH 5–7, and 5% 2-mercaptoethanol. 25 μl of sample containing 180,000 c.p.m. and approximately 10μg protein were loaded in the gel in the first dimension (IEF). The gel in the second dimension (SDS) was a 9.25 to 14.4% exponential acrylamide gradient. At an exposure of 825 h, it is possible to count 1000 spots on the original autoradiogram. (From O'Farrell, 1975.)

agarose. To facilitate the loading of the IEF gel rod, Imada (1978) has described a cell in which one of the two glass plates has been bent at an angle of 6.5°. When multiple gel rods have to be cast on a stack of SDS gels, the two systems of Anderson and Anderson (1978a,b) can be utilized. This set-up for multiple, parallel IEF–SDS runs has been termed ISO–DALT system, to indicate that separation is based on ISOelectric focusing (charge) in the first dimension and on molecular mass (DALtons) in the second. This has spurred Dean (1979) to describe an apparatus

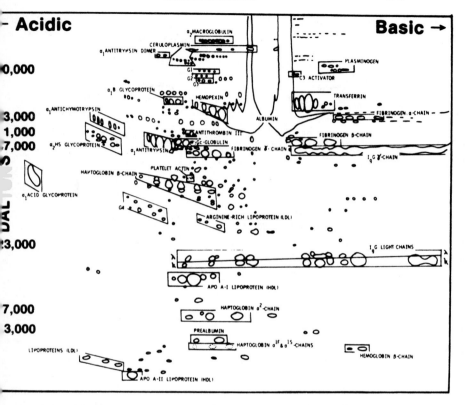

Fig. 3.32. Map drawn from a stained 2-D gel, and labeled to indicate positions of known plasma proteins. This is a slightly corrected version of the map published in Proc. Natl. Acad. Sci. (USA), *74*, 5421 (1977). The fibrinogen α and γ chain labels were reversed in the earlier map. (From Anderson and Anderson, 1977.)

for simultaneous processing of eight polyacrylamide gel slabs complementary to the standard eight-tube disc electrophoretic cell. A micro 2-D technique, utilizing gel slabs of approximately stamp size (ca. 1 cm) has been developed by Rüchel (1977). Miniaturized systems for pore gradient electrophoresis in slabs have also been adopted by Matsudaira and Burgess (1978), by Ogita and Market (1979) and by Poehling and

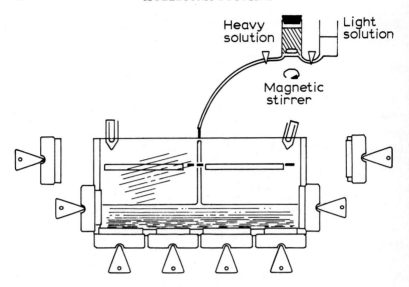

Pouring of an ultrathin-layer exponential gradient gel

Fig. 3.33. Set-up for casting a concave exponential polyacrylamide gel gradient from 4% T to 22.5% T. The cell is from the LKB Multiphor apparatus and the U-gasket is cut out of 3 layers of parafilm (360 μm gel thickness). Mixing chamber: 5 ml of 22.5% T, 4% C, 55% glycerol; reservoir: 5 ml of 0% acrylamide, 0% glycerol; both solutions buffered with 375 mM Tris–Gly, pH 8.8. (From Görg et al., 1981.)

Neuhoff (1980) (see also §3.2.17). An example of the separations obtained by these 2-D techniques is shown in Fig. 3.31: in the *E. coli* lysate, more than 1,000 spots could be revealed. Detection and quantitation of protein species representing as little as 10^{-6} to 10^{-7} of the total cell proteins could be achieved. An example of a 2-D map, as applied to the analysis of human plasma proteins, is given in Fig. 3.32 (Anderson and Anderson, 1977).

(b) *Horizontal systems.* They represent an interesting alternative to the previous method (Görg et al., 1980, 1981). The second dimension run is performed in a concave exponential gel from 4% T to 22.5% T, barely 360 μm thick (although today we prefer a more comfortable 0.5 mm thickness) cast against a silanized polyester foil (Fig. 3.33) (see also

Ultrathin – Layer 2D electrophoresis

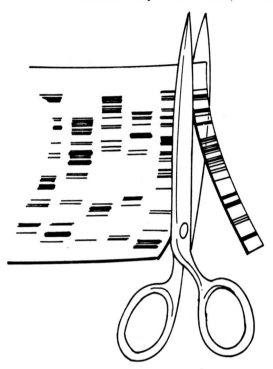

Cutting off the focusing gel strip

Fig. 3.34. Handling of the IEF gel slab. After IEF, the protein zones are fixed and stained in Coomassie Brilliant Blue G-250. For transfer to the 2-D matrix, a gel strip as wide as the trench in the 2-D gel is cut away. (From Görg et al., 1981.)

§3.2.13). The first dimension is also performed on a gel slab, 240 μm thick, bound to a plastic support. At the end of the IEF run, the proteins are fixed and stained with Coomassie G-250, thus providing a permanent record pattern. When ready to perform the 2-D run, a sample track is cut away with scissors (Fig. 3.34), the gel strip equilibrated for 2–4 min in SDS buffer and then loaded onto the SDS-gel slab. Figure 3.35 shows how this step is performed: the IEF gel strip is lifted at the

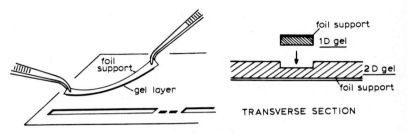

Transfer of the focusing gel strip to the
electrophoresis gel

Fig. 3.35. Interfacing the 1-D with the 2-D gel. The stained IEF strip, rinsed in
water, equilibrated in SDS-buffer and blotted, is transferred, gel layer facing down,
to the trench of the 2-D matrix. The operation is greatly facilitated by the plastic
backing of the gel. On the right, we can see a transverse section of the gels, indicating that there is a gel floor in the trench dug in the 2-D gel. (From Görg et al.,
1981.)

two extremes with tweezers which hold onto the two protruding edges
of the supporting plastic film and lowered, gel layer facing down, into
the trench preformed in the SDS slab. Several of the advantages of this
technique are: the 1-D to 2-D gel contact is optimized due to the flat joint;
transfer of proteins from the first to second dimension is complete, due
to the extreme gel thinness; gel equilibration, staining, destaining and
drying are performed in a fraction of the time needed in conventional
systems.

(c) *Markers.* When looking at a 2-D map with more than 2,000 spots
scattered over it one cannot avoid the impression of being confronted
with a starry sky on a clear night. As astronomers have long known, it
is imperative to have reference points for orientation. For the IEF
dimension, perhaps the simplest and most ingenious procedure is the
production of 'carbamylation trains' described by Anderson and Hickman (1979) and Tollaksen et al. (1981). It is known that, when a protein is heated in a solution of urea, its amino groups are progressively
carbamylated through the production of cyanate. Since the loss of a free
amino group below pH 8.5 results in a unit change in the charge of a
molecule, the products of this reaction appear, in the SDS gel, as a row
of spots at roughly constant molecular mass, spaced apart by about

0.1 pH units. Carbamylation of human Hb β-chain gives a row of 12 such spots corresponding to the blockage of 11 Lys and 1 α-amino group. A carbamylation train of carbonic anhydrase (from bovine erythrocyte) gives a string of 20 spots while a train of creatine phosphokinase (rabbit muscle) is as long as 30 wagons. For the SDS dimension, a series of cross-linked polymers of one protein of known $\bar{M}_r$ could be used as a $\bar{M}_r$ standard (Inouye, 1971), however, they might exhibit anomalous behavior in SDS gels. For this reason, Giometti et al. (1980) have suggested the use of rat heart whole homogenate as an internal reference standard in the DALT dimension. In the SDS run, this homogenate generates 80 lines, of which 12 are major spots distributed at convenient intervals along the gel pattern covering $\bar{M}_r$ increments from as little as 300 daltons (about 3 amino acid residues) up to 10,000 daltons at the upper end of the gel. This homogenate, when extracted in SDS, dispensed in small aliquots in 0.7% agarose and stored frozen at -20 to $-70\,^{\circ}$C, gives fully reproducible patterns for over 12-month periods of use. As a simple, tracking colored protein for the IEF–SDS gels, cytochrome c has been suggested (Leader, 1980), while, as a complete set of pI-M_r standards, bacteriophage T4 coat proteins have been proposed by Kurian et al. (1981).

(d) *Densitometry.*   Part of this topic has been covered in §3.3.3. In addition to the references reported there, an outline must be given of recent computer programs developed for analysis of 2-D electropherograms, especially in connection with silver stain. In fact, while 2-D electrophoresis has replaced a single variable with a pair of orthogonal variables representing the net charge and molecular mass of component proteins, the silver stain process of Sammons et al. (1981) has extended this pair to three: charge, mass and color. In addition to a characteristic position on a gel and an intensity, the spot formed by a given protein exhibits upon silver staining a peculiar hue as well, usually of various shades of yellow, red, green and blue. In order to quantify this 'third dimension' Vincent et al. (1981) have developed a computer program for multispectral digital image analysis of color in two-dimensional electropherograms. To this purpose, these authors have drawn on a large body of generalized image-processing software developed for analysis of multispectral image data from remote sensing satellites, such as LANDSAT.

Thus, when I stated that today's biochemist should have a training as an astronomer (see §3.3.3) I only said part of the truth: they should also have an astronaut's training. Another computer system for analysis of 2-D gels called 'GELLAB', has been developed by Lemkin and Lipkin (1981). Other systems for matching and stretching 2-D gel patterns have been described by Taylor et al. (1981) and by Kronberg et al. (1981). We can only hope that, out of this large body of software available, somebody will think of distilling a single, simple yet succulent program of universal use in any lab around the world.

(e) *Applications.*   Given the incredible explosion of 2-D techniques, it would be foolish of me to attempt to summarize this field. I will only briefly survey here some recent applications. Extensive sections on 2-D techniques can be found in the proceedings of recent meetings (Righetti, 1975; Radola and Graesslin, 1977; Catsimpoolas, 1978; Radola, 1980; Allen and Arnaud, 1981). Anderson and Anderson (1977, 1978a,b) have begun the systematic mapping of an estimated 30,000 to 50,000 human gene products. They include: mapping of human erythrocyte proteins (Edwards et al., 1980), saliva (Giometti and Anderson, 1980), urinary proteins (Tollaksen and Anderson, 1980), peripheral blood lymphocytes (Willard and Anderson, 1980; Anderson 1981). Applications in clinical chemistry (Thorsrud et al., 1980), and to the analysis of human lipoproteins (Zarnnis and Breslow, 1980), of virus infected cells (Drzenieck et al., 1980), of human fibroblasts (Singh et al., 1978; Burghes et al., 1981), of liver cytosol proteins (Elliott et al., 1978), of nuclear chromosomal proteins (Hurley et al., 1978), of basic cellular proteins (Sanders et al., 1980; Willard et al., 1979), of erythrocyte membrane proteins (Harell and Morrison, 1979), of mitochondrial ribosomal (Czempiel, 1979) and microsomal (Knüfermann, 1977) membrane proteins, of hair roots (Singh et al., 1981), of cystic fibrosis saliva (Bustos and Fung, 1981) and of human amniotic fluid proteins (Burdett et al., 1981) have been described. Unfortunately, by the time you will be able to read this, the list will be largely outdated.

3.5. Transient state IEF (TRANSIEF)

Catsimpoolas (1973c–f) has developed some elegant techniques for continuous analytical scanning IEF which allow estimates of the first and second moments of the concentration profile throughout the course of the IEF experiment. This method gives quantitative information on parameters of the experimental aspects of IEF and gives valuable insight into other physico-chemical properties of amphoteric molecules. He coupled a scanning device to an on-line digital acquisition and processing system. This allows continuous monitoring and recording of peak positions, peak area, segmental pH gradient and pI. Unlike IEF, which is a steady-state system, TRANSIEF is essentially a kinetic-method which is assumed to consist of three stages: initial focusing (IE), defocusing (DF) and refocusing (RF). The first stage involves the electrophoretic migration of amphoteric molecules to their pI with the attainment of steady-state focusing. The second stage (DF) is concerned with analyses of diffusion in the absence of current and the third stage (RF) studies the reapproach to the steady state when the electrical field is reapplied.

TRANSIEF gives valuable information about the minimal focusing time for a protein zone. It also provides quantitative data on the shifts or instability of the pH gradient, such as the so-called 'plateau phenomenon' or cathodic drift (see p. 299). In agreement with previous findings (Righetti and Drysdale, 1971), Catsimpoolas (1973c) found that the plateau phenomenon usually involved a progressive migration of the more basic ampholytes toward the cathode while the anodic zones remained essentially unaltered. Catsimpoolas has used the DF stage to estimate diffusion coefficients of proteins. In following the rate of zone spreading in the absence of an electric field, the measurement of variance (σ) of the diffusing zone as a function of time and gel concentration yields a linear relationship in which the slope corresponds to the apparent diffusion coefficient (D) of the protein. Measurements of diffusion coefficients given by this method are, however, substantially higher than those given from ultracentrifugation data, perhaps because of interaction with the gel matrix. It is also possible to obtain information by TRANSIEF about the effective molecular radius and molecular mass of proteins by using the retardation coefficient measured at different gel strengths. The computer program for TRANSIEF data evaluation has been published by Catsimpoolas and Griffith (1973). Figure 3.36 summarizes the various parameters measurable by TRANSIEF (Catsimpoolas, 1975b). Unfortunately, as Catsimpoolas lamented: 'the kinetic theory and available instrumentation allow us to do the necessary

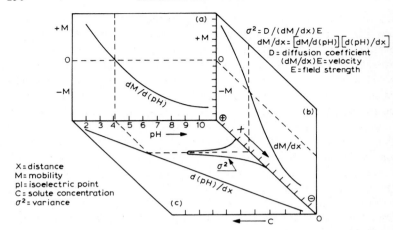

Fig. 3.36. Schematic diagram of the relationship among pH, mobility, distance and protein concentration in isoelectric focusing. The migration of a protein towards its isoelectric point is a function of its pH-mobility curve, $dM/d(pH)$ (panel a) and the pH gradient curve $d(pH)/dx$ (panel c). The observed mobility dM/dx (panel b) is the product of the above two parameters. X, distance; M, mobility; pI, isoelectric point; C, solute concentration; σ^2, variance; $\sigma^2 = D/(dM/dx) E$; $dM/dx = [dM/d(pH)]$ $[d(pH)/dx]$; D, diffusion coefficient; $(dM/dx)E$ = velocity; E, field strength. (From Catsimpoolas, 1975b.)

measurements, but the compiling of correction factors due to non ideal effects, stemming primarily from the present imperfection of carrier ampholytes, renders the method impractical. Dynamic development of new methodology and instrumentation coupled with a selective extension of the kinetic theory and the much-needed synthesis of second generation ampholytes will suggest new avenues of application'. Will the immobilized pH gradients (see §1.11.7) be the answer to the quest for the Holy Grail?

3.6. Detection of neutral mutations

The vast majority of spot mutations so far detected are point variants affecting a charged amino acid. For instance, in the case of human hemoglobin (in October, 1973), 174 point mutations had been described (Hunt and Dayhoff, 1974), distributed as follows: 61 in α-chains, 99 in

β-chains, 8 in γ-chains and 6 in δ-chains. Of those, only 35 (i.e., ca. 20%) are neutral to neutral mutants (especially Pro $\leftrightarrows$ Leu, which account for about 30% of all neutral mutations) which probably have not been detected by electrophoretic techniques but through altered functional properties and chemical analysis (tryptic digestion, fingerprinting and amino acid analysis of anomalous peptides). If we generalize these findings, given the total frequency of charged amino acids in proteins, this means that only about 40% of all possible mutations will result in charged mutants detectable by electrophoretic techniques. A good 60% will be 'silent' mutants, involving neutral amino acid substitutions, undetectable by any electrophoretic technique. There are hints that careful use of IEF can indeed resolve even these silent mutants. A very interesting example comes from a recent article by Whitney III et al. (1979). When working with mouse hemoglobins (Hb), they were able to detect by IEF several neutral substitutions in the α-chain. The resolved haplotypes, carrying mutations in three positions (α^{25}, α^{62} and α^{68}), were — Hb_1: Gly^{25} Val^{62} Asn^{68}; Hb_2: Gly^{25} Val^{62} Ser^{68}; Hb_3: Gly^{25} Val^{62} Thr^{68}; and Hb_4: Val^{25} Ile^{62} Ser^{68}. In going from Hb_1 to Hb_4 the isoelectric points of the focused, native tetramers, would progressively decrease, while the overall hydrophobicity would concomitantly increase (see Fig. 3.37). In fact, from the hydrophobicity scale of Nozaki and Tanford (1971), I have calculated a $\Delta f_t = -500$ in going from Hb_1 to Hb_2 and a $\Delta f_t = -800$ for the transition Hb_2 to Hb_3 (the overall hydrophobicity of Hb_4 is more difficult to calculate as not enough data are available). I hypothesize that hydrophobicity increments in the α-chain bring about pI decrements in the tetramer via loss of positive charges. This is probably not an isolated case, but could be a more general phenomenon. Thus, in horseradish peroxidase, when the prosthetic group protoheme IX binds to the imidazole of a His residue in the protein chain (Mauk and Girotti, 1974), the pK of this group is lowered from its usual value of six to about zero (Dunford and Stillman, 1976). When the protoporphyrin IX (which contains four imidazoles, of which two are protonatable with a pK of ca. 6.5) is dissolved in neutral micelles of Triton X-100, a pK_{app} of ca. 0.9 is found (Savitskii et al., 1978). I suggest that, in the native mouse hemoglobin, there could be a positively charged group (a most likely candidate would be Lys) lying within a few angstroms distance from the amino acids in positions 25, 62 and 68 of

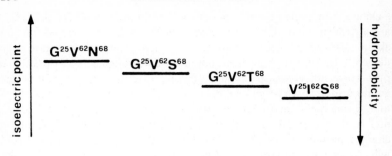

Fig. 3.37. Separation of mouse hemoglobins carrying neutral to neutral spot mutations, The mutants are (from left to right): $Gly^{25}Val^{62}Asn^{68}$; $Gly^{25}Val^{62}Ser^{68}$; $Gly^{25}Val^{62}Thr^{68}$; $Val^{25}Ile^{62}Ser^{68}$. Notice how the isoelectric points decrease as the hydrophobicity of the mutations increase. Drawn from IEF data of Whitney et al. (1979). (From Righetti et al., 1980.)

the α-chain. As the hydrophobicity of these amino acids is increased, and the dielectric constant of the environment progressively decreased, concomitantly the pK of the basic group would be lowered, down to a total loss of 1 proton at pH = pI in going from Hb_1 to Hb_4 (Righetti, 1979). That in the native tetramer neutral mutations can produce subtle, but detectable, pI changes, is not surprising. However, we have been able to separate neutral mutants even when working with fully denatured globin chains, where conformational transitions cannot account for pK shifts. We found that out when trying to separate by IEF β- from γ-globin chains, for thalassemia screening (Righetti et al., 1979; Comi et al., 1979; Saglio et al., 1979; Guerasio et al., 1979). When the IEF gels were run in 8 M urea and 3% NP-40, the γ zone was split into two bands, a pI 6.95 chain, corresponding to G_γ (glycine) and a lower pI band (pI 6.85) corresponding to A_γ (alanine) globin (Fig. 3.38). These chains differ by having either Gly or Ala at position 136 and are the products of two non-allelic loci, closely linked to α- and β-globin genes. To account for the fact that two neutral mutants, in fully random configuration, do exhibit different pI values in the presence of a neutral detergent, Righetti et al. (1980) suggested that the NP-40 micelle would bind preferentially to the A_γ-chain in the hydrophobic stretch of the mutation, going from Met^{133} to Leu^{141}, masking in this process Lys^{132}, thus producing a charge-shift (loss of one proton unit) in this phenotype.

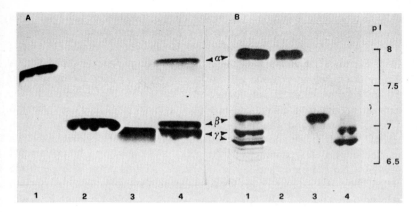

Fig. 3.38. Separation of human globin chains by IEF in the absence (A) and in the presence (B) of NP-40. Lanes 1, 2 and 3 of A and lanes 2, 3 and 4 of B show chromatographically pure α-, β-, and γ-globin chains, respectively. Lanes 4 of A and 1 of B show unfractionated globins. (From Saglio et al., 1979.)

In the model proposed, we had hypothesized that this peptide segment in the A_γ-chain had been sorbed in the Stern layer of the micelle, however an alternative model could depict it embedded in the hydrophobic nucleus of the micelle, as demonstrated in the case of hemin (Simplicio et al., 1975) or hydrophobic membrane proteins (Hackenberg and Klingenberg, 1980). The case of $A_\gamma-G_\gamma$-chains does not seem to be a unique phenomenon, as a similar type of charge-shift has been demonstrated even in the case of histones. If similar neutral substitutions prove to be equally detectable in other proteins, then it appears that the use of IEF will enable the detection of twice as many genetic variants as have been found through the use of standard electrophoretic techniques. This could in fact already be the case (see §3.3.6 and Fig. 3.25).

3.7. Titration curves

This chapter has been so demanding that I should like to end up with some fireworks in celebration. That is why I am ending it with this 2-D technique, which should belong to §3.4. The first report on this method was presented in 1976, at a meeting in Hamburg, when Rosengren et al.

(1977) described 'a simple method for choosing optimum pH conditions for electrophoresis', which was in fact a direct display in a polyacrylamide gel slab of the titration curves of all the proteins present in a mixture. Figure 3.39 shows how these pH-mobility curves are generated. A 2-mm thick polyacrylamide gel slab is cast with a trench in the middle, 10 cm × 1 mm, which can be loaded with up to 150 μl of sample (kits for generating titration curves are now available from LKB Produkter AB [No 2117-801] and Bio Rad Labs; they allow casting 1.5 mm thick slabs). The first dimension consists of sorting electrophoretically the carrier ampholytes contained in the gel, thus generating a stationary pH gradient. No sample is applied at this stage. At this point, the electrode strips, with the respective gel layers underneath are chopped away with a long knife (Fig. 3.39 B). This step is essential, otherwise much heat will be generated in these regions in the second dimension, due to the presence of 1 M acid and base. New electrodes are then applied perpendicular to the first run and the trench filled with the sample to be analyzed (Fig. 3.39C). Now electrophoresis perpendicular to the stationary pH gradient is run, usually at 600 V/12.5 cm and for periods of 10 up to 45 min (Fig. 3.99D). Here are examples of fields of application of this technique:

(a) *Genetic mutants.*   It is possible to perform 'differential' titration curves by running a protein and its genetic mutants in a mixture. The shape of the respective titration curves should reveal which charged amino acid had been substituted in the mutant phenotype (Fig. 3.40). For instance, in the case of Lys mutants, the two curves should meet around pH 11, while for Glu or Asp mutants the confluence point should be around pH 3. Double charge mutants (e.g., Lys → Glu) or same charge replacements (e.g., Arg → His) can also be detected. These theoretical titration curves were experimentally verified by running normal human adult hemoglobin (HbA) mixed with any of the following mutants: HbS, HbC, Hb Suresnes and HbG Philadelphia (Righetti et al., 1978; Righetti and Gianazza, 1979).

(b) *Macromolecule–ligand interaction.*   Krishnamoorthy et al. (1978) have been able to run titration curves of met-Hb-inositol hexaphosphate (IHP) and met-Hb-inositol hexasulphate (IHS) complexes. The pH ranges of stability of these liganded states are pH 4.5–6.0 for met-Hb-IHP and

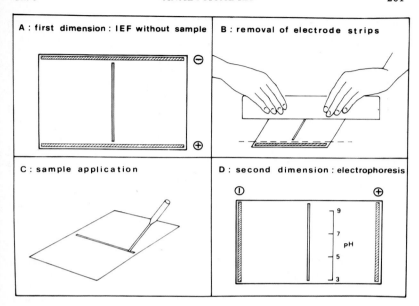

Fig. 3.39. Experimental procedure for generating titration curves by IEF-electro-phoresis. The pH gradient is first formed by focusing the carrier ampholyte mixture (A); the electrode strips are then removed (B); the sample is applied in a trench cut perpendicular to the pH gradient (C); the second dimension run is started perpen-dicular to the first dimension axis (D). (From Righetti and Gianazza, 1979.)

3.7–6.0 for met-Hb-IHS. Both complexes appear to have a dissociation constant (K_d) of the order of μM, and survive under these experimental conditions up to 8–10 min of electrophoresis. When the Hb-2,3-diphos-phoglycerate complex (known to have a K_d greater than mM) was tit-rated, only the titration curve of free Hb could be demonstrated within the pH 3–10 range, indicating that the complex was immediately split by the current. It should be possible, however, to detect very labile intermediates by performing the experiment at -15 to $-20\,^{\circ}$C, as des-cribed by Perrella et al. (1978, 1979). By using the same 2-D technique, Constans et al. (1980) have demonstrated the binding of vitamin D_3, and its derivatives, to the human serum vitamin D-binding protein. It has been shown that the binding of ligand induces conformational changes in

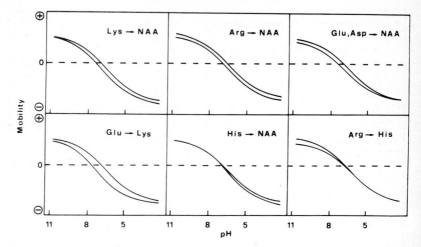

Fig. 3.40. Theoretical titration curves of hemoglobin and its genetic mutants in the presence of a given amino acid substitution. NAA: neutral amino acid. The broken lines represent the zero mobility plane (application trench). The arrows indicate the direction of amino acid mutation. (From Righetti and Gianazza, 1979.)

the apoprotein. The stoichiometry of the complex, and the pH range of stability of the holoprotein could be investigated.

(c) *Macromolecule–macromolecule interactions.* Righetti et al. (1978b) have exploited this 2-D technique to study the interaction between cytochrome b_5(Cyt b_5) and met Hb. While the two proteins, when run singly in the gel, develop the classical sigmoidal shape of a titrated macromolecule, when run in a mixture they exhibit strongly distorted patterns, above the pI in the case of Cyt b_5 and both, above and below the pI for met Hb. The maximum interaction appears to occur in the pH 8.0–8.3 range, and is consistent with a predominant role of Lys residues of met Hb in the binding to acidic amino acids of Cyt b_5. As an extension of this work Lostanlen et al. (1980) have demonstrated the interaction of NADH-cytochrome b_5 reductase and cytochrome b_5 by a direct enzyme titration curve. Direct detection of titration curves by zymograms or immunoprints has been independently reported by

Gianazza et al. (1980). The interaction between hemoglobin and hapto-globin (Hp), as well as the binding of intact α- and β-globin chains to Hp, have also been studied. While Hp—Hb complexes were completely stable in the pH range 3—10, indicating a predominant role of non-ionic interaction, free β-chains did not appear to bind to Hp while free α-globins exhibited an intriguing, pH-dependent pattern, indicating a mixed type of interaction (unpublished experiments with R. Krishna-moorthy, D. Labie and M. Waks).

(d) *Titration curves in 8 M urea and detergents.* Since pH-mobility curves of macromolecules, under native conditions, do not allow direct titration of all ionizable groups, but only of surface groups accessible to solvent not engaged in subunit contacts, or other interaction, Righetti et al. (1979a) have developed electrophoretic titrations in denaturing solvents, such as 8 M urea. In this system, many proteins will exist as random coils, subunits will be split apart, buried groups will be exposed to the solvent and the macromolecule will be stripped free of non-covalently bound ligands or co-factors. When running titration curves of heme-free, α- and β-globin chains in 8 M urea, a 'bird's-eye' view of the total amino acid composition of these two chains could be obtained. In fact, since α- and β-chains differ mostly in their acidic residues, they come very close below pH 3, where only one positive charge difference is left. Disturbing features are encountered in the fact that they indeed join around pH 3 (while they should not) and that each curve forks both below and above the pI, probably due to partial precipitation and aggregation in the pI neighborhood. However, if the titration curves are performed in 8 M urea and 1% NP-40, both of these disturbances disappear since most probably the detergent, by binding to hydrophobic stretches in the polypeptide chain, prevents inter-chain interactions, which would favor flocculation in proximity of the pI (Righetti and Gianazza, 1979b, 1980).

(e) *Titration curves in highly porous matrices.* Bianchi Bosisio et al. (1980) have attempted to run titration curves in highly porous media (acrylamide gels containing 15—50% cross-linker, either Bis or diallyl-tartardiamide (DATD)), which should allow almost unhindered migration of macromolecules in the multimillion $\overline{M}_r$ range. Unfortunately, DATD gels contain up to 80—90% unpolymerized DATD which reacts with proteins and produces gluey and highly stretchable matrices. On the other hand, highly cross-linked Bis gels, at 40 to 50% C levels, are too hydrophobic and produce a collapsed matrix which keeps exuding water. An acceptable compromise are 30% C_{Bis} gels, which are stable and allow practically unhindered migration of globular proteins up to 0.5×10^6 daltons.

(f) *Direct pK determinations from titration curves.* Righetti et al. (1979c) and Valentini et al. (1980) have developed a mathematical theory which would allow direct pK determination of ionizable groups from the shape of the pH-mobility curves. Equations have been derived

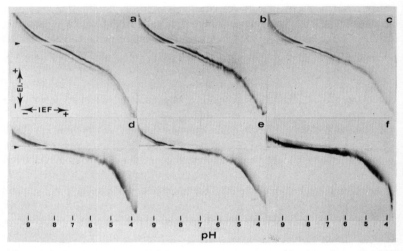

Fig. 3.41. Affino-titration curves of lectin from *Ricinus communis* seeds. The gel contained 6% T, 4% C_{Bis}, 2% Ampholine, pH 3.5–10, and 2 mM Glu, Asp, Lys and Arg. The amount of ally-α-D-galactose copolymerized in the gel matrix was: (a), control, no ligand; (b), $1 \cdot 10^{-5}$ M; (c), $4 \cdot 10^{-5}$ M; (d), $5 \cdot 10^{-5}$ M; (e), $7 \cdot 10^{-5}$ M and (f), $10 \cdot 10^{-5}$ M. The amount of protein loaded in all cases was 200 μg in 100 μl volume. 1-D, 80 min at 10 W constant. 2-D, 20 min at 700 V constant. In both dimensions the electrolytes were 1 M NaOH at the cathode and 1 M H$_3$PO$_4$ at the anode. The gel was cooled at 4 °C with a Lauda KR4 thermostat. The two arrow heads in (a) and (d) indicate the sample application trench (zero mobility plane). The two double arrows with positive and negative symbols represent the direction and polarity of isoelectric focusing (IEF) and electrophoresis (E1). (From Ek et al., 1980.)

linking cationic or anionic mobilities to the degree of ionization of
simple cations and anions and of uni-uni-valent amphoteric molecules.
For non-amphoteric ions, a direct determination of either pK_c or pK_a
can be made by measuring the pH ($pH_{\frac{1}{2}}$) corresponding to $1/2$ mobility
in the cathodic or anodic direction, respectively. For amphoteric species,
the $pH_{\frac{1}{2}}$ values will have to be corrected by a term accounting for the
influence of the degree of ionization of the opposite charge ion on the
mobility curve of the ion being measured.

(g) K_d *determinations from titration curves.* One of the most recent
extensions of pH-mobility curves is the possibility of determining dissociation constants (K_d) of ligands to proteins and their pH dependence. This is achieved by techniques developed for affinity electrophoresis (Horejsi, 1979). If the ligand is a macromolecule, it is simply
entrapped in the gel matrix, if it is a small molecule, it is covalently
bound to the gel fibers. In presence of increasing concentrations of ligand
the titration curve of the protein is progressively retarded, in a pH-

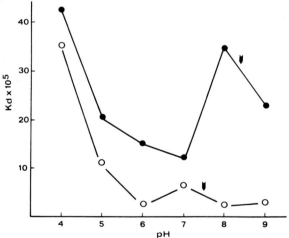

Fig. 3.42. Variation of the dissociation constants (K_d) as a function of pH for
Ricinus communis (o) and for *Lens culinaris* (•) lectins. The K_d values reported
here have been calculated from the retarded mobilities, at different pH values, in
affino-titration curves, such as the ones shown in Fig. 3.41. The two arrows indicate
the pI of each lectin. (From Ek et al., 1980.)

dependent fashion, and the mobility decrements, when plotted against the ligand molarity in the gel, can be used to calculate K_d values at any pH value. Ek et al. (1980) and Ek and Righetti (1980) have developed this technique for studying the binding of glycogen to phosphorylases a and b, of blue dextran to several dehydrogenases and of sugars to lectins. An example of these affino-titration curves is given in Fig. 3.41, which shows binding of Ricinus communis seed lectins to allyl-α-D-galactose copolymerized in the gel matrix. K_d values can be determined at any pH (usually at 6 different pH values, from pH 4 to 9, in 1 pH unit increments) and then plotted against pH (see Fig. 3.42) in order to evaluate the pH dependence of K_d.

As a general conclusion, it can be stated that titration curves hold a great potential for investigating several physico-chemical parameters of proteins and dynamical aspects of their interactions with ligands. For practical aspects on how to perform a titration curve see Ek (1981) and Righetti and Gianazza (1981). An interesting example on the use of titration curves is shown in Fig. 3.43. It has been known for a long time that some protein samples focus sharply if applied near the anode and give diffused bands if loaded at the cathode, and vice versa. No ready explanation has been available for this phenomenon, and in general interaction with carrier ampholytes has been suspected. This phenomenon can now be readily understood by inspection of Fig. 3.43, which shows the titration curves of isozymes of bovine carbonic anhydrase and, in the upper part, the equilibrium focusing patterns. Most of these isozymes have very steep pH/mobility curves around their pI values and thus focus sharply independently from the anodic or cathodic application point in equilibrium IEF. However, the two most basic isozymes have very flat titration curves above their pI values, while presenting a steeper slope in the portion below their pI. Consequently these two bands focus sharply when driven to equilibrium from their anodic side, but present diffuse zones, removed from their pI positions, if applied to the cathode, because their rate of approach to equilibrium is extremely small as compared with the other isozymes.

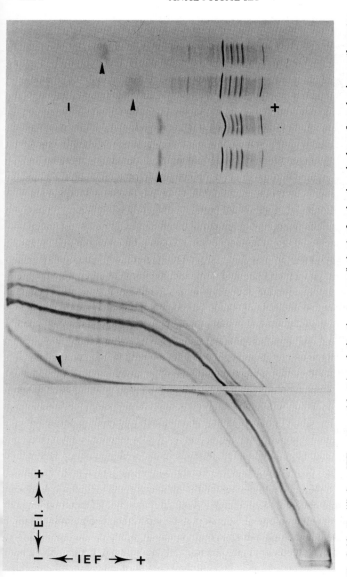

Fig. 3.43. Equilibrium IEF pattern (above) and titration curves (below) of carbonic anhydrase from bovine erythrocytes over the pH range 3.5 to 9.5. Conditions for the titration curve: 1st D: 10 W, 50 min; 2nd D: 700 V, 15 min. The arrow indicates the isozyme with a very shallow titration curve on the alkaline branch. This component, when applied at the cathodic side of the gel, focuses as a broad zone (see the two upper sample tracks in the equilibrium IEF pattern, marked by two arrow heads) while, when applied along the gel track at pH values below its pI, it focuses as a sharp zone (see the two lower sample tracks in the equilibrium IEF pattern, marked by an arrow head). In the lower left corner, the two double arrows with positive and negative symbols represent the direction and polarity of isoelectric focusing (IEF) and electrophoresis (El.). (From Ek, 1981.)

CHAPTER 4

General experimental aspects

4.1. Isoelectric precipitation

One great drawback of IEF in liquid media is the flocculation of less soluble proteins at their pI. Precipitates, once formed, aggregate into larger particles which may sediment slowly through the column. Thus, separations are often spoiled by precipitates sticking to the walls and disturbing the density gradient or by contaminating adjacent zones. Sometimes a precipitate will redissolve as it moves away from its pI, only to reprecipitate when forced back to its pI. Consequently, a quasi-equilibrium may occur with a protein existing in different physical states over a wide pH range. Such problems may be prevented by decreasing the protein load or by increasing the level of carrier ampholyte. Neither remedy is entirely satisfactory since both lead to an economically unfavorable protein–Ampholine ratio. A better approach is to use additives (see §4.2). Should this fail, there are some methods to circumvent problems of isoelectric precipitation in vertical, density gradient stabilized columns. One such method is to select the electrode polarity in such a way as to ensure that the precipitate will focus and collect near the bottom of the column, while the protein of interest will focus away from it, and possibly close to the top of the column. In this situation, sample collection by column drainage will be a difficult proposition, since the precipitate may clog the drain tube and the sample of interest may diffuse considerably. A better solution is to withdraw the sample from the column top. This operation, which is easily done in the two LKB columns, is schematically represented in Fig. 4.1. When electrofocusing is completed, the power supply is switched off and the bottom valve

268

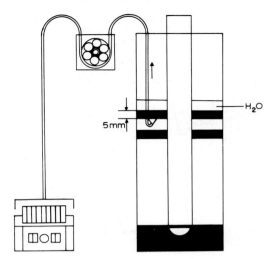

5mm

H_2O

(a)

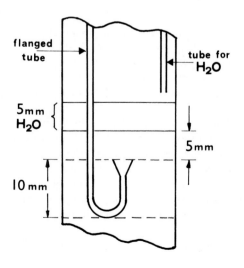

flanged tube

tube for H_2O

5mm H_2O

5mm

10 mm

(b)

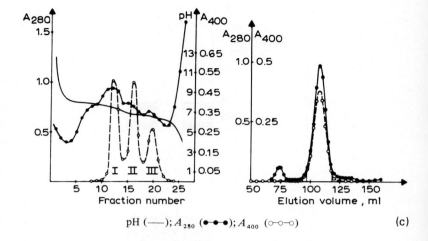

pH (——); A_{280} (●–●); A_{400} (○–○–○)   (c)

Fig. 4.1. Gradient elution from the Ampholine Electrofocusing Column in the presence of precipitates. It is assumed that the protein of interest focuses away from the precipitate. The column is then emptied from the top, through nipple (10) (see Fig. 2.2) by a stepwise procedure. The end of a capillary tubing is flanged (60°) and U-bent. This capillary is introduced 5 mm below the surface of the gradient and a 5 mm water layer is layered on top of it with the aid of a second tube (b). As shown in (a), this layer is eluted with the aid of a peristaltic pump operated at a flow rate of 100–120 ml/h. This stepwise procedure is then sequentially repeated. (c) Examples of the separation achieved are given in the two graphs. (Courtesy of LKB Produkter AB.)

connecting the dense electrode solution to the central platinum wire is closed (see Fig. 2.2). From the upper nipple, used to fill the density gradient into the column, a capillary tube is inserted 1 cm below the upper electrode and the light electrode solution is pumped out. A second tube is now used for sample removal. Before insertion, the tubing is flanged, with the aid of a 60 degree flanging tool, by gently heating the end of the capillary with a match or lighter. By the same heating procedure, a U-bend is formed in the tubing 1 cm from the flange. This tube is now inserted in the column, 5 mm below the surface of the liquid to be eluted. To ensure that all sample is washed off the column walls, a 5 mm layer of distilled water is introduced, with the aid of the first capillary,

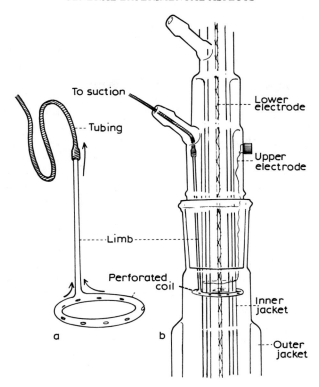

Fig. 4.2. Removal of an isoelectrically precipitated sample from the Ampholine Electrofocusing Column. In this case it is assumed that the precipitate is the sample itself. When mounting the column, a perforated glass coil (a) is placed either near the upper electrode or in the predetermined vicinity of the precipitate (b). Once a precipitate is formed, it is removed by suction via a tubing attached to the limb of the perforated coil. (From Rathnam and Saxena, 1970.)

before each fraction is removed. Then the U-bent capillary is used to withdraw the 5 mm thick sample layer and the 5 mm water layer overlying it. This is done with the aid of a peristaltic pump at a flow rate of 120 ml/h. Then the U-shaped tube is lowered another 5 mm, a water layer of 5 mm is layered with the first capillary and the operation repeated sequentially until all the fractions of interest are eluted. For further details, see the LKB instruction manual I-8100-EO4.

By this method, Janson (1972) has been able to focus 7.3 g protein in the 440 ml LKB column and to recover from the top, well separated, three isozymes of α-glucosidase from *Cytophaga johnsonii*. Interestingly, in this case, IEF was used as the first preparative step in the purification procedure and yet, in this single step, a 300-fold purification (based on removal of inactive proteins) was achieved with a yield of 20% in α-glucosidase activity. Sometimes, however, one is faced with the opposite problem, i.e., the removal not of a protein soluble at its pI, in the presence of precipitates, but of the protein itself precipitated at its pI. Rathnam and Saxena (1970) have devised an ingenious solution to this problem. The protein bands precipitated during IEF are removed by the use of an indwelling perforated glass coil (see Fig. 4.2a). The perforations in the coil directly opposite the limb have greater diameter and their diameter decreases in size as they approach the limb. This arrangement permits equalization of suction at each perforation around the circumference of the coil. The limb is attached to the coil at a slightly obtuse angle (see Fig. 4.2b) which facilitates vertical upward and downward motion of the coil in a horizontal plane. As shown in Fig. 4.2b, the coil is placed near the upper electrode or, better, in the predetermined vicinity of the precipitate before the assembly of the column. Following the appearance of the precipitate, this is removed by suction via a tube connected to the coil. This operation can be performed, with due precaution, while the column is under voltage, with no disturbances for other parts of the gradient. Should the upper electrode solution be lowered below the upper platinum electrode, more can be pumped in with the aid of a second tubing. Actually, the two operations can be performed simultaneously.

Isoelectric precipitation is not so critical in horizontal flat beds stabilized by capillary systems, such as Sephadex, polyacrylamide, agarose or Pevikon beads (see §2.2.3) since the granulated bed packing renders the precipitate zone gravitationally stable. It must be borne in mind, however, that at the very low ionic strength prevailing in IEF, and at the high protein loads compatible with IEF, proteins might start to precipitate ca. 1/2 pH unit below and 1/2 pH unit above their pI values, thus producing a smear which might impair resolution. Moreover heavy protein precipitates might clog the gel pores resulting in uneven current distribution and distortion of the focused zones. It might be possible to reduce isoelectric precipitation by performing IEF at high salt concentration (up to 0.1 M NaCl) as suggested by Righetti and Chrambach (1978) but this would require large electrode reservoirs in order to ensure a constant background transport of the electrolytes. It has also been suggested (Hjelmeland et al., 1981) that polyacrylamide gels could be co-polymerized with zwitterionic derivatives, such as MAPS [(3-sulfopropyl) dimethyl (3-methacrylamidopropyl) ammonium inner salt] but it does not seem that this compound can indeed increase ionic strength or equalize conductance in IEF. The best way to increase ionic strength in a controlled fashion, however, and to minimize isoelectric precipitation even at considerable protein loads, is to use immobilized pH gradients (see § 1.11.7) (Bjellqvist et al., 1982a,b).

4.2. Additives

Isoelectric precipitation can often be prevented or lessened by the proper use of additives (Vesterberg, 1970). In addition to stabilizing or solubilizing focused bands, some additives have the added advantage of reducing differences in osmolarity along the gradient caused by focused ampholytes. Also, by increasing the osmolarity, additives reduce the risk of an unacceptable large film of water. Urea and formamide up to concentrations of 4 M have been used to solubilize proteins. However, these substances may markedly alter the physico-chemical nature of proteins, particularly multimeric proteins and should be used with caution. Nonionic detergents have also been successfully used, in concentrations ranging from 0.1 to 5%. Detergents have proven especially useful for maintaining the solubility of focused membrane bound proteins. Many such detergents are now available under various trade names, e.g., Tween 80, Emasol, Brij 39, Triton X-100 (Sigma Chemical Co.). Serum lipoproteins have been maintained in solution by using 33% ethylene glycol (Kostner et al., 1969), or tetramethyl urea (Gidez et al., 1975). Wadström (1975) has suggested the combined use of additives in some instances. For instance, he found a mixture of urea and Triton X-100 to be very useful in the analysis of protein components of the red blood cell membrane. The same results have also been reported by Bhakdi et al. (1975). Wadström (1975) has also stressed the importance of using highly purified nonionic detergents, since many preparations contain impurities which adversely affect the separation. In the case of Triton X-100, he suggested the scintillation grade product from Serva. Today, in fact, it is a common procedure to use a mixture of 9M urea and 2% Nonidet P-40 or Triton X-100 when analyzing complex mixtures of cellular or membrane proteins, especially during the first dimension of a 2-D separation (O'Farrell, 1975; Whalen et al., 1976; Friesen et al., 1976; Zechel, 1977; Schmidt-Ullrich and Wallach, 1977). Urea dissociates aggregates and subunit assemblies and unfolds polypeptide chains, thereby releasing non covalently bound cofactors, while non ionic detergents disrupt hydrophobic bonds.

As reported by Jacobs (1971), additional problems may arise from the oxidation of cysteine and methionine to cysteic acid and methionine sulphoxide. This can be prevented by performing IEF in the presence of antioxidants, such as thiodiglycol or ascorbic acid. The latter seems to be more effective, and also prevents possible modifications of tyrosine and arginine residues (Jacobs, 1973). In the case of sulphydryl-dependent enzymes, their activity can be preserved by working in the presence of a low concentration (about 10^{-4} M) of thiol compounds (Wadström and Hisatsune, 1970a,b). For this purpose, 2-mercaptoethanol, 2,3-dimercaptopropanol and dithiothreitol (or its isomer, dithioerythritol) (Cleland, 1964) have been effectively used. Working with milk-fat globule membranes and with human erythrocyte membranes, Allen and Humphries (1975) found it useful to incorporate in the sucrose density gradient zwitterionic surfactants, such as:

$$
\begin{array}{c}
CH_3 \\
| \\
R-N^+-CH_2-COO^- \qquad (R = C_{10} \text{ to } C_{16}) \\
| \\
CH_3
\end{array}
$$

of the alkylbetaine type (Empigen BB, supplied by the Marchon Division of Albright and Wilson Ltd.) or:

$$
\begin{array}{c}
CH_3 \\
| \\
R-N^+-CH_2-CH_2-SO_3^- \qquad (R = C_{12}) \\
| \\
CH_3
\end{array}
$$

of the sulphobetaine type (sulphobetaine DLH, supplied by Texilana Corp.). These compounds are very effective in solubilizing membranes at low concentrations (between 5 to 30 mM; 0.1 to 0.5% w/v) and under very mild conditions. Enzymatic activities are not, in general, destroyed by the action of these surfactants (an exception is erythrocyte acetylcholinesterase which is inactivated by Empigen but not by sulphobetaine DLH). This retention of biological activity with zwitterionic surfactants is in contrast to the denaturant effect of anionic surfactants such as sodium dodecyl sulphate (SDS). Zwitterionic detergents are compatible with IEF: they are usually required in only low concentrations, they are electrically neutral, and the pK values of the charged groups lie on either side of the pH range of 3 to 10 which is usually employed. Alper et al. (1975) have used a similar approach in the analysis by gel IEF of the polymorphism of the sixth component of complement. They incorporated 0.2 M taurine in the gel. Taurine ($NH_3^+-CH_2-CH_2-SO_3^-$) is zwitterionic and has no net charge within the pH range of the run; it therefore serves to raise the osmotic pressure of the gel without contributing to its electrolyte concentration or to the viscosity of the liquid phase. More recently, a mixture of non-ionic and zwitterionic detergents has been suggested for complete solubilization of membrane proteins. Thus Hjelmeland et al. (1979) have solubilized liver microsomes in 1% Triton X-100 and 1% SB_{14} (N-tetradecyl-N,N-dimethyl-3-amino-1-propane sulfonic acid or zwittergent TM3-14 from Calbiochem.) and focused the dissolved proteins in gels containing 0.05% of each surfactant. While this detergent mixture is able to fully solubilize membrane proteins, it does not seem to preserve their biological activity as well, contrary to previous reports (Allen and Humphries, 1975). It has been suggested that these last authors have used too low SB_{14} concentrations, inadequate to effect full solubilization of membranes. For this purpose, according to Gonenne and Ernst (1978) a detergent:protein ratio in excess of 10:1 is required. The denaturing properties of SB_{14} could be representative of the general class of n-alkyl zwitterions, and could be related to the flexible nature of the hydrocarbon tail in these molecules (Hjelmeland et al., 1978). Figure 4.3 lists the most common non-ionic and zwitterionic detergents compatible with IEF (Hjelmeland and Chrambach, 1981). The Tritons have in general a rather low critical micellar concentration (CMC) (0.075 mM for Triton N-101 and 0.240 mM for Triton X-100), a high aggregation number

Structural Formula	Chemical or Trade Name
Anionic Detergents	
	Sodium dodecylsulfate (SDS)
	Sodium cholate
	Sodium deoxycholate (DOC)
	Sodium taurocholate
Cationic Detergents	
	Cetyltrimethylammonium bromide
Zwitterionic Detergents	
	Zwittergent 3-14
	CHAPS
	Lysophosphatidylcholine
Nonionic Detergents	
	Lubrol PX
	Triton X-100
	Triton N-101
	Ammonyx LO
	Digitonin
	Octyl glucoside

(140 to 150 monomers in the micelle) and thus a high micellar mass (90,000 daltons for Triton X-100) and a large Stoke's radius (5 nm for the pure Triton micelle).

Thus, when focusing in 9 M urea and 2% Triton X-100 (or Nonidet P-40) one has to remember that the solution viscosity will be extremely high and the approach to the pI position quite slow, so that considerably longer focusing times will have to be applied (at least twice as long as when focusing in the absence of these additives). In addition to that, strongly hydrophobic membrane proteins can be fully incorporated into the Triton micelle, creating complexes of remarkably large Stoke's radius. Figure 4.4 gives an example of such a case (Hackenberg and Klingenberg, 1980). The ADP–ATP carrier protein (a dimer) from heart mitochondria is seen to be fully enveloped in an ellipsoid micelle of ca. 150 Triton monomers, which exhibits a Stoke's radius of 6.5 nm. Such a large oblate ellipsoid might have a hard time to reach its pI position, if ever, especially in sieving matrices. Moreover, these detergents have been found to bind to carrier ampholytes, especially in the alkaline region (Gianazza et al., 1979) so that gel staining requires a lengthy pre-washing step in TCA-isopropanol to disrupt and wash-out the detergent micelle. The fast-stain technique (colloidal dispersion of Coomassie G-250 in 12% TCA) cannot be used in detergent gels, because a tertiary complex dye-detergent-carrier ampholyte is irreversibly precipitated in the gel matrix (Righetti and Chillemi, 1978). It might be of interest to explore other surfactants, such as octyl glucoside, which have a very small micellar size (2,000–3,000 daltons) (Helenius and Simons, 1975; Tanford and Reynolds, 1976).

In the case of tenacious complexes, other disaggregating agents have been successfully tried: 50% dimethylsulphoxide (DMSO), 50% dimethylformamide (DMF), 50% tetramethyl urea (TMU) and up to 90% formamide (FA) (Righetti et al., 1977). Except for FA, the concentration limit in IEF appears to be about 50%, since higher levels will precipitate Ampholine. If they are added to a gel, DMF and TMU should be used at

Fig. 4.3. Structures of some common zwitterionic and non-ionic detergents compatible with isoelectric focusing. On the left the chemical formula and on the right the chemical or commercial name. CHAPS: 3-[(3-cholamidopropyl)dimethyl ammonio]-1-propane sulfonate. (From Hjelmeland and Chrambach 1981.)

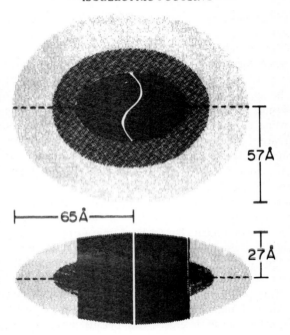

Fig. 4.4. Geometrical model of the carboxyatractylate (CAT)-protein-Triton complex. It is supposed that the CAT-protein is completely enveloped in an ellipsoid micelle composed of *ca.* 150 Triton and *ca.* 16 phospholipid molecules. Due to the insertion of the protein along the small axis, the Stoke's radius of the pure Triton micelle is increased from 5.3 nm to 6.5 nm. (From Hackenberg and Klingenberg, 1980.)

levels of about 30–35%, as higher levels hamper gel polymerization. In the case of Ampholine-dye complexes, the most powerful disaggregating agents are DMF and TMU and the weakest FA (on an equimolar basis) (see Fig. 4.5). FA could be a useful additive, since it can be used up to 95% without inhibiting gel polymerization or precipitating Ampholine; unfortunately it is rapidly hydrolyzed at the cathode, giving free ammonia and formic acid, which severely interfere with pH gradient formation and sample focusing.

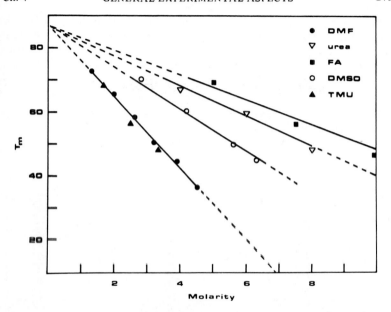

Fig. 4.5. Hyperchromic thermal transitions. Melting curves were measured with a Gilford Model 2400 automatic spectrophotometer with a programmed temperature increase of 1.0°C/min at 585 nm. Benzo New Blue 5BS (30 μg/ml) was dissolved in 50 mM acetate buffer (pH 3.85) containing increasing molarities of each of the five dissociating agents tested. The melting point (T_m) of the Ampholine–dye complex, extrapolated to zero additive concentration, is 87°C. DMF: dimethyl formamide; FA: formamide; DMSO: dimethyl sulphoxide; TMU: tetramethyl urea. The most powerful disaggregating agents are DMF and TMU. (From Righetti et al., 1977.)

4.3. Sample application

I have already discussed at length sample application in gel rods (see §3.2.10), in gel slabs (see §3.2.14) and, briefly, also in vertical columns stabilized by sucrose density gradients (see §2.1.1). In this last case, more methodological details will be given here. The sample can be added to the column either before or after the pH gradient has formed. In the former case, the sample can be uniformly distributed over the entire column length or added in a narrow zone where it is expected to focus.

Often this last method is preferred, since it allows shorter focusing times and avoids harmful contact of the sample with anolyte and catholyte. In this last case, the following procedure should be used. Fill the density gradient in the column until it has reached the level where the sample is expected to focus. Now, fill three test tubes with the following volumes of density gradient: 0.5 ml (1st tube); 3.0 ml (2nd tube); 0.5 ml (3rd tube) for the 110 ml column, or 1.0 ml (1st tube); 10.0 ml (2nd tube); 1.0 ml (3rd tube) for the 440 ml column.

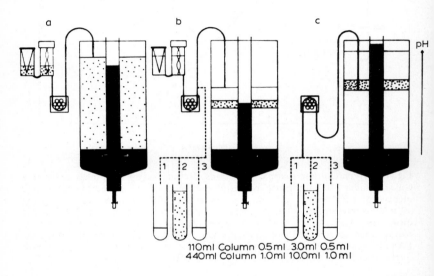

110ml Column 0.5ml 3.0ml 0.5ml
440ml Column 1.0ml 10.0ml 1.0ml

Fig. 4.6. Sample load in preparative IEF. (a) The sample is uniformly distributed throughout the density gradient. (b) The sample is loaded in a narrow zone, close to its expected pI, during the formation of the pH gradient. In this case, at the chosen level in the column, three test tubes are filled with solution from the gradient mixer (0.5 ml, 3.0 ml and 0.5 ml respectively for the 110 ml column or 1.0 ml, 10 ml and 1.0 ml respectively for the 440 ml column). The sample is dissolved in the middle tube and its density adjusted as explained in the text. The sample solution is then layered in the density gradient and the column filled with the remaining gradient solution. (c) The sample is loaded after the formation of the pH gradient. Upon focusing the Ampholine in the density gradient, harvest the desired amount of gradient solution as explained in (b). Add the sample to the middle tube, adjust its density, as described in the text and pump it back into the column. (Courtesy of LKB Produkter AB.)

Add the sample to the second tube, mix gently and then dissolve few sucrose crystals in it until the density of the solution in the second tube is between the higher density of the first tube and the lower density of the third tube. The density of the second tube is satisfactory when a droplet of solution from the first tube will sink in it, while a droplet of solution from the third tube will float on its surface. At this point, the sample containing-solution can be pumped in the column and layering of the density gradient terminated (see Fig. 4.6). It is often desirable, however, to add the sample after the pH gradient is formed, to minimize contact time with Ampholine, for instance in the case of metallo-enzymes or in the case of components which are inactivated or unstable at their pI, e.g., when focusing intact cells. To facilitate this, Sherbet and Lakshmi (1973) have modified the LKB 8100-1 column by fusing a side arm with a special septum to the middle of the column. After prefocusing, the sample is injected into the middle of the column and the separation takes place in only 2 to 3 h (see also Chapter 5). In home-made columns, Boltz et al. (1978) have devised a sample inlet capillary inserted and fused to the walls of the inner cooling finger (see Fig. 4.7). The cell suspension is pumped through four outlets into the gradient column at a height selected by vertical positioning of the movable cooling finger. Alternatively, a simpler method can be used. After the pH gradient has been formed, insert a capillary tubing, via the upper nipple of the column, to the level in the column where it is desired to introduce the sample. Withdraw 4 ml of gradient solution from the 110 ml LKB column and 12 ml from the 440 ml LKB column. Divide the solution into three tubes and add the sample to the second tube, just as described above. After the proper density adjustments in the sample tube, this solution is pumped back into the column and the capillary tubing removed (Fig. 4.6). Focusing is then continued until termination of the experiment.

When focusing in gel slabs, in addition to the sample application techniques reported in Chapter 3 (see Fig. 3.20), the preferred method today is to use surface application strips made of silicone rubber (or polyester foil) and containing slots or circular holes (Fig. 4.8). The advantage of these strips over paper applicators is that no proteins or enzymes are lost by adsorption; they are also to be preferred to pockets precast in the gel, since in this last case the conductivity in the gel is not uniform due to the variation of thickness. In general, these applicators

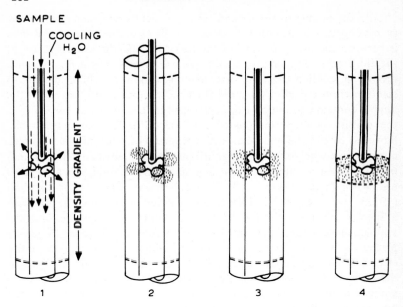

Fig. 4.7. Sample inlet system for isoelectric focusing of cells. Two concentric glass cylinders form the upper half of the column cooling finger. Cell suspensions are admitted at the top of the innermost cylinder which has four outlets that penetrate the cooling finger and allow cell suspension to enter the gradient column at a height selected by vertical positioning of the movable cooling finger. (From Boltz et al., 1978.)

are applied 10–20% of the separation distance away from the electrodes, in any case in a gel zone which does not interfere with the expected focusing pattern. The applicator strips can be washed with 1 N alkali, rinsed in water and used repeatedly. When focusing in gel tubes, where the sample is usually applied as a dense liquid layer at one gel extremity, Cantrell et al. (1981) have recently reported that, when the sample is applied at the cathodic extremity, the amount of protein in it affects the pH gradient and can decrease it by as much as one pH unit. Several explanations have been given for this protein load effect on alkaline pH gradients, but none is quite convincing to me. Perhaps these authors have

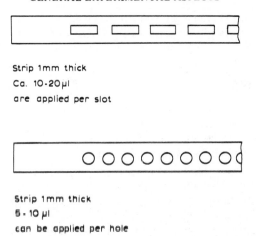

Strip 1mm thick
Ca. 10-20 µl
are applied per slot

Strip 1mm thick
5 - 10 µl
can be applied per hole

Fig. 4.8. Surface sample application strips for gel IEF, made of silicon rubber and containing slots or circular holes. (Courtesy of Desaga.)

overlooked the most plausible possibility: the sample contains 5% 2-mercaptoethanol (which corresponds to a ca. 0.5 M solution) which has a pK of ca. 9.5. It is conceivable that this amount of thiol compound overcomes the buffering capacity of carrier ampholytes (which are barely 20 mM!) and lowers the pH of the cathodic gel end closer to its pK. When running denatured proteins containing 2-mercaptoethanol, I am usually careful enough to apply them at the anodic gel side and I have never seen any pH gradient disturbance or decay.

4.4. Choice of pH gradient. Production of narrow pH gradients

It is well known, from Rilbe's equation (Svensson, 1962a) on the resolving power in IEF, that good resolution is obtained with compounds having a low diffusion coefficient and a high pH mobility slope at their pI, as well as by increasing the field strength (V/cm) and by decreasing the slope of the pH gradient. This last proposition is, in practice, achieved by utilizing the so-called 'narrow pH ranges', which create pH gradients of 1.5 to 2 pH units over the separation distance. When higher resolution is needed, even shallower pH ranges are available, such as pre-cast gel slabs in the pH 4—5 range for the phenotyping of α_1-AT. Available pH ranges, or home-made batches, can be further subfractionated by the IEF process (Righetti and Drysdale, 1971; Gianazza et al., 1975), down to narrow cuts encompassing about 0.4 to 0.5 pH units. As demonstrated by Charlionet et al. (1979), this can result in the resolution of protein species differing in pI by as little as one thousandth of a pH unit. Another way of modifying the pH gradient slope in IEF is the 'separator' technique, described by Caspers et al. (1977) and Brown et al. (1977). By adding amphoteric substances (separators or 'pH-gradient modifiers') (Beccaria et al., 1978) to an Ampholine pH gradient, its slope is altered and usually flattened in a pH region corresponding to, or in proximity of the pI of the added separator. The following chemicals have been used as separa-tors: tetraglycine (pI 5.2), proline (pI 6.3), β-alanine (pI 6.9), 5-amino-valeric acid (pI 7.5), 4-aminobutyric acid, 7-amino caprylic acid, histidine (pI 7.6) and 6-aminocaproic acid (pI 8.0). All of them, except His, are poor carrier ampholytes (pI-pK values of 3 or 4) and thus focus as broad plateaus in the Ampholine gradient. Therefore, in order to be able to modify the pH gradient, they have to be used at rather high concentra-tions, usually about 0.3 to 0.6 M. On the other hand, good ampholytes, such as His (or the dipeptide His-Gly, successfully used for the separation of HbA from HbA$_{1c}$, a minor glycosylated Hb) (Beccaria et al., 1978) are already effective at concentrations of 10—50 mM and probably act by interposing themselves between the two species to be separated. pH-gradient modifier IEF adds another dimension to IEF by further im-proving separations which are at the resolution limit of the technique. The successful separation of HbA from the glycosylated component HbA$_{1c}$, utilized in diabetes screening, by the use of β-alanine as separator, has also been reported by Jeppson et al. (1980). Mixture of different 'separators' can also be used to improve the resolution in a given pH region. Thus Cossu et al. (1982) have used an equimolar mixture of β-alanine and 6-amino caproic acid for the separation of HbA, HbF and HbF$_{ac}$ in umbilical cord blood for thalassemia screening in newborns. Separator IEF has also been recently applied to agarose matrices. Thus Qureshi and Punnett (1981) have described the phenotyping of α_1-anti-trypsin in agarose gels containing N-(2-acetamide) 2-aminoethane sulfonic acid (ACES).

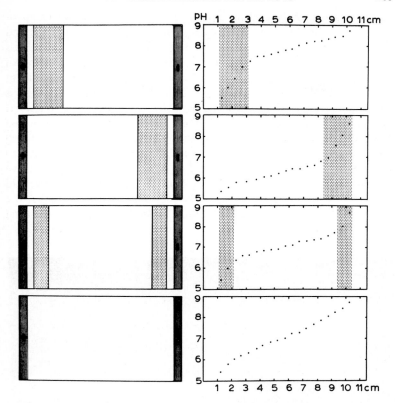

Fig. 4.9. pH gradients produced by different locations of 2 mm thick gel strips on a 350 μm thin base gel containing 2% (w/v) Servalyte carrier ampholytes in the pH range of 6–8. The locations of the gel strips are indicated by the shadowed areas. Running conditions: 16°C, 4 h at 7.5 mA (constant current), 1000 V (limit). (From Altland and Kaempfer, 1980.)

Yet another variant for pH gradient flattening is the 'strip' technique recently reported by Altland and Kaempfer (1980). It is based on the assumption that modifying the profile of the gel along the pH axis should be accompanied by changes of pI distances along that axis. Such a modification of the gel thickness along the pH axis can be achieved simply by overlaying a gel strip of a defined thickness and width on the gel at any

desired location. In the electric field the overlaid gel strip delivers ampholytes of a distinct pI interval into the strip-free part of the base gel and
removes from the latter some ampholytes present in the system; this
process results in a flattening of the pH gradient in the IEF gel slab. An
example of this technique is given in Fig. 4.9, which indicates focusing in
a pH range of 6–8 with strips added at different locations. If the strip is
placed at the anode, this results in a flat pH 7.2–8 gradient; if overlaid at
the cathode, conversely, a flat pH 6.0–7.0 gradient will be generated
while, if placed symmetrically at both anode and cathode, the central
portion of the gradient (pH 6.5–7.5) will be expanded. That this is also
the case in practice is illustrated in Fig. 4.10, which shows progressive
increments of separation among HbA_2, HbA and HbF in a gel slab overlaid
with two wedge strips placed at both electrodes. For best results, the
volume of the overlay gel strip should be at least equivalent to the total

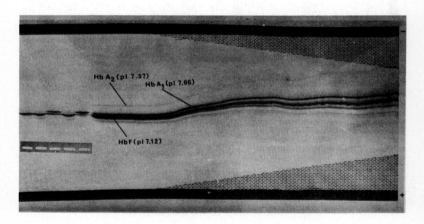

Fig. 4.10. Continuous flattening of the pH gradient to increase the separation of
hemoglobins. The 350 μm thin base gel contained 2% Servalyte pH 6–8. The wedge-
like 2 mm thick strips had a width of 2 cm each at their base (right edge of the
figure) and covered the shaded areas throughout the total run. Samples (from left
to right): 1, 2 and 4 = 15 μl RBC lysate from an adult human individual; 5 = same,
but from a human newborn; 3 = 1:1 mixture of 1 and 5; 6 (long band) = 3:1 mixture of samples 1 and 5. Running conditions: 16°C, prerun, 30 min, 7.5 mA (constant current), 1000 V (limit); run, 5 h, 1000 V (constant). (From Altland and
Kaempfer, 1980.)

volume of the base gel slab, while the area occupied on its surface should not exceed 20–25% of the electrode distance. For these reasons, ultrathin base gel slabs are preferred (200–300 μm) so that the thickness of the overlay gel can be kept in the range 1.5–2 mm. It would be of interest to see if this technique, in conjunction with 'separator' IEF, can further flatten pH gradients thus incrementing the resolution obtainable. The Altland and Kaempfer (1980) technique has also been 'discovered' by Låås and Olsson (1981) who have named it 'thickness modified pH-gradients'. These last authors have described another way for altering the course of a pH gradient on IEF: 'concentration modified pH gradients'. The principle of these two methods is shown in Fig. 4.11A,B. The last technique is a bit more cumbersome, since it requires polymerizing different gel sections containing variable amounts of carrier ampholytes. In any event, the mechanism of both methods is similar in that, over the gel discontinuity, there exists a much higher voltage gradient as compared with the surrounding gel regions, which also results in a sharpening of the protein bands over this gel zone. It might also be possible to combine these three different methods ('separators', thickness and concentration modified pH gradients) to achieve even greater resolution for difficult separation problems.

4.5. Measurement of pH gradients

The pI of a protein determined by IEF also represents its isoionic point in the absence of complex-forming ions. By definition, the isoionic point is a measure of the intrinsic acidity of a pure protein, as it is defined as that pH which does not change on addition of a small amount of pure protein (Sorensen et al., 1926). This definition is also applicable to a protein in IEF, as the pH of the isoelectric zone does not alter on addition of more protein; pI values estimated by IEF are temperature dependent and usually decrease with increasing temperature (Vesterberg and Svensson, 1966). The difference in pI for the same protein, measured at 25 °C and 4 °C, could be as high as 0.5 pH units, the higher value being obtained at the lower temperature. This difference is usually more pronounced in alkaline regions and when a protein has a pI value close to the pK of some of its functional groups. In fact, apart from carboxyls, the protolytic groups in proteins have rather large standard heats of ionization

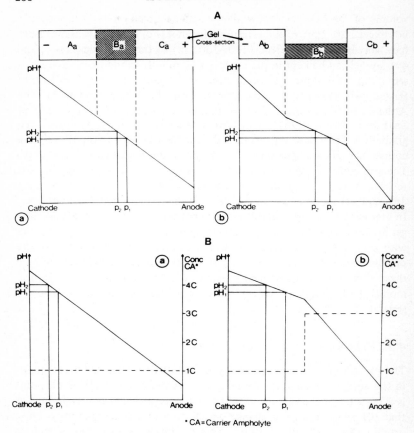

* CA = Carrier Ampholyte

Fig. 4.11. (A) Schematic illustration of *thickness modified pH gradient.* (a) illustrates the 'normal' experiment with even gel thickness, whereas (b) shows the expected pH gradient when the thickness of the indicated part is decreased. The volume within the shaded regions occupies the same percentage of the total gel volume and the same carrier ampholyte species in the two cases. Also indicated are the expected positions of two imaginary proteins P_1 and P_2 with isoelectric points pH_1 and pH_2 respectively. (B) Schematic illustration of *concentration modified pH gradient.* (a) pH gradient in the normal case with even concentration of carrier ampholyte over the whole plate. (b) Expected pH gradient when half the plate has three times as high carrier ampholyte concentration as the rest. Also indicated are the positions of two imaginary proteins P_1 and P_2 with isoelectric points pH_1 and pH_2 respectively. (From Låås and Olsson, 1981.)

TABLE 4.1

Values of pK^i_{int} and ΔH^o_i used in the calculation of pI shifts for model proteins (from Fredriksson, 1977)

Ionizable group	pK^i_{int} (25 °C)	ΔH^o_i (kcal/mole)	pK^i_{int} (4 °C)[a]	ΔpK^i_{int}
Carboxyl	4.50	1	4.56	0.06
Imidazole	6.50	7	6.89	0.39
α-Amino	7.80	10	8.36	0.56
Phenolic	9.80	6	10.13	0.33
ϵ-Amino	10.00	12	10.67	0.67

[a] Calculated from the van't Hoff equation and the ΔH^o_i given.

H^o_i = standard heat of ionization

pK^i_{int} = intrinsic dissociation constant

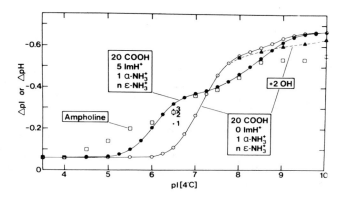

Fig. 4.12. The pI shifts (ΔpI) accompanying the increase of temperature from 4 to 25°C vs. pI at 4°C for a number of model proteins. The assumed protolytic compositions of the model proteins are given within the squares. The number (n) of ϵ-amino groups in each model protein was not specified in advance but was adjusted to the (generally non-integer) value that made the protein isoelectric at a pre-chosen pH (4.0, 4.5, etc.) at 4°C. The points numbered 1–3 refer to model proteins having pI = 6.5 at 4°C and containing 1, 2, and 3 imidazole groups per molecule, respectively. The figure also contains the pH shifts (ΔpH) accompanying the warming up from 4 to 25°C of Ampholine carrier ampholyte fractions being isoelectric at 4°C (calculated from data of Davies, 1970). (From Fredriksson, 1977.)

TABLE 4.2a
pH markers for isoelectric focusing

Protein	pI at 25°C	Reference	pI at 25°C	Reference	pI at 4°C	Reference
Cytochrome c	9.28 ± 0.02	Radola (1973a)				
Ribonuclease	8.88 ± 0.03	Radola (1973a)				
Myoglobin (sperm whale)						
major component	8.18 ± 0.02	Radola (1973a)	8.18 ± 0.04	Nakhleh et al. (1972)		
minor component	7.68 ± 0.02	Radola (1973a)				
Myoglobin (horse)						
major component	7.33 ± 0.01	Radola (1973a)	7.45 ± 0.04	Nakhleh et al. (1972)	7.58 ± 0.02	Bours (1973d)
minor component	6.88 ± 0.02	Radola (1973a)	7.15	Nakhleh et al. (1972)	7.22 ± 0.05	Bours (1973d)
Bovine hemoglobin A	6.80	Conway-Jacobs and Lewin (1971)			6.18 ± 0.02	Bours (1973d)
Carbonic anhydrase (bovine)						
Conalbumin	5.88 ± 0.02	Radola (1973a)			5.45 ± 0.02	Bours (1973d)
β-Lactoglobulin B	5.31	Radola (1973a)			5.35	Bours (1973d)
β-Lactoglobulin A	5.14 ± 0.01	Radola (1973a)				
Bovine insulin	5.32 ± 0.02	Conway-Jacobs and Lewin (1971)				
Albumin (bovine)						
Cohn fraction V	4.90	Conway-Jacobs and Lewin (1971)			4.95 ± 0.02	Bours (1973d)
Ovalbumin	4.70	Conway-Jacobs and Lewin (1971)				
Horse spleen ferritin I	4.50 ± 0.02	Radola (1973a)				
Horse spleen ferritin II	4.38 ± 0.02	Radola (1973a)				
Horse spleen ferritin III	4.23 ± 0.03	Radola (1973a)				

TABLE 4.2b
pH markers for isoelectric focusing

Dye	pI at 25°C	Reference
Tris(5-OH)iron(II)[a]	7.15	Nakhleh et al. (1972)
Bis(5-OH)-4-OH iron(II)[b]	6.82	Nakhleh et al. (1972)
Bis(4-OH)-5-OH iron(II)[c]	6.24	Nakhleh et al. (1972)
Congo red	5.80	Conway Jacobs and Lewin (1971)
Tris(4-OH)iron(II)[d]	5.45	Nakhleh et al. (1972)
Evans blue	5.35	Conway-Jacobs and Lewin (1971)
Methyl blue	3.60	Conway-Jacobs and Lewin (1971)
Fast green FCF		
(major component)	3.05	Conway-Jacobs and Lewin (1971)
Patent blue V	3.00	Conway Jacobs and Lewin (1971)

[a] Tris(5-hydroxy-1,10-phenanthroline)iron(II).
[b] Mixed complex between tris (5-hydroxy-1,10-phenantroline)iron(II) and tris(4-hydroxy-1,10-phenanthroline) iron(II).
[c] Mixed complex between tris(4-hydroxy-1,10-phenanthroline)iron(II) and tris(5-hydroxy-1,10-phenanthroline)iron(II).
[d] Tris(4-hydroxy-1,10-phenanthroline)iron(II).

(ΔH_i^o in the range 6–14 kcal/mol) (see Table 4.1). Therefore the variation of intrinsic pK (pK_{int}^i) in the temperature interval 4–25°C, while being very small for carboxyls ($\Delta pK = 0.06$) is quite large for an imidazole ($\Delta pK = 0.39$) and very large for the ϵ-amino group of Lys ($\Delta pK = 0.67$) (Tanford, 1962; Bull, 1971). Therefore, it is to be expected that neutral and basic proteins will exhibit large dpI/dT increments. As shown by Fredriksson (1977), in fact, the dpH/dT and dpI/dT curves for carrier ampholytes and proteins, respectively, coincide only below pH 5 and greatly diverge in the pH range 5 to 9, especially as a function of imidazole and Lys residues of a protein. As shown in Fig. 4.12, due to the fact that the different curves cross-over at pH 7.4, it was this mere coincidence that allowed Vesterberg and Svensson (1966) to state that the pI values of horse myoglobin determined at 4°C and 25°C were identical. For the greater majority of proteins, this statement will not hold true and it is to be expected that pI values at 4°C will be higher than pI values at 25°C. In general, for a strongly acidic protein, dpI/dT should be ca. −0.005 pH units per degree around 4°C, whereas for a strongly basic protein it should be ca. −0.03 pH unit per degree. Fredriksson (1978) has published tables which allow the pI, taken by pH measurements at 25 °C, of a protein focused at 4 °C, to be converted into a true value at 4 °C. This is made possible by measuring the pH shifts which accompany the increase of temperature from 4 °C to 25 °C of focused Ampholine fractions in sucrose density gradients.

Ideally, a good mixture of pH markers, of known pI values at 4 °C and 25 °C, could avoid tedious pH measurements after IEF. Tables on the use of protein and dye markers have been reported (Bours, 1973d; Radola, 1973a; Righetti and Caravaggio, 1976; Conway-Jacobs and Lewin, 1971; Nakhleh et al., 1972). Protein markers should not be heterogeneous or at least have a major, easily identifiable, band in order to avoid confusion in pH assessments. Dyes would appear to be very promising, except that most of them, especially sulphonated derivatives, interact strongly with carrier ampholytes, originating a multitude of bands representing specific ampholyte-dye complexes (Righetti et al., 1977). Table 4.2a,b lists a series of such pI markers. They are now commercially available from several manufacturers (Pharmacia, BDH etc.). For pH measurements after IEF in density gradients, Jonsson et al. (1969), Secchi (1973) and Strongin et al. (1973) have described flow cells which allow continuous pH monitoring of the eluate. The pH curve can be superimposed on the UV profile in the same chart, thus providing easy reference for pH assessment. Gelsema and DeLigny (1977) and Gelsema et al. (1977, 1978) have stressed that, when using additives (especially in gradients) several correction factors should be introduced to correct for the apparent isoelectric point (pI_{app}) determined. Their conclusions can be summarized as follows:

(a) pI_{app}–pI values, i.e., the errors made in the measurement of isoelectric points by the conventional performance of density gradient IEF, depend upon the acidity, basicity and the chemical type of the ampholyte and upon the solvent composition;

(b) these errors are within ±0.1 pH unit for Ampholine (and presumably proteins) having $3 < pI < 9$ in gradients of 0–60% sucrose;

(c) the temperature coefficient of pI values of Ampholine (and presumably proteins) is in good agreement with that of pK values of carboxylic and amino groups;

(d) the primary medium effect (pI^*–pI) of sucrose, glycerol and ethylene glycol on pI values of Ampholine is in good agreement with that (pK^*–

pK) on pK values of carboxylic acids and alkyl substituted ammonium ions;

(e) pI_{app}–pI values holding for Ampholine can be used for the correction of pI_{app} values of proteins determined by density gradient IEF.

When using 6M urea as an additive in sucrose density gradient IEF, Ui (1971) suggests an overall correction factor of 0.42 pH unit, to be substracted from the apparent pI of the protein in this medium. A correction factor ranging from 0.7 to 0.9 pH unit has been recommended by Josephson et al. (1971) in 7 M urea solutions, while Gianazza et al. (1977) find that the urea effect varies with the various Ampholine pH ranges, from 5 to 7.5 hundredths of a pH unit/unit of urea molarity in going from acidic to alkaline pH ranges. In agreement with these last findings, Gelsema and DeLigny (1977) report that the use of a unique correction term irrespective of the pI values of the ampholytes studied is incorrect. For pH measurements in gel tubes, it was customary to section the gel in 2 to 5 mm segments, which were eluted in 200 to 300 μl of 10 mM KCl prior to pH readings. Recently, this operation has been automated with continuous pH readings on an intact gel rod (Chidakel et al., 1977). A contact microelectrode (Ingold No. E547310-1), mounted on a mechanical drive, moves along the surface of a gel cylinder resting on a trench in a plastic block. The gel is scanned at a rate of 1 cm/1.5 min and the pH is automatically recorded on a chart. In gel slabs, the pH can be measured directly on the gel surface with a flat membrane electrode, such as LOT type 403-30-M8 from Ingold (with a 6 mm membrane diam.) while punching 1–2 mm reference holes on the gel as markers for pH readings (Righetti and Caravaggio, 1976). The original antimony electrode of Beeley et al. (1972) has been mostly abandoned because of its low reproducibility and lack of 'Nernstian' properties. Recently, a microelectrode of iridium (a needle, with a 0.2 to 0.5 mm tip) has been described (Papeschi et al., 1976; Gianazza et al., 1977). This is a true Nernstian electrode, with linear pH/volts and pH/temperature responses and seems to be the only microelectrode able to perform pH readings at 4 °C, the usual temperature for IEF experiments (with all other microelectrodes, the gel has to be warmed up to room temperature before pH readings).

As a last comment on pI measurements, a note of caution concerning the comparison of pI data obtained after IEF in density gradients or in gel-stabilized systems. It has been known for a long time that for a number of alkaline proteins, e.g., cytochrome c (Radola, 1973a; Bobb, 1973), chymotrypsinogen (Salaman and Williamson, 1971), ribonuclease (Delincée and Radola, 1975) and subtilopeptidase (Wadström and Smyth, 1975), the pI values measured in gel-layer IEF were consistently lower (by as much as 1 to 2 pH units) than the values determined in density gradient IEF. Delincée and Radola (1978) have found this to be due to CO_2 absorption by the open gel surface in thin-layer gels. The interference of CO_2 is quite strong above pH 8.2 to 8.3. Thin-layer IEF in a CO_2 free atmosphere followed by pH measurements also in a CO_2 free atmosphere yields for alkaline marker proteins pI values in excellent agreement with those found by the density gradient technique.

4.6. Removal of carrier ampholytes after IEF

After IEF, it is often desirable to remove the carrier ampholytes from the protein fraction. Since most ampholytes have a mean $\bar{M}_r$ between 600 and 900 daltons (Bianchi Bosisio et al., 1981), it would appear to be a simple matter to remove them by dialysis. However, working with $[^{14}C]$ - Ampholine, Vesterberg (1970) found that a complete removal by dialysis requires at least 32 h, even when dialyzing against 0.1 M phosphate buffer, pH 7, in 0.5 M NaCl. Nilsson et al. (1970) and Li and Li (1973) obtained a very efficient removal of ampholytes by precipitation and subsequent washings of the protein with ammonium sulphate, in concentrations ranging from 60 to 100%. The protein precipitate is then washed three times with the same $(NH_4)_2SO_4$ solution. It appears that all ampholytes are soluble even at 100% salt saturation, so that their removal is quantitative. Quast and Vesterberg (1968) obtained a good separation of proteins from ampholytes by gel filtration in Sephadex G-50 fine. The separation is performed in columns equilibrated with 0.1 M phosphate buffer, containing 0.5 M NaCl, or 0.1 M ammonium bicarbonate or 0.1 M imidazole acetate. These buffers are usually kept at pH of about 7. The last two buffers have the advantage of being volatile, which allows them to be removed by lyophilization. Whichever method is used, it is important to have a high ionic strength in the buffer (0.1 M, or 0.5 M salt) to disrupt weak complexes which might form between Ampholine and proteins by electrostatic interaction. Alternatively, Brown and Green

(1970) have proposed the use of a mixed-bed ion-exchange resin (Bio Rad AG 501-X8) for separation of proteins and carrier ampholytes. This seems to be the method of choice, as also confirmed by Baumann and Chrambach (1975), who found by using [^{14}C]ampholyte, that less than 0.005 mol of Ampholine would remain bound to 1 mol of protein after chromatography on the bifunctional resin. This is tantamount to quantitative removal.

Ampholine can also, of course, be removed by electrophoresis. This is the method used by Suzuki et al. (1973) and by Stathakos (1975) to recover proteins from polyacrylamide gels after IEF on a preparative scale. The protein, with carrier ampholytes, is electrophoresed out of the gel into a chamber built with a dialysis membrane on the floor, through which Ampholine escapes (see also §2.2.4 and §2.2.5). A very high protein recovery (in a concentrated form and Ampholine-free) is thus achieved (see Fig. 4.13). Separation of carrier ampholytes from proteins could also be achieved by ultrafiltration, or by the use of the hollow-fiber technique (Bio Rad). If small amounts of carrier ampholytes (of the order of micrograms) are suspected to contaminate protein zones, they can be detected by thin-layer chromatography as suggested by Bloomster

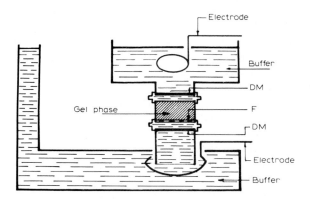

Fig. 4.13. Schematic drawing illustrating the electrophoretic recovery of focused bands. DM, dialysis membrane; F, filter membrane. The protein is recovered into the small chamber between F and DM, while the carrier ampholytes escape through the dialysis membrane. (From Stathakos, 1975.)

and Watson (1981). The sample proteins are spotted on cellulose thin-layer sheets and developed in 10% TCA. Proteins and large $\bar{M}_r$ species are precipitated at the origin, while contaminant carrier ampholytes migrate as a diffuse ninhydrin-positive layer with an R_f greater than 0.50. These authors have described another method for Ampholine removal: electro-dialysis for 4 h in acetate-buffered saline at pH 4.0. It might also be possible to remove carrier ampholytes by hydrophobic interaction chromatography on phenyl- or octyl-Sepharose, as described by Vester-berg and Hansen (1978) for concentration and desalting of body fluid (e.g., urine) proteins. Ampholytes are highly hydrophilic and would flow through the resin bed. While this could work for proteins, it does not seem suitable for separation of peptides from Ampholine, as demonstrated by Gelsema et al. (1980b).

4.7. Possible modification of proteins during IEF

Jacobs (1971, 1973) first reported on the partial modification of Cys and Met to cysteic acid and methionine sulphoxide upon prolonged IEF, as well as partial loss of Tyr and Arg from bovine ribonuclease. Both these phenomena could be largely suppressed by removal of O_2 from the column, and by addition of antioxidants, such as thiodiglycol and ascorbic acid. Satterlee and Snyder (1969) and Quinn (1973) have described the reduction of ferric myoglobin (metMb) to oxyMb during IEF. In view of the natural tendency of Mb to auto-oxidize and of the known oxidizing effects of persulphate in gels, this reduction process is quite unexpected. Funatsu et al. (1973) have also described the loss of glutamic acid from reduced ricin D during the process of IEF. Their hypothesis is that either the preparation is contaminated by traces of proteases, or that one of the two chains of ricin D is a protease itself, thus auto-digesting the molecule during IEF. Autolytic phenomena have also been documented by Needleman et al. (1975) during IEF of acidic lysosomal hydrolases (acid phosphatase, aryl phosphatase, β-glucuroni-dase, β-galactosidase and β-N-acetyl-hexosaminidase). Autolysis was minimized by adding to the IEF column 0.1% nitrophenyl oxamic acid, an inhibitor of lysosomal neuraminidase and cathepsin D. Perhaps one of the most controversial arguments in the field of IEF is whether or not artifactual heterogeneity can be elicited in a protein by binding of carrier ampholytes. If this were the case, indeed, it would shed doubt on most cases of microheterogeneity found by IEF and not readily detectable by other techniques. An example of such controversy comes from experiments with bovine serum albumin (BSA). It must be emphasized, however, that BSA represents an extreme case, since BSA is known to present a high affinity for a multitude of ligands (Steinhardt et al., 1972). By IEF in the pH range 4–6, Kaplan and Foster (1971) have presented evidence that a low pI peak in BSA represents albumin molecules strongly bound to a minor constituent or an impurity in the ampholyte mixture. Similar evidence for complex formation between BSA and ^{14}C-labeled carrier ampholytes has been presented by Wallevik (1973). On the other hand, these results have not been substantiated by Salaman and Williamson (1971) and by Spencer and King (1971) when focusing BSA both in the presence and in the absence of 6 M urea. Actually, in these last two publications, the possibility of artifactual heterogeneity is rejected. Another example of protein–ampholyte complex has been described by Frater (1970) when working with an acidic, low sulphur protein component from wool. In the light of the present information, it appears that evidence for protein–Ampholine interaction is rather scanty and some strong evidence to the contrary is now available. Thus, Baumann and Chrambach (1975a), working with human growth hormone and ovine prolactin, in the presence of ^{14}C-labeled Ampholine, have been able to exclude any interaction between protein and Ampholine. Protein binding to excess Ampholine was negligible. Ampholine binding to excess protein, if any, was below the detection limit of 0.2 mol Ampholine per mole protein. Similar results have been obtained by Dean and Messer (1975) working with purified proteins (albumin, ferritin and β-glucuronidase) and with serum proteins. Thus, in the light of these recent findings, it appears that artifacts in IEF due to ampholyte–protein interactions are at worst a rare occurrence. Often, sample modification might only be apparent and might be due to unfavorable pH and/or ionic strength conditions at the application site. For instance, as shown by Wadström and Smyth (1975b) in the case of bacterial cytoplasmic protein extracts, they are strongly adsorbed at the applicator site when applied to the cathode, while they are unaffected if applied at the anode. Usually, addition of 4 M urea to the gel and to the applicator minimizes this effect. Often, in crude extracts, the presence of large amounts of nucleic acids can give rise to blurred patterns and to pattern diversity for anodic and cathodic-species. This can often be prevented by pretreatment with DNase (Wad-ström and Smyth, 1975b).

Sometimes, the nature of the sample is such that meaningful results can only be obtained by focusing it either from the anode or from the cathode. Thus, working with zeins isolated from maize, Soave et al. (1975) found that, while the anodic pattern was highly heterogeneous, the cathodic pattern hardly showed any protein band at all. The two patterns could not be reconciliated even when performing IEF in 6 M urea or when applying the sample in a free solution, in a pocket on the gel surface. It was then found that at the cathode, due to unfavorable pH conditions, the zeins aggregated and precipitated at the application point, even in the presence of 6 M urea in the sample and in the gel. Thus, in this case, meaningful results could only be obtained by applying the sample at the anode.

4.8. Power requirements

Since the introduction of the technique of electrophoresis, supporting media, buffer systems, electrophoretic cells and cooling methods have all undergone extensive changes. Little attention, however, has been paid to the source of electrical potential. Among typical power sources used in electrophoresis, the only differences in output are in the range of power provided and in the type of regulation: constant voltage or constant current.

Recently, a second generation of power supplies has appeared, capable of operating either in the constant voltage, or in the constant current mode, and with automatic switch-over from one mode of operation to the other. With the advent of IEF, there has been a need for a third generation of power supplies, operating at constant wattage. Since the migration rate and the sharpness of focused bands at their pI values is proportional to the applied voltage, it is desirable to have the applied field strength as high as possible, while maintaining the joule heating (which is proportional to the square of the applied voltage) at an accept-able level. This maximum tolerable power, defined as optimum power by Schaffer and Johnson (1973), will depend on the conductivity of the system and on its efficiency to dissipate joule heating, and can be determined by inserting a thermic probe in the separation column. Once this optimum power, for a given system, has been determined, that system will run on an isotherm for the duration of the experiment.

Schaffer and Johnson (1973) have built a regulator which transforms an unregulated DC power supply into a constant wattage power unit. Söderholm and Lidström (1975) have built a constant wattage power pack, capable of delivering a maximum of 3000 V, 300 mA and 300 W, by applying a wattage regulator, which feeds the signal to a reversible synchronous motor (5 rev./min), used as a servomotor, to a constant voltage power supply.

There are now several constant wattage power supplies available on the market (e.g., from LKB, Pharmacia, Bio Rad) which are able to deliver up to 3,000 V, 200 mA and 150 W and are thus adequate for most IEF applications. With the availability of constant power supplies it is now possible to perform high speed IEF by the use of higher wattages, under thermally controlled conditions. For instance, we have already seen that, in the case of the 110 ml and 440 ml LKB columns, separations often required 48 to 72 h by applying wattages of 5 W and 10 W, respectively (see p. 114). By using a constant wattage power supply, Lundin et al. (1975) have shown that it is possible to obtain equilibrium conditions in 15—20 h by applying a maximum of 15 W and 1,600 V to the LKB 8100-1 column and a maximum of 30 W and 2,000 V to the LKB 8100-2 column. Under these conditions, the temperature was quite constant and only a few degrees above the coolant temperature, as measured with a pin point thermistor at different levels in the two columns. When using these higher wattages, Lundin et al. (1975) recommend that the experiment should terminate within 20 h, to avoid deformation and possible breakdown of the pH gradient upon prolonged focusing. In the case of thin-layer equipments, wattages as high as 20—25 W can be applied with no distortion of the pH gradient and no heat damage to the protein. In this last case, the experiment should be terminated within 2 h to avoid excessive cathodic drift.

4.9. The cathodic drift

From the early days of IEF in gels, it has been apparent that the steady-state envisaged by Svensson (1961, 1962a,b), with stationary carrier ampholyte distributions and protein patterns, is subjected to rapid decay with concomitant transport of samples and Ampholine mainly towards the cathode. This pH gradient instability was called 'plateau phenomenon' (Finlayson and Chrambach, 1971; Chrambach et al., 1973) when linked to the observation of a flattening in the centre of the pH gradient, or 'cathodic drift' (Righetti and Drysdale, 1971, 1973) when linked to the main direction of the flow. A dramatic representation of this phenomenon can be seen in a series of illustrations published by Davies (1975), covering an 8 h IEF experiment with hemoglobins. The cathodic drift also occurs when Ampholine is replaced by buffers (Nguyen and Chrambach, 1976) or when strongly acidic and basic electrolytes are replaced by buffer electrolytes (Nguyen and Chrambach, 1977a). Even under conditions where pH gradient stabilization is claimed to be achieved, the residual drift is in the cathodic direction (Nguyen and Chrambach, 1977b,c). Seven hypotheses (Haglund, 1975; Righetti and Drysdale, 1976) have been advanced to explain pH gradient decay: (1) electrophoretic migration of isoelectric carrier ampholytes; (2) electroendoosmosis; (3) formation of a zone of pure water in the neutral region of the pH gradient; (4) isoelectric focusing of water (considered to be an ampholyte) which results in its accumulation in the neutral region and back-flow towards the periphery of the gel; (5) selective deficiency or progressive destruction of ampholytes; (6) progressive gain or loss of charged ligand by Ampholine; (7) diffusion of Ampholine out of and/or electrolytes into the gel. Experimental evidence in favor of different models has been given by Fawcett (1975d) and Baumann and Chrambach (1975). Gianazza et al. (1979) and Arosio et al. (1978) have given as an additional hypothesis the interaction of carrier ampholytes among themselves and the presence of 'poor' amphoteric species, which could generate a cathodic drift. Delincée (1980) has blamed the drift on electroendoosmosis and CO_2 adsorption at the cathodic gel end. A valuable survey of literature data has been published by Rilbe (1977), who concluded that 'most observed forms of pH gradient instability can be explained by electroosmotic water flow and hydration water transport together with secondary phenomena'. However, the most generally accepted view, today, is the one put forward by Chrambach's group (Nguyen and Chrambach, 1975, 1977a–c; An der Laan and Chrambach, 1980) who appear to be able to stop cathodic drift by progressively increasing the pH of the anolyte (i.e., decreasing the proton molarity at this electrode). This finding is compatible with the computer simulations of pH gradient of Murel et al. (1979) which suggest that pH gradient stability may depend on a balance between mobilities and concentrations of protons and hydroxyl ions migrating in opposite directions. In other words, according to these authors, the 'proton cloud' generated at the anode travels towards the cathode faster than the '$-OH^-$ cloud' moving in the opposite direction, hence a net drift towards the cathode. If this hypothesis were correct, equalization of the product (H^+ molarity $\times$ H^+ mobility) with the product ($-OH^-$ molarity $\times$ $-OH^-$ mobility) should stop the cathodic drift, but this does not seem to be the case (Delincée, 1980). Since none of these mechanisms worked satisfactorily, we have set out to find an effective remedy to the cathodic drift. We have found that, indeed, polyacrylamide matrices are negatively charged due to: (a) trace impurities of acrylic acid in the gel; (b) covalent incorporation of catalysts (persulphate and riboflavin 5'-phosphate) as terminal groups in polyacrylamide chains; and (c) hydrolysis of amide groups to acrylic acid in the gel layer underneath the cathodic filter paper strip (Righetti and Macelloni, 1982). Thus, the only way we found effective in blocking pH gradient decay in IEF was the production of 'balanced matrices', i.e., the covalent incorporation of tertiary or quaternary amines in stoichiometric amounts as compared with the gel negative charges. To this end we have utilized 3-dimethylamino propyl methacrylamide (DMAPMA) or methacrylamido propyltrimethyl ammonium chloride (MAPTAC), as shown below:

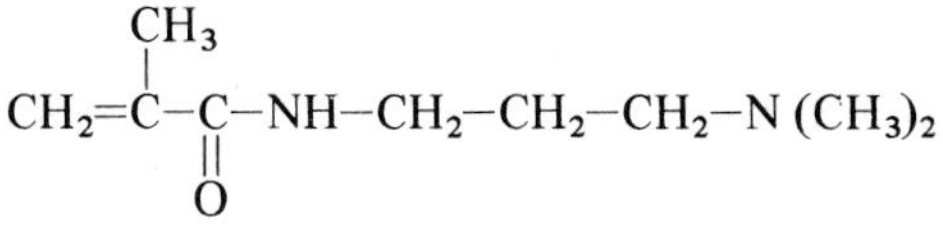

$$\underset{\displaystyle O}{\overset{\displaystyle CH_3}{CH_2=C-C-NH-CH_2-CH_2-CH_2-N(CH_3)_2}}$$

DMAPMA

$$\underset{\displaystyle O}{\overset{\displaystyle CH_3}{CH_2=C-C-NH-CH_2-CH_2-CH_2-N^+(CH_3)_3Cl^-}}$$

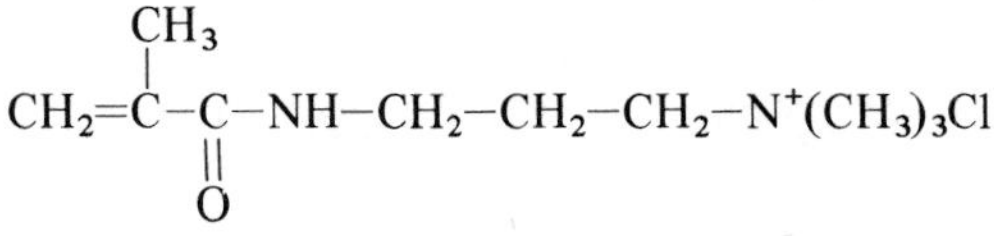

MAPTAC

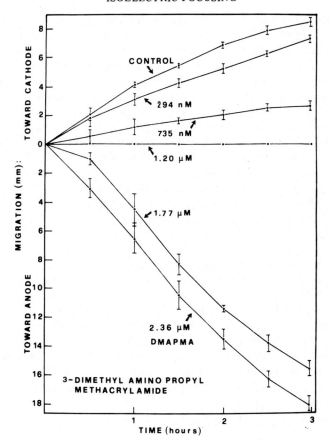

Fig. 4.14. Effect of a tertiary base on the cathodic drift. The gel was polymerized either as such (control, no DMAPMA added) or in presence of increasing amounts of 3-dimethylaminopropylmethacrylamide, from a minimum of 294 nM up to 2.36 μM. Notice that, above the equivalence point (1.20 μM DMAPMA added) the focused hemoglobin zones start drifting towards the anode (drift reversal). The gel was a 6% T, 4% C Bis matrix, 360 μm thick, and contained 2% Ampholine pH 3.5–10. Focusing was at 10°C, 20 min pre-run, followed by 1 h sample run (normal human adult hemoglobin). The time period marked on the abscissa begins after this total time of 80 min. All cathodic (or anodic) drift measurements were made at a constant voltage of 1000 V (100 V/cm). (From Righetti and Macelloni, 1982.)

Incorporation of DMAPMA in increasing amounts in polyacrylamide gels progressively quenches the cathodic decay down to 'zero drift' conditions at the equivalence point. When an excess of DMAPMA is incorporated, the drift reverses its polarity and becomes anodic (see Fig. 4.14). Thus, the fact that, according to the net charge incorporated into the gel, a cathodic drift, zero drift (balanced gels) or anodic drift can be generated, speaks in favor of electroendoosmosis being the major cause of pH gradient decay in IEF in gels. According to Bjellqvist, however (personal communication), some cathodic drift is due to a basic structural property of carrier ampholytes, whereby their cationic species have a mobility 5% higher than their anionic counterparts, so that the net result is a slow pH decay toward the cathode. 'Balanced gels', however, and the recently invented immobilized pH gradients (Bjellqwist et al., 1982a,b), appear to be the final answer to the vexatious problem of cathodic drift.

4.10. Trouble shooting (by E. Gianazza)

Problem: poor polymerization, long gelling time (polyacrylamide gels)
Cause and remedy:
(a) wrong composition of the mixture (too low $\%T$, or $\%C$);
(b) insufficient amount of polymerization initiator (persulphate, ribo flavin) or catalyst (TEMED): however, concentrations higher than $1.5\times$ (for persulphate) and $2\times$ (for TEMED) above the standard values should be avoided (Righetti and Macelloni, 1982);
(c) altered initiator: persulphate decomposes with time, solutions should be made fresh at least every week and stored at $4\,^{\circ}C$;
(d) wrong initiator: riboflavin is almost inactive at alkaline pH;
(e) too low temperature: for persulphate-gels, however, do not exceed $45\,^{\circ}C$ to $50\,^{\circ}C$;
(f) wrong lighting: for riboflavin-gels use UV, not visible-light source, i.e., fluorescent tubes, not tungstenum bulbs; do not use UV-absorbing material for the cassette, i.e., use thin glass, not thick plastic apparatus;
(g) presence of inhibitors: atmospheric oxygen: degas thoroughly and reequilibrate with nitrogen, pour in the casette immediately; alcohols, mercaptans; dirt: never leave plastic and rubber apparatus components soaking in soap;
(h) contact with air during polymerization: results in a wavy top of the

gel: completely fill an air-tight cassette, or carefully overlay the top with water.

Problem: amperage too high
Cause and remedy:
Temperature too high during run: excess of salt, either in the gel (wrong formula, excess of persulphate) or in the samples (if the defect develops only after their application; dialyze them against ampholytes [glycine, carrier ampholytes of appropriate pH range] or low molarity buffers [prefer $(NH_4)_2CO_3$, $CH_3COOH/Tris$, $Gly/Tris$ and like]) if conductivity is normal at the beginning but fails to fall with time, the electrodes could have been inverted: test the pH at the electrode strip − no remedy; if conductivity suddenly rises during the run, the apparatus is short-circuiting (see below, burning): turn it off immediately and drain excess liquid on the gel slab sides.

Problem: burning
Cause and remedy:
(a) if there is evaporation from the whole gel: wattage too high or temperature too high;
(b) if the gel burns along the edges: electrode strips are too long (they must be a few millimeters shorter than the gel); presence of water around the gel − after checking for the strips and the water, always cut and remove the burned gel side;
(c) if the gel burns close to the cathodic strip: formation of a zone of water: inadequate catholyte (choose a less basic one) − extend the pH gradient towards the basic region, adding some 6−8 or 3−10 carrier ampholytes; hydrolysis of the matrix: very long runs − if the sample application takes a long time, apply the electrode strips just before the start of the run;
(d) if the gel burns somewhere in the middle: sample with high salt concentration.

Problem: waviness
Cause and remedy:
(a) if the waves are present at the anodic side: carbonated catholyte − use fresh solutions and keep them in air tight bottles − the problem is

alleviated by adding Glu and/or Asp (2–5 mM) or some 3–10 ampholytes to the gel formula;

(b) if they develop around single samples: too high salt concentration if they correspond to the sample application site: remove the pads after ca. 1/2 h, or keep the slots filled with diluted ampholytes.

Problem: sample precipitation
Cause and remedy

(a) distinct lines: isoelectric precipitation: lower the concentration, choose a very soft gel formulation; interaction with carrier ampholytes: this has only been demonstrated for polyanions: use disaggregating agents;

(b) streaking at the application point: high molecular mass components: use soft polyacrylamide gels, or agarose, or granulated gels; water in-soluble components: use additives (detergents, urea; for some globulins, glycine); adverse effect of pH higher (or lower) than pI: add sample at the anode (or at the cathode) − it is advisable as a first trial to apply a sample in different positions on the gel: this would be a check both for the problem above and for the minimum focusing time; interacting (aggregating) components: add detergents or urea to the gel, if compatible (urea is compatible with most proteins at a concentration ca. 2 M) or thiols to the sample, or apply in another position on the gel (if ionic pairing is suspected).

Problem: pH instability (cathodic drift)
Cause and remedy:

(a) charges fixed to the gel: excess of polymerization initiators; presence of acrylic acid: recrystallize acrylamide, pass acrylamide solutions through an ion-exchange resin, incorporate positive charges into the gel (Righetti and Macelloni, 1982) − the problem is partially cured by increasing the viscosity of the medium, with sucrose, glycerol or urea (the most effective, but not always compatible with sample components): a reduction of $\%T$ is in this case advisable; or else a pH gradient extended towards alkaline values could be used; it is important in these cases to determine the minimum focusing time, and not to exceed it.

(b) a pH gradient different than expected is also obtained where the samples contain high amounts of specific amphoteric compounds, or of weak acid and bases.

Problem: irregularities of the gel surface

Cause and remedy:

(a) liquid on the surface: incomplete polymerization: carefully but completely remove it with tissue paper: keep moving it to avoid sticking;

(b) liquid at the cathode: gradient drift (see above): have a filter paper overlaying the gel 3—4 mm inside the cathodic strip and hanging free 2—3 cm outside;

(c) pads for sample application stuck to the surface: completely wet them with water, and wait a few seconds, remove excess water afterward;

(d) urea crystals: running (or storage) temperature too low (minimum for 8 M urea: 10 °C), or drying: apparatus has too large an open space around the gel: have a glass plate on top of the electrodes and moistened pads at the sides; few crystals may be dissolved with drops of water; the effect of urea crystallization, anyway, is usually less dramatic than its appearance;

(e) foaming around the electrodes, in presence of detergents: it is due to water transport (gradient drift) and to H_2 and O_2 bubbling (electrolysis products) — water must be removed (see above) to prevent short circuiting; a stack of two electrode strips should be used.

4.11. Artefacts: a unified view

With the widespread use of IEF, it has been accepted that microheterogeneity is a phenomenon of wide prevalence, whose causes and physiological implications are not fully understood, while homogeneity of a protein preparation seems to be an impossible dream. This had led many biochemical journals, in the early seventies, to reject papers dealing with IEF methodology, on the grounds of artefactual separations, a rejection crisis which has now largely, and luckily, subsided. Several explanations can be given for microheterogeneity:

(a) conformers: the components may represent alternative stable conformational substates;

(b) the separated bands could be due to possible oligomeric combinations of different subunits;

(c) in glycoproteins, the isolated IEF zones could correspond to nonuniform distribution of saccharide units between different molecules;

(d) varying degrees of enzymatic phosphorylation, methylation or acetylation could be responsible for the IEF distribution, as often is the case with nuclear proteins;

(e) there always remains the possibility of proteolytic damage, probably as an artefact of the preparation;

(f) partial deamidation of asparagine and glutamine residues;

(g) tightly bound co-factors, or inhibitors, to a protein moiety, further complicated by different charge states of these ligands, such as a redox state.

Many of the possibilities for charge heterogeneity have been discussed by Epstein and Schechter (1968), Kaplan (1968), Williamson et al. (1973) and Righetti and Drysdale (1976). A few reports on a possible artefactual heterogeneity, elicited by the IEF technique itself have been published (see §4.7) but no clear evidence was ever presented. Indeed, we have found that there are classes of compounds which display an artefactual heterogeneity upon IEF fractionation, but they belong to limiting structures and sequences, which bear no relationship with those commonly found in proteins. I will list these cases below, together with a proper interpretation of these phenomena.

(1) *IEF of nucleic acids.*   In 1972 Drysdale and Righetti reported the first fractionation of tRNA species by IEF. Subsequently, several reports appeared on the IEF separation of α- and β-globin mRNAs as well as of ribosomal 18 S and 28 S RNAs from rabbit reticulocytes (Shafritz and Drysdale, 1975; Drysdale, 1977). These 'isoelectric fractionations' are unexpected on theoretical grounds, since a method of 'pure' charge fractionation, such as IEF, is indifferent to polydispersity in molecular mass where charge-to-mass ratio is constant. Moreover, a true theoretical pI for tRNA (and probably for most nucleic acids) is centered around pH 2.6 (Morris, 1978), while all RNAs tested exhibited 'apparent pI values' (pI_{app}) in the pH range 3.5–5.2. A close scrutiny of the published material reveals the following:

(a) for tRNA, the number and relative distribution of the IEF bands depends on the mode of sample application and on the relative tRNA–Ampholine ratio;

(b) the IEF profile of tRNA is strongly sensitive to 6 M urea in the gel (the number of bands is drastically reduced and the pI_{app} considerably lowered);

(c) for tRNA, isolated fractions redistributed when refocused, though to a lesser extent than in the first fractionation;

(d) ribosomal RNAs banded at pI_{app} values of 4.8 (18 S) and 5.2 (28 S), but moving down from the cathode (sample application site) in a 'size-dependent' rather than a 'charge-dependent' fashion. In fact, the tails of the two peaks are smears which move down from the cathodic gel end. While both peaks appear to have barely entered the gel, 4 S RNA, in the same experiment, moves considerably down the gel toward the anode and bands around pI_{app} 4.0. There is no difference between a typical size fractionation in sucrose gradients of these species and their apparent charge fractionation by IEF;

(e) by the same token, α-globin mRNA migrates slightly faster (pI_{app} 4.7–4.8) than β-globin mRNA (pI_{app} 5.1–5.2), which is known to have molecular dimensions somewhat greater than the former.

Notwithstanding claims to the contrary (Drysdale, 1977), all these observations indicate that 'isoelectric' separation of nucleic acids is definitely 'non-isoelectric'. Indeed Galante et al. (1976), on the basis of variations of melting points of tRNA in presence of Ampholine, have demonstrated that the mechanism of IEF fractionation of nucleic acids is a strong complex formation between them and the carrier ampholytes used to generate and stabilize the pH gradient in IEF. The complex is strongly pH-dependent: it is practically non-existent at neutral pH (pH 7.4), rather weak at pH 5.4 and very strong at pH 4.2 and lower, i.e., in the typical pH range of apparent isoelectric fractionation of tRNAs.

(2) *IEF of heparin.*   Nader et al. (1974) and McDuffie et al. (1975) reported that heparin could be fractionated by IEF into 21 components, with pI_{app} values between pH 3.2 and 4.2. The basis for the fractionation was associated with the polydispersity in $\bar{M}_r$, the 'low pI' components exhibiting the smallest size (3,000) while the 'high pI' components had the highest $\bar{M}_r$ (37,500), with graded intervals of ca. 1,500–2,000 daltons in mass. Heparin fractionation by IEF is even more puzzling than nucleic acid separation, since while at least the latter are amphoteric, the former is not and is a simple polyanion, with no positive counter-charges. It was later noted that different heparin preparations with mutually exclusive $\bar{M}_r$ values gave a number of common isoelectric bands (Johnson and Mulloy, 1976), suggesting that the IEF profile of heparin was not specifically related to $\bar{M}_r$, despite the fact that very low $\bar{M}_r$ samples did seem

to migrate further down the pH gradient to the 'low pI' range.

The mechanism of this fractionation was unraveled by Righetti and Gianazza (1978) and Righetti et al. (1978). On the basis of spectra taken in solution and in the focused gel, of repeated runs of isolated, focused bands, of IEF fractionations performed in the presence and in the absence of urea, at variable Ampholine–heparin ratios and with heparins of varying degrees of carboxylation and sulphation, they demonstrated the IEF heparin profile to be artefactually elicited by interaction with carrier ampholytes. The 21 IEF fractions indeed represent 21 different complexes of the same macromolecule with 21 specific Ampholine molecules. The complexes differed markedly in binding strength, the high pI_{app} species forming the weakest interactions, which could be extensively split by Toluidine blue, while the lowest pI_{app} species exhibited the strongest aggregates, which could not be disaggregated by excess basic dye. It was then found that not only the binding strengths, but also the stoichiometries of these complexes could vary considerably, depending on the relative Ampholine–heparin ratios in solution. In an excess of Ampholine, aggregation is perpendicular to the heparin chain, with the end ammonium charge of each Ampholine molecule neutralizing one negative charge along the heparin molecule (at this stoichiometry, heparin would bind an amount of Ampholine twice as much its own mass, giving a complex of ca. 37,000 daltons, starting from its original $\bar{M}_r$ of 12,600). In heparin excess, the bound Ampholine is aligned parallel to the heparin molecule, so that on the average one Ampholine species neutralizes approximately three negative charges. As shown in Fig. 4.15, these two models represent two limiting stoichiometries, with any possible ratios in between. In both models, unbound, protonated nitrogens in the Ampholine molecule can act as cross-links or bridges among different heparin helices, thus favoring the formation of huge macromolecular aggregates.

(3) *IEF of polyanions.* It is now understood (Gianazza and Righetti, 1978) that the anomalous fractionations of nucleic acids and heparin were just particular cases of a more general phenomenon, by which all polyanions would elicit complex formation with carrier ampholytes in the pH range 3.2–5.0 (see Fig. 4.16). As previously reported, the affinity of Ampholine for polyanions is strongly pH dependent, at pH 3 being ca.

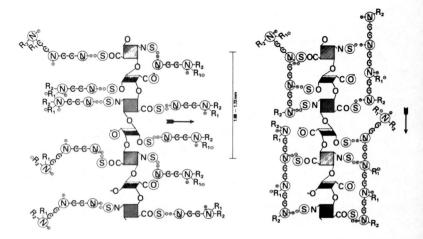

Fig. 4.15. Model for the interaction between segments of heparin and segments of Ampholine molecules. The left side represents the aggregation of ampholytes on heparin at Ampholine excess, e.g., a ratio of one Ampholine molecule to one heparin charge. The right side shows the complex formed in heparin excess, e.g., one Ampholine species per three charges on heparin at pH = 3.5. Notice how in going from one extreme to the other, the vectorial orientation of the ampholytes varies from perpendicular to parallel to the heparin helix. (From Righetti et al., 1978.)

10,000-fold greater than at pH 7. This seems to be due to the fact that at pH 3 Ampholine species would exhibit a stretch of 4 to 5 protonated nitrogens (each two methylene groups apart), while most of its carboxyls will bear no negative charge, thus allowing for strong, cooperative binding to polyanions. At pH 7, the net charge will be reversed, and the anionic Ampholine will be repelled by the negatively charged polyanion (Righetti et al., 1978). Binding is affected by the type of negative charge, its density and spatial orientation on the polyanion. On the basis of the type of negative charge, the binding strength decreases in the following order: polyphosphate > polysulphate > polycarboxylate. Given the same type of negative charge, the binding is dependent on charge density and its space orientation: thus polyglutamic forms stronger complexes than polygalacturonic acid. The minimum length of the polyanion eliciting a measurable binding appears to be of the order of about six negative charges, as demonstrated with hexametaphosphate.

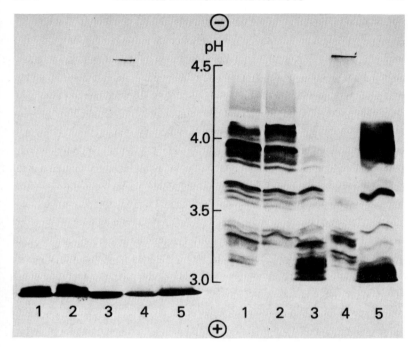

Fig. 4.16. IEF of polyanions in 5% acrylamide gel, 2% Ampholine pH 2.5–5 in the absence (right side) and in the presence (left side) of 8 M urea. 1: heparin A; 2: heparin B; 3: carboxyl-reduced heparin B; 4: polygalacturonic acid; 5: polyglutamic acid. Staining with toluidine blue. (From Righetti et al., 1978.)

(4) *IEF of dyes.* The same types of phenomena reported with all polyanions were then encountered when focusing acidic dyes (mostly mono- to tetra-sulphonated species) (Righetti et al., 1977). This was quite puzzling and unexpected and not quite easily correlated with the other types of artefacts. Yet we now understand that the same underlying mechanism applies also in this case, except that here it is the Ampholine molecule itself which acts as a template to which as many as four to five dye molecules can bind simultaneously. Moreover, the dye can act as a cross-link, over several Ampholine species, forming a tri-dimensional

network dye-Ampholine which aggregates and precipitates in the focusing process. In fact, even in the test tube, simple addition of Ampholine buffered at pH 3 to any polyanions or to most of the dyes tested gives rise to turbidity and eventually to sample precipitation. Needless to say, this 'non-isoelectric precipitation' does not take place if the pH of the polyanion or dye solution is lowered to pH 3 in the absence of Ampholine. These complexes can only be disaggregated by changing the dielectric constant of the solvent or by high temperatures, or both. On an equimolar basis, the order of dissociating power of the most commonly used disaggregating agents is: dimethylformamide − tetramethylurea > dimethylsulphoxide > urea > formamide (see Fig. 4.5). We have postulated a two-step binding model: first a strong, polydentate and undissociated Ampholine−dye salt, followed by hydrophobic dye−dye interaction.

(5) *IEF of polycations.* The next logical step was to see if the same type of phenomena would hold true at the other extreme of the pH scale. In fact, when we ran poly-Arg (Gianazza and Righetti, 1980), the complex was so strong that the polycation, applied at the anode to a prefocused gel slab, swept away on its wake the focused Ampholine molecules and precipitated as a smear in the pH range 7−8, where the carrier ampholytes would behave as anions. This 'perturbed Ampholine pattern', is a good indication of complex formation, provided the applied sample is salt-free, since excess salt in the sample could distort the pattern of focused Ampholine. However, this phenomenon seems to be limited to polyArg, since very little, if any, binding occurs with polyLys and no binding at all could be demonstrated with polyHis. Moreover, histones run through the entire gel length collecting at the cathodic filter paper strip, with no indication of complexes with Ampholine.

All these artefacts have been confirmed by theoretical calculations of Cann and Stimpson (1977), Stimpson and Cann (1977) and Cann et al. (1978). However, they used only a bimodal model (a single peak splitting into two upon interaction with carrier ampholytes). It was only when I confronted Cann with our 21 heparin zones in IEF that I convinced him the most likely chance for artefacts in IEF was a multimodal distribution, since the macromolecule, once it interacts with carrier ampholytes, will complex with several different species, not just a single pI component. This view was later accepted (Cann, 1979). In addition to all the classes of artefacts I have listed, there appears to be yet another group of artefacts: when focusing mammalian cells, there is strong evidence that carrier ampholytes can coat the cell surface by binding to sialic acid residues via the protonated nitrogens of their oligo-amino backbone (McGuire et al., 1980). This phenomena will be dealt with more extensively in the next chapter (see Fig. 5.6).

Except for these artefacts generated by peculiar structures, there is practically no evidence that carrier ampholytes will elicit the same multimodal distribution by interacting with proteins. Even an earlier claim of complex formation between an acidic, low sulphur protein fraction and Ampholine (Frater, 1970) has been dismissed by Marshall and Blagrove (1979) who have successfully fractionated by IEF the same protein. The only real artefact on IEF of a protein has been described by Jonsson et al. (1978) working with rat incisor phosphoprotein (RIP). RIP binds Ampholine, generating a series of bands with apparent pI values in the pH range 2.5–3.1, while its true pI is 1.1. This is not unexpected: RIP is an extremely acidic protein, with a charge density similar to that of heparin, so that at all practical effects it behaves as a polyanion. Perhaps lipoprotein lipase also behaves in a similar fashion (Bengtsson and Olivecrona, 1977), but in the vast majority of cases the heterogeneity detected in a given sample upon IEF will be inherent to the protein itself, and the scientist will have to find a way to live with it.

CHAPTER 5

Some applications of IEF

There are at least 4,000 articles describing applications of IEF to a host of biochemical, genetical and biological problems. If you think you will find them all reviewed here, then you are reading the wrong book, or you are the wrong reader, or both (to be honest there could be another possibility: I am the wrong writer). I have listed in §1.4 books and reviews on IEF and I am afraid you will have to leaf through them to find your field of interest. In this chapter, I will briefly survey some applications to a few, selected topics which can show the great progress made by the proper use of this technique in the understanding of modern biochemical processes. Some of the topics have been selected as examples of pitfalls and limitations encountered when using IEF for the wrong purification process.

5.1. IEF of peptides

Due to its very high resolving power and to the sharpening effect of sample zones in their pI position, IEF appears to be particularly useful for the analysis of peptides, either natural or obtained by fragmentation or by synthetic processes, since these substances have a rather high diffusion coefficient as compared with proteins, and usually produce diffused zones in conventional separation techniques. Analytical IEF of peptides was not feasible, however, due to their very close similarity and common reactivity with carrier ampholytes. The classical techniques used for peptide detection (ninhydrin stain, Folin reagent, microbiuret, permanganate oxidation, starch–iodine reaction), in fact work wonders for the detection of focused Ampholine in the gel matrix. The first attempts were reported as early as 1973 (Kopwillem et al.), when active fragments of the human growth hormone (hGH), synthetized by the solid-phase method of Merrifield (1969), were analyzed by IEF. The detection of the focused zones was made possible by the presence of a chromophore (a dinitrophenyl-His) in the polypeptide chains, but this method was not generally applicable. In the case of amphoteric anti-biotics of low solubility in water, Righetti and Righetti (1974) performed IEF in gels containing 50% dimethylsulphoxide (DMSO). After focusing, the gel is extruded into distilled water: the DMSO is rapidly leached out and the focused bands precipitate at their pI as opalescent zones which can be quantitated by a scan at 600 nm and recovered fully active. Even this method, however, is of limited applicability.

A breakthrough came in 1978, when Righetti and Chillemi (1978) and Bibring and Baxandall (1978) independently described the IEF separation of peptides and their detection by general protein stains, usually a colloidal dispersion of Coomassie Brilliant Blue G-250 (or R-250) in 12–15% trichloroacetic acid (TCA). The micellar dye is in a leuco form (a transparent, pale greenish solution for CBB G-250) and only upon uptake by the peptide (or protein) band it turns to the blue color. The focused gel is bathed directly in this solution, whereby the carrier am-pholytes are leached out from the gel matrix while the peptides, which are only partly soluble in 12–15% TCA, are further fixed by the dye molecules, which probably act as cross-links over different chains, thus forming a macromolecular aggregate which can be trapped in the random meshwork of the gel fibers. The gel matrix remains unstained, probably because the dye micelles cannot diffuse within the gel, so that destaining in acid–alcohol (which could remove bound dye and solubilize the precipitated peptides) is not required. By this method, the minimum critical peptide length for fixation and staining appears to be around 12--14 amino acids (ca. 1,500 daltons) (see Fig. 5.1). Dye adsorption onto the peptide is favored by the presence of basic amino acids and by a relatively high content of hydrophobic amino acids. Besides binding to basic groups the dye can interact with hydrophobic residues, particularly if they are in a close sequence. A possible mechanism for peptide insolu-bilization in the staining step is shown in Fig. 5.2 (Righetti and Chillemi, 1978): it is hypothesized that the dye binds two chains simultaneously (via its $-SO_3^-$ groups) thus acting as a bridge or a cross-link between chains

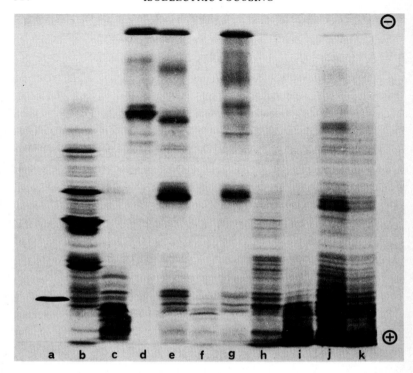

Fig. 5.1. IEF of peptides in the length range 8−54 amino acids. The gel slab was 0.7 mm thick and contained 7% acrylamide, 2% Ampholine pH 3.5−10 and 8 M urea. 10−15 μl of sample (10 mg/ml) were applied in filter paper strips at the anode after 1 h prefocusing. Total running time, 4 h at 10 W (1,000 V at equilibrium). The gel was then dipped in a colloidal dispersion of Coomassie G-250 in 12% TCA. The samples are the following synthetic fragments of the human growth hormone (hGH): a = hGH 31−44; b = hGH 15−36; c = hGH 111−134; d = hGH 1−24; e = hGH 166−191; f = hGH 25−51; g = hGH 157−191; h = hGH 1−36; i = hGH 96−134; j = hGH 115−156 and k = hGH 103−156. Shorter fragments (octa-, nona-, deca- and dodecapeptides) were neither fixed in the gel nor stained. (From Righetti and Chillemi, 1978.)

and forming a macromolecular aggregate that is trapped in the gel fibers (the dye formula is given by Diezel et al., 1972).

Even though the present method represents an important step forward

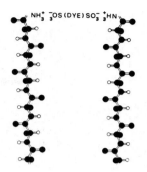

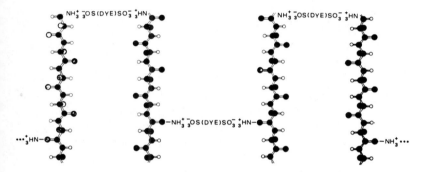

Fig. 5.2. Proposed dye-peptide binding model. A peptide with no basic residues (upper part), will only bind the dye via its NH$_2$ terminus, thus forming a dimeric complex that readily splits because it is extremely unstable in H$_2$O. The lower part shows a peptide with an additional binding site in the chain (an ϵ-NH$_2$ group of Lys). In this case, a macromolecular aggregate is formed, that is trapped in the gel matrix and is insoluble in water. This model should be seen not as linear, as depicted, but as a three-dimensional array of chains. The aggregate is further stabilized by stacking among aromatic rings of different dye molecules, made possible by the ionic interactions of the two sulphate groups of the dye with the peptide amino groups.
(From Righetti and Chillemi, 1978.)

in peptide analysis, it unfortunately leaves out the class of oligopeptides (from di- to tetradeca-residues) where many biologically active substances are located. By exploiting the ultrathin gels described by Görg et al. (1978), Gianazza et al. (1979b) have been able to close this gap partially.

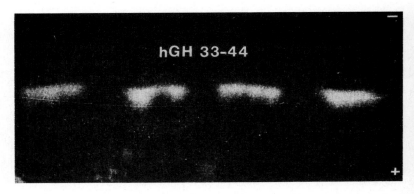

Fig. 5.3. IEF of hGH 33–44 fragment (150 μg/sample) on a polyacrylamide gel slab. The spots were visualized by a short (a few seconds) exposure to iodine. The gel was then left in an air stream until the background faded, the bands were cut, eluted in 80% acetic acid, and the eluate was then hydrolyzed and run on an automatic amino acid analyzer. (From Gianazza et al., 1980b.)

300 μm thin acrylamide gels are cast, and used as a medium for oligopeptide focusing. Rather high voltages can thus be applied and thin bands obtained. After IEF, the gel is pasted to a sheet of filter paper and immediately dried in an oven at 110 °C, whereby the gel layer is reduced to a film of vanishing thickness. The gel-paper sheet can now be sprayed, like a typical paper chromatogram, with any of the specific stain solutions for Arg, Met, Cys, Tyr, Trp and His. In this way, 30% of all the possible amino acids in oligopeptides can be detected, with a sensitivity level of only a few micrograms of peptides per IEF zone. A method for direct His staining of focused peptides, in a wet, instead of dried gel, has also been independently described by Faupel and Von Arx (1978). There is now evidence that these developments allow the display of a peptide map, in the single IEF dimension, with great saving of time and materials when compared with the classical, two-dimensional fingerprints (Gianazza and Righetti, 1979; Righetti et al., 1980c).

As a recent development, Gianazza et al. (1980b) have been able to find a general stain, independent from peptide length. When a focused and dried ultrathin-layer gel is exposed to iodine vapors in a jar, the carrier ampholytes adsorb the iodine molecules quickly, giving a uniformly brown stain in the gel-paper sheet. By a mechanism still not completely understood, the focused peptides inhibit this adsorption, thus being detected by negative staining, as white zones on a brown background. The staining is fully reversible, since no chemical reaction occurs, and is developed within a few minutes of exposure to iodine. In order to obtain a uniform brown color, rather than a striped background, 4% carrier ampholytes are used in the focusing gel, instead of the usual 2% levels, so that the gaps in between the 'stripes' of focused Ampholine are smoothed. A drawback of the present method is its low sensitivity, a minimum of 30 μg peptide/zone being required for detection. However, this technique can be exploited for small scale preparative applications: the peptides can be eluted from the gel in 80% acetic acid with recoveries up to 85% (Fig. 5.3). As an alternative to all these staining procedures, Stock et al. (1980) have tried, after IEF of peptides, to cross-link them in 5% formaldehyde and then detect them by common Coomassie stains in acid—alcohol. This method, however, does not seem satisfactory because carrier ampholytes are also fixed and stained. IEF of peptides is now in widespread use, especially in the field of hormone research (Noble et al., 1977; Biscoglio De Jimenez Bonino et al., 1981; Schwartz et al., 1981; Kohnert et al., 1973).

Notwithstanding the success of analytical IEF of peptides, its preparative counterpart is still not quite feasible. This is mainly due to the fact that carrier ampholytes and peptides have quite similar physico-chemical properties, both in terms of net charge and mass, so that their separation, after IEF, by gel chromatography, dialysis or ion-exchange is an extremely complicated proposition. Notwithstanding the fact that carrier ampholytes generally have a molecular mass ($\bar{M}_r$) not exceeding 1,000, their separation even from higher $\bar{M}_r$ peptides is complicated by the fact that they tend to aggregate in solution (Bianchi Bosisio et al., 1981). Recently, Gelsema et al. (1980b) have studied the feasibility of separating carrier ampholytes from peptides by hydrophobic interaction chromatography on Octyl Sepharose. However, one cannot expect this method to be generally applicable, because most peptides below 5,000 in molecular mass are not adsorbed to amphiphilic gels (Hjertén, 1974), which is in agreement with the authors' conclusion: 'the separation of peptides from carrier ampholytes by hydrophobic interaction chromatography in the presence of salt is in general not feasible for peptides with molecular weights in the most interesting range (say 2,000–8,000)'. As an alternative to this method, one could try to focus against a background of small $\bar{M}_r$ carrier ampholytes (Ampholines have an average $\bar{M}_r = 750$). This approach has been used by Gasparic and Rosengren (1975) who have synthesized small $\bar{M}_r$ Ampholine (average $\bar{M}_r = 300$) but these compounds have never been tried for actual separation of peptides and are not available on the market. Righetti and Hjertén (1981) have tried to solve this problem at the other end of the $\bar{M}_r$ scale: instead of using smaller and smaller ampholytes, they have synthesized extremely large carrier ampholytes (see also §1.7.4). A giant polyethylene imine ($\bar{M}_r$ range 40,000–60,000) is mixed in a linear gradient of acrylic acid in a flow-through system and allowed to react at 80°C for 70 h. Giant carrier ampholytes ($\bar{M}_r$ range 50,000–90,000) are thus obtained. Although these compounds interact strongly among themselves, especially in acidic pH ranges, giving rise to aggregates and precipitates, they interact only weakly and reversibly with proteins and no interactions are apparent with model dipeptides (His-Ser, His-Met, His-Phe and His-Lys) which could be recovered with ca. 85% yields from preparative runs.

All these data indicate that, albeit with some difficulties, analytical and preparative separations of peptides are feasible. Even these difficulties might be totally eliminated with the use of the recently introduced technique of immobilized pH gradients (Immobiline, LKB Produkter AB).

Recent data from our group (Gianazza and Righetti, unpublished) indicate that, analytically, all focused peptides can be revealed by spraying with ninhydrin, or fluorescamine, or any reagent for primary amino groups, since the grafted buffering groups do not interfere with any of these reactions. Preparatively, considerable amounts of peptides can be loaded onto immobilized pH gradients and recovered uncontaminated because the buffering groups are covalently linked to the gel matrix. I predict that the field of peptide chemistry will greatly benefit from this newly developed technique.

5.2. IEF of cells, subcellular particles, bacteria and viruses

IEF of cells was first introduced by Sherbet et al. (1972) and Sherbet and Lakshmi (1973) by using stationary, linear gradients of sucrose (10–55%), glycerol (10–70%) or Ficoll (5–30%) containing 1% Ampholine in a standard LKB 110 ml column. In these early systems, the cells were subjected to the IEF process for as long as 30 h. Subsequently, Just et al. (1975a,b) and Just and Werner (1977a,b) introduced the method of continuous-flow IEF separation of cells, in a modified Hannig apparatus. In this system, the residence time of the cells in the flow-chamber was only between 7–10 min. Cell focusing has also been described in several other reports. Leise and LeSane (1974) reported the IEF separation of peripheral lymphocytes of human and rabbit origin in a gradient of dextran-40, while Hirsch and Gray (1976) analyzed rat peripheral lymphocytes in dextran gradients in isotonic sucrose. Boltz et al. (1977) reported the IEF fractionation of Chinese hamster fibroblasts in a linear Ficoll density gradient made isotonic throughout by sucrose and glucose; Manske et al. (1977) have described the IEF analysis of Ehrlich–Lettré mouse ascites tumor cells, as well as rat hepatocytes, in Ficoll–sucrose gradients; even boar (Moore and Hibbit, 1975) as well as bull and rabbit (Hammerstedt et al., 1979) spermatozoa have been isolated by IEF in Ficoll gradients. In the field of procaryotic cells, the IEF of bacterial cells has been reported by Sherbet (1978), and by Langton et al. (1975). According to Talbot (1975), Mandel (1971), Chlumechka et al. (1973) and Rice and Horst (1972) even viruses seem to be amenable to isoelectric fractionation.

In an extensive treatise on electrophoresis and isoelectric focusing of cells, Sherbet (1978) has given evidence that IEF can be used as a probe for characterization of ionizable groups on the cell surface, for measuring cell surface charge densities, pK values of ionizable groups on the cell membrane and for following chemical modifications of charged groups on the cell envelope (see Table 5.1). According to Sherbet (1978), IEF can probe the electrokinetic zone of shear of cells down to a much greater depth than zone electrophoresis. Conventional electrophoresis can probe the surface to a depth of about 1.4 nm, while it appears that the isoelectric zone extends at least five times as much, to a depth of about 6–7 nm below the surface. Moreover, it appears that the isoelectric point (pI) values of cells, as determined by IEF, can be used to calculate their surface charge density (σ), as well as their electrophoretic mobility (a list of cell pI values is given in Table 5.2). In turns, charge densities (expressed as number of charges/cm^2 of cell surface) can be linearly correlated to the electrokinetic potential (ζ potential), provided the

TABLE 5.1

Isoelectric equilibrium analysis of E. coli cells after different treatments and of viruses[a]

Cell type[e]	Observed pI	Probable ionogenic groups in isoelectric zone[d]		Calculated pI[b]	No. of net negative charges per cell $\times 10^{-6}$
		Present	Excluded		
E. coli cells (untreated controls)	5.60 ± 0.10	COOH; NH_2pK $= 7.5$	Acidic groups of pK < 3.2; ϵ-NH_3^+ of lysine	5.38	0.37
AW-EI-E. coli	8.55 ± 0.04	NH_2pK $= 7.5$	NH_2 groups of phospholipids; phenolic—HO; guanidyl groups	8.7	$+0.41$
HCHO-E. coli	3.85 ± 0.06	COOH groups[g]	Acidic groups with pK < 3.2	3.5	0.84
E. coli-CPDS	4.28 ± 0.05	As in E. coli control	Thiols[f]		
E. coli AW-EI-CPDS	7.47 ± 0.18	As in AW-EI cells	Thiols[f]		
Foot-and-mouth disease virus	6.9 and 4.3				

[a] All values from Sherbet and Lakshmi (1975) except for foot-and-mouth disease virus, taken from Talbot (1975).

[b] pI values are calculated from surface pK of ionogenic groups corrected using the Hartley–Roe equation, pI $= -\log (\Sigma k'a \times k'w / \Sigma k'b)^{1/2}$ where $k'a$ and $k'b$ are corrected dissociation constants of the strongest anionogenic and cationogenic groups, respectively, $k'w = 6.81 \times 10^{-15}$ at 20°C.

c   The number of net negative charges on the cell surface was calculated as described by Sherbet et al. (1972) using the equation $Q = PDr^2 K \times 3.3 \ 10^{-3}/e$, where Q is the number of net negative charges per cell. The potential, P, on the surface of the cells is estimated as the potential difference between a solution at neutral pH and a solution isoelectric with the cell surface and is given by $(7 - 1) \times 2.303 \ RT/F$, where R is the gas constant (8.315 Joules/°C, T the absolute temperature, F the Faraday (96,500 C). The value of $2.303 \ RT/F = 0.592$ at 25 °C. D is the dielectric constant of water (78.54 at 25 °C), r the radius of the cell in cm, K the Debye–Hückel function ($0.327 \times 10^8 I^{1/2}$) at 25 °C where I is the ionic concentration due to Ampholine assumed to be 0.01 M, and e the electronic charge 4.8×10^{-10} e.s.u.

d   The isoelectric zone is estimated to extend to a depth of 6.0 nm below the cell surface.

e   AW-EI-*E. coli*, acid-washed ethyleneimine-treated *E. coli* cells treated with formaldehyde; AW-EI-CPDS, AW-EI cells treated with 6,6'-dithiodinicotinic acid (CPDS); *E. coli*-CPDS, *E. coli* cells treated with CPDS.

f   The total number of thiol groups on the *E. coli* surface equals the number of net negative charges introduced by reacting them with 6,6'-dithiodinicotinic acid, which is approx. 0.35×10^6 groups. Only a proportion of these (less than 3%) are dissociated at physiological pH.

g   COOH groups of glucuronic and neuraminic acids, β-COOH of aspartic acid and γ-COOH of glutamic acid.

TABLE 5.2

Isoelectric points of animal cells, bacteria and viruses determined by isoelectric equilibrium method[e]

Type	pI	Reference
HeLa (Human carcinoma of the cervix)	6.85	
γ-Globulin-treated HeLa	6.36	
Yoshida ascites sarcoma	6.35	Sherbet et al. (1972)[a]
Ehrlich ascites sarcoma	5.6	
Py cells	6.4	
Normal liver cells	6.5	
Butter yellow-induced rat hepatoma	6.75	Sherbet and Lakshmi (unpublished)[a]
3T3 mouse fibroblasts	4.6	Sherbet and Lakshmi (unpublished)[b]
SV-3T3 cells	4.8	
SV-CHK	4.66	Sherbet and Lakshmi (1976)[b]
SV-TRK	4.6	
Py-3T3	4.78	
Human astrocytoma Grade I	5.15	
Grade II	5.0	
Grades III and IV	4.48	Sherbet and Lakshmi (1974)[b]
Human meningioma	4.73	
Human fetal brain cells	4.38	
Escherichia coli	5.6	Sherbet and Lakshmi (1973, 1975)[a]
Human erythrocytes	5.6	
Mouse erythrocytes	5.8	Just et al. (1975)
Rabbit erythrocytes	6.0	
Boar spermatozoa	6.5	Moore and Hibbit (1975)
Turnip yellow mosaic virus	4.0	
Cowpea chlorotic mottle virus	4.1	
Qβ bacteriophage	4.1	Rice and Horst (1972)[c]
Satellite of tobacco necrosis virus	4.3	
Brome mosaic virus	6.8	
Poxviruses	4.0	Hill et al. (1963)[d]

[a] pI determined at 25 °C using 0.008 M Ampholine.
[b] pI determined in 0.009 M Ampholine.
[c] pI measured at 2.0% Ampholine.
[d] pI determined by electrophoresis (= pIE).
[e] From Sherbet (1978).

latter does not exceed a value of 27 mV. I have tried to plot the values given by Sherbet, in order to see how accurate the data are. Figure 5.4 shows the correlation between pI and surface charge densities for eleven

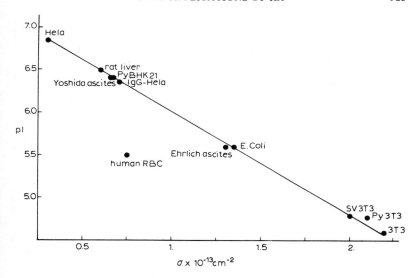

Fig. 5.4. Plot of isoelectric points (pI) of cells, as determined by IEF, vs. their respective charge densities (σ). Notice the extremely good fit for all cell lines and the huge deviation of red blood cells. Drawn from data tabulated by Sherbet (1978). (From Righetti et al., 1980.)

mammalian cells. The fit is remarkably good, as practically 10 cell lines, ranging in pI values from ca. 4.5 to ca. 7.0, fall on the linear plot (which gives decrements of σ of $0.85 \times 10^{-13}/cm^2$ for increments of 1 pI unit). However, there is one cell type, human RBCs, which falls completely out of the linear relationship. Given a pI of 5.5, as reported by Just et al. (1975a), the charge density should be about twice as much. If the σ value is correct, then the pI should be about 6.3. Given the fact that RBCs are perhaps the most extensively characterized mammalian cells, this very poor fit is quite suspicious. If we now plot the ζ potential against charge density, a linear relationship is again obtained (Fig. 5.5), however with a considerable scatter of points (50% of the points deviate considerably from linearity).

Notwithstanding this large body of experimental data, IEF of cells still has not been quite established. For instance, Catsimpoolas and

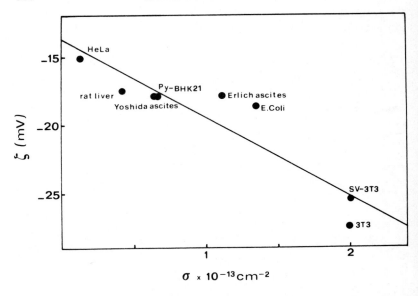

Fig. 5.5. Plot of electrokinetic potential (ζ potential) of different cell lines vs. their respective charge densities (σ). Drawn from data tabulated by Sherbet (1978). (From Righetti et al., 1980.)

Griffith (1977), upon focusing mouse spleen lymphocytes, have reported that as soon as the cells are focused they start to defocus and are finally lysed. They hypothesize that, in the pI region, changes in the membrane occur that could alter the pI towards a more acidic value so that the cells have to seek a new pI position. This may occur repetitively until extensive damage to the membrane causes lysis of the cells. It is a fact that, even though many workers (Hirsch and Gray, 1976 and Manske et al., 1977) have claimed high cell viabilities (up to 90%) after IEF, as measured by dye exclusion tests, in reality viabilities in terms of plating efficiencies are usually low (Boltz et al., 1977), or recultured cells after IEF remain dormant for very long periods (Longton et al., 1974). This has led many workers to suspect that carrier ampholytes might have some degree of cytotoxicity. Therefore, the question of whether membrane stability and hence cell viability can be maintained under electrofocusing conditions should be analyzed critically. There also remains the question of whether the measured cell pI values are 'true' isoelectric points, representing a balance between positive and negative groups in the cell surface. For instance, the pI of *E. coli* is given as 5.6 (Sherbet, 1978), practically the same as the pI of RBCs (see Fig. 5.4). However, in *E. coli*, the ratio of negative to positive charges is 2:1 ($2.899 \times 10^{-13}/cm^2$ carboxyl groups versus $1.433 \times 10^{-13}/cm^2$), while in RBCs the ratio of negative to positive groups is 25:1 (Seaman, 1975), i.e., enormously higher. How these two cell types can exhibit the same pI remains a mystery.

Additionally, three other parameters that further increase the uncertainty of the data should be considered: (a) at the steady-state, focused carrier ampholytes represent a medium of very low 'ionic strength' (now known to be in the range 0.5 to 1 mg ion/l for 1% Ampholine) (Righetti, 1980); (b) ampholytes for IEF are capable of chelating doubly positively charged metal ions (Galante et al., 1975); (c) they have been demonstrated to form complexes with polyanions, such as nucleic acids (Galante et al., 1976) and sulphated and carboxylated polysaccharides (Righetti and Gianazza, 1978; Gianazza and Righetti, 1978), including polyglutamate and polyaspartate. In the light of these data, in fact, McGuire et al. (1980) have recently demonstrated a pH-dependent binding of carrier ampholytes to the surface of RBCs: the binding is very strong at pH 4, weak but still appreciable at pH 5 and abolished at pH 6 and above. The same phenomena could be reproduced with PEHA, a polyamine which could typically represent the backbone of carrier ampholytes. On the basis of these observations, McGuire et al. (1980) have drawn a model depicting a segment of an Ampholine molecule bound, via a stretch of four protonated nitrogens (each two methylene groups apart), to four negative charges (possibly sialic acids or sulphate residues attached to the carbohydrate side chains) on the cell surface (see Fig. 5.6). The bound Ampholine might change the surface charge characteristics, thus altering the cell stability and contributing to the decrease in cell viability. It is a fact that, even in early studies by the Sherbet group (1978), all cells focusing around pH 5 (4.70–5.30) (Yoshida ascites, Ehrlich ascites, Py cells, HeLa cells, normal liver cells), where binding of Ampholine is still appreciable, were in fact all non-viable cells. These data are further corroborated by Hammerstedt et al. (1979), who found that the apparent isoelectric point (pI_{app}) of sperm cells is not altered when the cells are

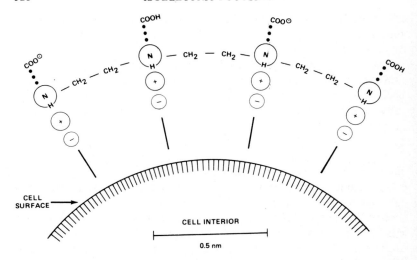

Fig. 5.6. Hypothetical model of the interaction of one Ampholine species with the cell surface. A segment of an Ampholine molecule, represented by a stretch of four protonated nitrogens, is depicted bound to four negative charges (possibly sialic acid residues) on the cell surface. The carboxyls in the carrier ampholyte are drawn facing away from the plasma membrane. At pH 4, these weak carboxyls (average $pK = 4$) would be 50% protonated. The curvature of the cell surface, over a distance of 1 nm, has been grossly exaggerated for easy visualization. In reality, over such a short distance, it should be almost planar. (From McGuire et al., 1980.)

inserted in a prefocused column at either pH 7 or pH 8. However, if cells were added to the prefocused gradient at a low pH, the pI_{app} was 2 pH units lower, suggesting either extensive binding of acidic carrier ampholytes, or cell surface damage due to low pH, or peeling-off of basic proteins adsorbed onto the cell glycocalyx region. Actually, binding of Ampholine is more consistent with an increase of pI_{app}, rather than with its decrease, and could explain the rather high pI_{app} of most cells, reported to be in the range 6.0–6.8. On the other hand, when focusing is performed in citrate buffer (Boltz et al., 1978), most of the cells investigated appeared to have pI values in the pH range 3.5–4.7. This type of focusing should be further explored, as citrate allows higher medium ionic strengths, prevents cell aggregation and, being oligoanionic, should not bind to the cell surface (Pittz et al., 1977). IEF of cells in the presence of 'poor' zwitterions, such as glycine, taurine or trimethylaminopropane sulphonate, able to maintain a constant osmolarity in a given pH interval, has also been suggested (McGuire et al., 1980).

Given these highly controversial aspects, I feel that IEF, as a method of cell separation and characterization, should be taken with great caution. It might turn out that its pitfalls and limitations outweigh any of its possible advantages.

5.3. IEF of ferritins

Ferritin is a ubiquitous iron-storage protein present in every mammalian tissue, in high amounts in iron-deposit tissues such as liver, spleen and the reticuloendothelial system, and at lower levels in heart, kidney and spleen. Small, but significant amounts of ferritin are also found in serum (Drysdale, 1977b).

Ferritin is composed of a proteinaceous shell (apoferritin) and of a core of precipitated iron (III) hydroxide. The protein envelope has an $\bar{M}_r$ of 450,000 and is shaped like a hollow sphere. Its outer diameter is of 12 nm, while the central cavity has a diameter of 7.0 nm (Hoare et al., 1975). Since the inner iron core is completely shielded from the outer environment, it does not contribute to the surface properties of ferritin, such as electrophoretic and immunological properties (Harrison et al., 1974). The protein is composed of 24 subunits, once believed to be identical and of similar size (19,000 daltons) (Bryce and Crichton, 1971) but recently demonstrated, by IEF-SDS analysis, to be of different mass and, possibly, slightly different amino acid composition (Drysdale, 1977; Arosio et al., 1978b). X-Ray analysis has demonstrated that six channels connect the outer surface with the inner cavity: each channel has a maximum width of 1.6 nm and a minimum width of 0.7 nm (Banyard et al., 1978). Ferritin also displays catalytic properties, since it can adsorb and oxidize iron, thus acting as a redox system (Stefanini et al., 1976; Melino et al., 1978). Its inner core can be totally devoid of iron or be saturated with up to 4,000 iron atoms (Niitsu and Listowsky, 1973). Ferritin also displays a marked size heterogeneity, due to its tendency to form dimers, trimers and higher aggregates, which are readily separated by gel filtration or electrophoresis (Suran and Tarver, 1975; Righetti et al., 1971).

IEF analysis has been instrumental in leading to the present understanding of the structure and function of tissue ferritins. By IEF each ferritin has been demonstrated as microheterogeneous, and to be composed of a closed family of isomers with a narrow pI distribution (Drys-

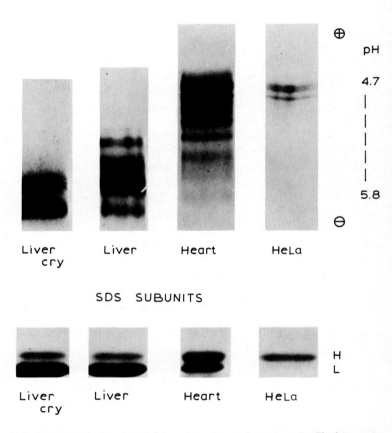

Fig. 5.7. IEF profile of isoferritins from liver, heart and HeLa cells. The lower part shows the subunit distribution of the same tissue isoferritins as determined by SDS electrophoresis (cry = crystallized). (From Drysdale, 1977b.)

dale, 1974). The IEF profiles are tissue specific and they all present a certain degree of heterogeneity. As shown in Fig. 5.7, the most acidic and least heterogeneous is HeLa ferritin, while the most alkaline is liver ferritin. This microheterogeneity is not due to different levels of iron in the molecule, nor is it an artefact of the IEF technique, since the bands do not redistribute upon refocusing (Urushizaki et al., 1971; Ishitani et al., 1975). Indeed the heterogeneity is due to structural differences, since all ferritins analyzed so far are resolved into two subunits, whose ratio varies

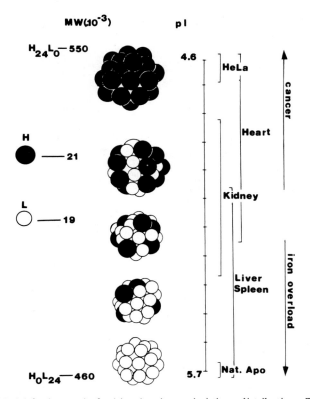

Fig. 5.8. Model for human isoferritins showing typical tissue distributions. The two subunits H (heavy) and L (light) are assembled in the protein shell in different ratios. The size differences in the multimeric structures are exaggerated for emphasis. (From Drysdale, 1977b.)

in different tissues and is related to their isoferritin content (Arosio et al., 1978). A model has thus been derived showing how isoferritins are composed of variable proportions of two subunit types: an L subunit (light, $\bar{M}_r$ 19,000) and an H protomer (heavy, $\bar{M}_r$ 21,000). As shown in Fig. 5.8, the low pI species (pI $\simeq$ 4.6, e.g., HeLa ferritin) is composed prevalently of H subunits, while the high pI bands (pI $\simeq$ 5.7, e.g., natural apoferritin or serum ferritin) are made almost exclusively of L protomers (Drysdale, 1977b; Arosio et al., 1978). H and L subunits appear to have also different immunological reactivity since antibodies produced against ferritins of the H-type give only a weak cross reaction with L-type molecules (Arosio et al., 1976). There also appears to be a different physiological role of the various isoferritins: according to Wagstaff et al. (1978) acidic ferritins (H-rich species) display a much faster iron uptake than higher pI ferritins (L-rich bands) (see Fig. 5.9). Moreover, L-type ferritins usually increase

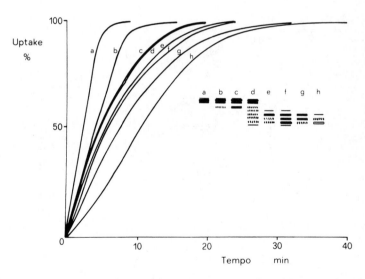

Fig. 5.9. Iron binding curves of different human apoferritins. The insert shows the IEF profile of the different tissue ferritins. The most acidic isoferritins (a to d) (which are also the ones richer in the heavier subunit, H) exhibit the fastest iron uptake. (From Wagstaff et al., 1978.)

under iron overload, while H-rich ferritins are most abundant in neoplasia and during iron-deficiency (Drysdale, 1979).

Serum ferritin, recently discovered by radioimmunoassay (Addison et al., 1972), has today assumed an important clinical and diagnostical role. Its structure is quite similar to that of tissue ferritins, but its composition in isoferritins, as revealed by IEF, changes in different pathological conditions. Moreover, some minor but important differences suggest that serum ferritin has an origin and physiological role different from its tissue counterparts. These differences are: (a) serum ferritin contains much less iron (5%) than tissue ferritins (av. 20%), which suggests a possible similarity with native apoferritin (Arosio et al., 1977) (Fig. 5.10).

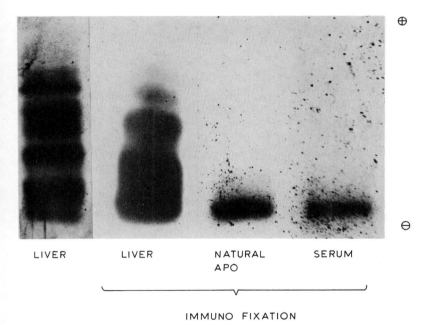

IMMUNO FIXATION

Fig. 5.10. Comparison of the IEF profiles of serum ferritin (iron overload), native apoferritin and liver ferritin (as detected by Coomassie stain, far left, and by immuno fixation). Serum ferritin and native apoferritin appear to be present as a main, more alkaline band, probably representing the homopolymer of the L subunit.
(From Drysdale et al., 1977.)

Its iron content is increased in some pathological conditions, such as liver necrosis; (b) serum ferritin contains carbohydrate while the tissue components do not (Worwood et al., 1977); and (c) from a point of view of immunological reactivity, in healthy individuals it cross-reacts with higher affinity with liver ferritin antisera, while in cases of neoplasia it has higher affinity for antisera against heart ferritin. The serum isoferritin profile resolved by IEF follows the same pattern: it resembles liver components in physiological conditions and heart isoferritins during neoplastic processes (Halliday et al., 1976). In general, serum ferritin levels are markedly increased in carcinomas, leukemias, myelomas and hepatomas (Niitsu et al., 1975). According to Gropp et al. (1978) serum ferritin can be a good tumor marker, similar and complementary to the carcino embrionic antigen.

5.4. IEF of hemoglobins

IEF of hemoglobins (Hb) (for recent reviews see Bunn, 1977 and Basset et al., 1982) has now become a standard tool in hematology departments for mass screening of blood samples and for the diagnosis of hemoglobinopathies. As of 1976, a data collection of Hb mutants included 269 variants with the following distribution: 80 in α-chains, 166 in β-chains, 14 in γ-chains and 9 in δ-chains (Bunn et al., 1977). Today, over 300 variants have been described. Fortunately, the majority of known variants are not associated with any apparent clinical sequelae. In addition, nine changes in chain length are described. These include six variant β-chains resulting from the deletion of from one to five adjacent amino acid residues within the chain and one β-chain with ten additional residues at its carboxyl end. Since the Hb γ, δ and β-genes are contiguous to each other, fusion gene products consisting of both Hb δ and Hb β (Hb Lepore), Hb β and Hb δ (Hb Miyada) and Hb γ and Hb β (Hb Kenia) fragments can arise from unequal crossover during normal meiosis. Two variant chains with additional residues at the carboxyl end are also known.

Genetic mutants.   IEF has been instrumental in the mapping of genetic variants. Its use was first described in 1971 by Drysdale et al., utilizing gel cylinders. Figure 5.11A shows a typical IEF pattern of Hbs from patients homozygous for HbC (β 6Glu → Lys) and HbS (β 6Glu → Val) and from a double heterozygous ASI. All samples contain traces of HbA_2 ($\alpha_2\delta_2$) while the two Hb C and S contain discrete amounts of fetal Hbs F_I and F_{II}. Subsequently, Jeppsson and Berglund (1972) introduced thin-layer separations (TLIEF), which are today, by far, the most widely used methods. Fig. 5.11B shows the detection of a new variant, Hb Malmö, by TLIEF. Several new mutants have been detected and characterized by IEF: e.g., Hb Cochin-Port-Royal (Wajcman et al., 1975); Hb Wood (Taketa et al., 1975); Hb Bicêtre (Wajcman et al., 1976); Hb Pontoise (Thillet et al., 1977); Hb Rothschild (Gacon et al., 1977a) and Hb Djelfa (Gacon et al., 1977a). Rosa's group in Creteil has extensively used TLIEF for mass screening of Hbs (Monte et al., 1976; Basset et al., 1978). These authors have constructed a map of more than 80 human Hb

(A)

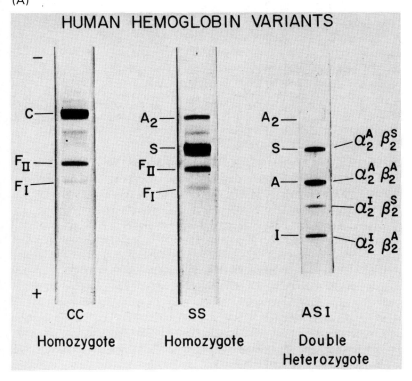

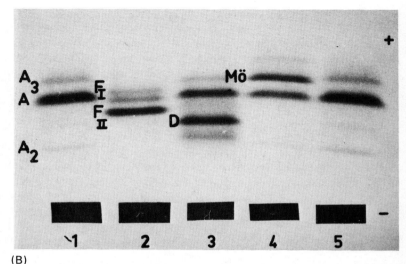

(B)

Fig. 5.11. (A) IEF of genetic Hb variants. IEF conditions: 5% acrylamide gels, 2% Ampholine pH 6–8. From left to right: HbC homozygote; HbS homozygote and HbASI heterozygote. (From Drysdale et al., 1971.) (B) IEF of human Hbs in the pH range 6–10. (1) and (5) normal adult blood; (2) cord blood; (3) blood of a patient heterozygous for HbD Punjab; (4) blood of a patient heterozygous for Hb Malmö. (From Jeppsson and Berglund, 1972.)

variants occupying nearly 50 distinct pI positions (Fig. 5.12). As seen in this map, most Hbs focus within less than 1 pH unit, from pI ca. 6.6 up to pI 7.5. Three clusters can be distinguished in this map: a group of 'C-like' variants, a second of 'S-like' and a third of 'A-like'.

The 'C-like' zone is the least crowded, and allows resolution of most of the mutants focusing in this region. The 'S-like' region is quite crowded, since it contains more than 30 other mutants with very similar pI values. Since Hb-S is the most widely distributed abnormal hemoglobin in the world and since the homozygous S/S state, which is responsible for sickle-cell anemia, could be confused with other, less critical disorders (e.g., double heterozygotes for HbS/D or HbS/P or HbS/G), the IEF analysis has to be supplemented with other tests. Also the 'A-like' region is over-crowded and contains several important pathological hemoglobins, like Hb Köln and Hb St. Etienne (which tend to lose their heme groups), or

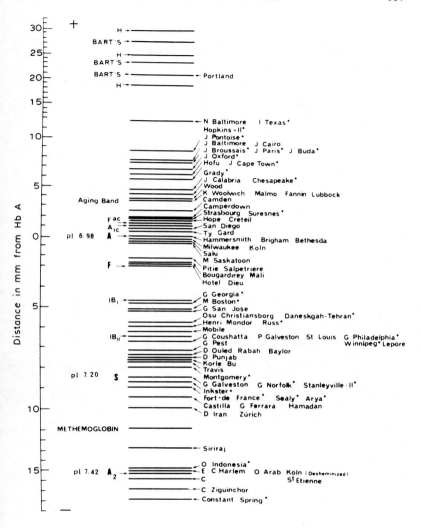

Fig. 5.12. IEF map of 79 human Hb variants exhibiting nearly 50 distinct pI positions. Asterisk (*) indicates α-chain mutations. IB$_I$ and IB$_{II}$ are the ferrous-ferric hybrids of HbA. (From Basset et al., 1978.)

the M hemoglobins, which cause congenital cyanosis, since they are readily oxydized to met forms, or the high-oxygen-affinity variants (e.g., Hbs Bethesda, Brigham, Olympia) which are usually accompanied by erythrocytosis or chronic hemolysis. Thus, even in this last case IEF analysis should be confirmed by complementary investigations. Nevertheless, the increment in resolution obtained by IEF runs as compared with conventional cellulose acetate electrophoresis at alkaline pH or citrate agar separations is impressive.

A comparison between the results obtained by TLIEF and by a classical electrophoretic system can be appreciated in Fig. 5.13. Four variants of HbA having the same type of amino acid substitution (a glycine replaced by aspartic acid) were admixed with Hb-A and run on TLIEF and on cellulose acetate strips. Isoelectric focusing distinguished the five expected bands, only two of which are seen upon electrophoresis (Basset et al., 1982).

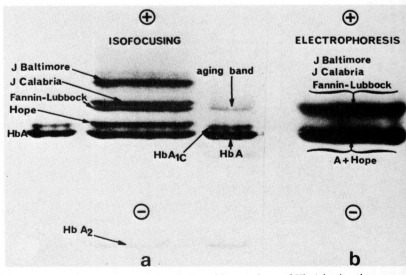

Fig. 5.13. Migration patterns of a mixture of four variants of Hb-A having the same type of amino-acid substitution: (a) by isoelectric focusing in a pH gradient range 6–9; (b) by electrophoresis on cellulose-acetate strips at alkaline pH. (From Basset et al., 1982.)

The Bohr effect. Perutz (1970) has shown that deoxyHb is held in a constrained or taut (T) structure by means of inter- and intra-subunit salt bonds. Upon the addition of ligands such as oxygen or CO to the heme groups, the salt bonds are broken and the molecule snaps into a relaxed (R) conformation, in which the affinity of the hemes for oxygen is markedly increased. The T structure has a relatively high affinity for protons and thus deoxyHb has a significantly higher pI than oxyHb. This can be easily demonstrated by IEF (Bunn and McDonough, 1974): the purple band of deoxy Hb has a pI of 7.07, which is about 0.12 pH units above that of the red carboxyhemoglobin band (pI 6.95). The Bohr effect could be exploited to enhance the separation between HbA and mutants of very close pI, provided the latter have a subnormal Bohr effect. This could in fact be obtained with Hb Syracuse (Jensen et al., 1975). This is a variant having very high oxygen affinity and a Bohr effect that is half the normal and it cannot be separated from HbA by conventional electrophoresis. Partial separation is obtained in the oxy form, which is greatly enhanced when the two deoxy species are run in IEF (Fig. 5.14).

IEF could also be used for an accurate measurement of the intrinsic alkaline Bohr effect of normal and mutant human Hbs (Poyart et al., 1981). This has been possible by measuring the differences between the pI values of different Hbs when completely deoxygenated and when fully liganded with CO [Δ pI (deox.–ox.)]. This value, in HbA is 0.17 pH units, which corresponds to a difference of 1.20 positive charges between the oxy and deoxy states of the tetrameric Hb. The ΔpI (deox.–ox.) values of mutant or chemically modified Hbs carrying an abnormality at the N- or C-terminus of the α-chains are decreased by 30% as compared with HbA. When the C-terminus of the β-chain is altered, as in Hb Nancy, the ΔpI value was decreased by as much as 70% as compared with HbA (Fig. 5.15). Moreover, even in Hbs exhibiting a normal $\Delta \log pI_{50}/\Delta pH$ value in solution, the ΔpI decreases with increasing pI (ox.), indicating the pH-dependence of the intrinsic Bohr effect. These IEF data are in close agreement with the estimated respective roles of the two major Bohr groups, Val-1α and His-146β, at the origin of the intrinsic alkaline Bohr effect (Kilmartin et al., 1973; Perutz et al., 1980). In mutant Hbs it was also demonstrated that the ΔpI could be decreased or abolished when the substitution leads to a destabilization of one of the two quaternary

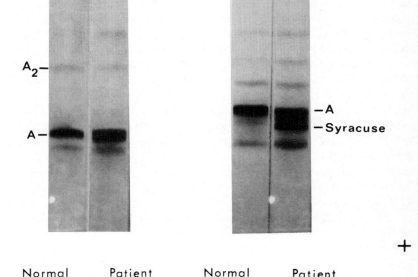

Fig. 5.14. Comparison of gel electrofocusing patterns of normal hemolysate and hemolysate of a patient with hemoglobin Syracuse. *Left:* Hemoglobin with CO. *Right:* Deoxygenated hemoglobin. The improved separation of Hb Syracuse from Hb A following deoxygenation is a reflection of the fact that Hb Syracuse has a decreased Bohr effect. (From Jensen et al., 1975.)

structures of the tetramer and/or to a low value of the L-allosteric constant (Poyart, 1981).

Red-ox state in the ligand.   Some mutant Hbs, which carry a mutation is proximal (F8) or distal (E7) histidines, tend to either oxidize spontaneously, or to lose their heme group. To the first class belong M Hbs: four of them correspond to the substitution of a Tyr residue for each of the 4 α- and β-histidines, giving rise to a permanent bond with the oxi-

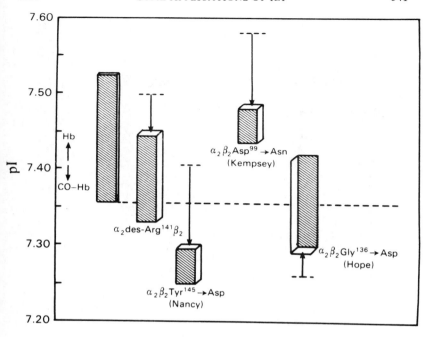

Fig. 5.15. Schematic representation of ΔpI (deox.–ox.) of various human hemo-globins as measured by isoelectric focusing. Compared with HbA$_0$ (left-hand block) all these haemoglobins have lowered ΔpI and also low Bohr coefficients in solution. The broken lines indicate the missing part of the ΔpI if HbA$_0$ is taken as the refer-ence. The arrows indicate which of the two pI values, either that of the deoxy or the oxy form, appears to be preferentially altered. (From Poyart et al., 1981.)

dized heme. To the second group belong unstable Hbs such as Gun Hill (Bradley et al., 1967) and St. Etienne (Beuzard et al., 1972) in which the proximal His of β-chains does not exist, being deleted in one case and substituted by glutamine in the other; consequently there is no heme bound to these β-chains. The different red-ox states of the iron in the heme group of these M Hbs can markedly improve the IEF separation of these species. Thus when Hb M Saskatoon (β 63 His → Tyr) (Bunn, 1977) or HbM Bicêtre (β 63 His → Pro) (Wajcman et al., 1976) are analyzed by IEF, they appear as a distinct brown band having a slightly higher pI than

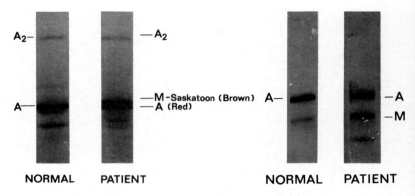

Fig. 5.16. Gel electrofocusing patterns obtained on hemolysate of patient with Hb M Saskatoon. Left, Pattern obtained on oxygenated specimens showing separations of red Hb A band and brown Hb M Saskatoon band. Right, Pattern obtained after ferricyanide, showing wider separation of Hbs A and M Saskatoon. (From Bunn et al., 1977.)

the red HbA zone. However, the two species are barely separated, allowing for a poor quantitation of the relative ratios by densitometry. If, on the other hand, both Hbs are oxidized with ferricyanide, a much wider separation of M and A hemoglobins is achieved (Fig. 5.16). Met HbA has a much higher isoelectric point (more positive surface charge) than Met HbM, since in the latter the Fe^{3+} atoms of the abnormal subunits are neutralized by an internal ligand.

Valence hybrids.   Not only mutant Hbs can be found under oxidized form, but also normal human adult Hb can slowly undergo spontaneous auto-oxidation in vitro. In vivo, however, in healthy individuals, Hb is usually in a reduced form, since, in order to fulfill its physiological role, the heme iron must remain in a ferrous state. Heme that has been oxidized to the ferric form (met Hb) is unable to bind oxygen. Despite the significant difference in charge between ferro Hb ($\alpha_2 \beta_2$) and its oxidized forms ($\alpha_2^+\beta_2$, $\alpha_2\beta_2^+$, $\alpha_2^+\beta_2^+$) these species have never been isolated by conventional electrophoresis. In contrast, gel IEF has permitted clear-cut separations of fully oxidized and partially oxidized intermediates. After HbA is partially oxidized with an agent such as ferricyanide, three new

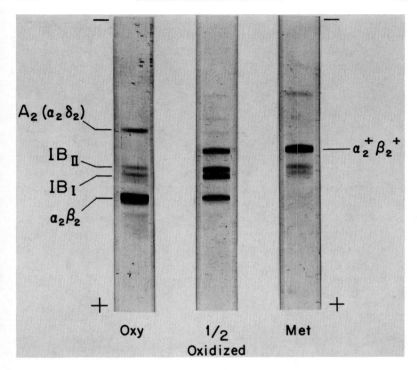

Fig. 5.17. Separation of oxidized forms of human hemoglobin. IEF fractionations were performed on hemolysates in 0.05 M phosphate buffer (pH 7.1). *Left*, no oxidant added; *middle*, hemoglobin partially oxidized by the addition of a half equivalent amount of ferricyanide; *right*, hemoglobin, fully oxidized by the addition of an excess of ferricyanide. Spectra of these specimens showed the presence of 0%, 50% and 100% ferrihemoglobin, respectively. (From Bunn and Drysdale, 1971.)

bands can be visualized (Fig. 5.17). The fully oxidized Hb ($\alpha_2^+\beta_2^+$) appears as a brown band with a pI of 7.20, 0.25 pH units above that of HbA (Bunn and Drysdale, 1971). Equidistant between fully oxidized Hb and unoxidized Hb ($\alpha_2\beta_2$) are two closely spaced bands, designated IB$_I$ and IB$_{II}$, which have been shown by Park (1973) to be $\alpha_2\beta_2^+$ and $\alpha_2^+\beta_2$, respectively. The fact that these two components are separable in this high-resolution system adds to a large body of evidence indicating differences in the heme environments of the α- and β-chains, inducing a slightly different net surface charge (conformational transitions) in otherwise identical total charge proteins.

As it turns out, the story is much more complex than that. In theory, there should be ten valence hybrids obtainable by partial oxidation of $\alpha_2\beta_2$ (Perrella et al., 1981). Since only four were found means that these are the only species stable under those experimental conditions. By quenching a Hb solution partially saturated with CO into a hydro-organic solvent containing ferricyanide, Perrella et al. (1981) were able to produce a population of partially oxidized and CO-bound Hb molecules. When these valence hybrids were analyzed by IEF at temperatures as low as $-23\,^\circ$C, all the intermediate species were resolved (Fig. 5.18). This is an excellent example of the use of IEF as a structural probe.

Subunit exchange.   Under physiological conditions (pH 7.4, 37 $^\circ$C) the oxygenated hemoglobin tetramer dissociates readily and reversibly into $\alpha\beta$ dimers:

$$\text{Oxy, Carboxy:}\quad \alpha_2\beta_2 \rightleftharpoons 2\alpha\beta \qquad K_{4,2} \simeq 2 \times 10^{-6}\,\text{M}$$

Therefore, in a solution containing two different Hbs (such as Hbs A and S), the hybrid Hb, $\alpha_2\beta^A\beta^S$ should be formed. Indeed, if equal amounts of Hbs A and S are present, 50% of the tetramers should be hybrids. However, these hybrids cannot be demonstrated by conventional electrophoresis or chromatographic techniques since, during the separation process the hybrid tetramer will dissociate into unlike dimers of different charge that will then sort with like dimers. With the use of gel IEF, hybrid species can be demonstrated due to the fact that deoxygenated (unliganded) Hb dissociates much less readily into dimers:

$$\text{Deoxy:}\quad \alpha_2\beta_2 \rightleftharpoons 2\alpha\beta \qquad K_{4,2} \simeq 3 \times 10^{-12}\,\text{M}$$

Bunn and McDonough (1974) have indeed demonstrated the existence of such asymmetrical Hb hybrids, by mixing deoxy Hbs A and C (Fig. 5.19). By following the decay of the hybrid band during prolonged focusing, they could also measure the first-order rate constant for the

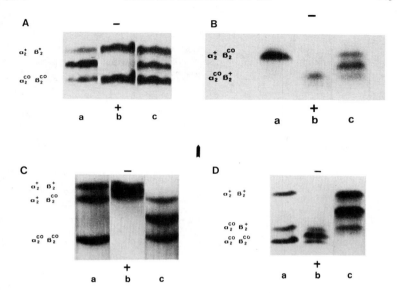

Fig. 5.18. Patterns obtained after isoelectric focusing at $-23\,°C$ of various hemo-globin species. The arrow indicates the direction of the increasing pa_H in the gradient. A, a, $\alpha_2^+\beta_2^+$ is mixed and incubated with $\alpha_2^{CO}\beta_2^{CO}$ for 5 min at $0\,°C$. The mixture is then quenched and separated at $-23\,°C$ as described in the text. b, $\alpha_2^+\beta_2^+$ is mixed with $\alpha_2^{CO}\beta_2^{CO}$ at $-23\,°C$, pa_H 8.5, and the mixture is incubated for 30 min under these conditions prior to focusing. c, the same as b, but at pa_H 7.5. B, a, $\alpha_2^+\beta_2^{CO}$; b, $\alpha_2^{CO}\beta_2^+$; c, $\alpha_2^+\beta_2^{CO}$ and $\alpha_2^{CO}\beta_2^+$ mixed and incubated at $0\,°C$ for 5 min prior to quenching at $-23\,°C$. C, a, $\alpha_2^+\beta_2^+$, $\alpha_2^+\beta_2^{CO}$, and $\alpha_2^{CO}\beta_2^{CO}$ mixed and incubated at $-23\,°C$, pa_H 8.5, for 5 min prior to focusing, b, $\alpha_2^+\beta_2^+$ and $\alpha_2^+\beta_2^{CO}$ mixed and incubated at $0\,°C$ for 5 min prior to quenching at $-23\,°C$. c, $\alpha_2^+\beta_2^{CO}$ and $\alpha_2^{CO}\beta_2^{CO}$ mixed and incubated at $0\,°C$ for 5 min prior to quenching at $-23\,°C$. D, a, $\alpha_2^+\beta_2^+$, $\alpha_2^{CO}\beta_2^+$, and $\alpha_2^{CO}\beta_2^{CO}$ mixed at $-23\,°C$, pa_H 8.5, and incubated for 5 min under these conditions prior to focusing. b, $\alpha_2^{CO}\beta_2^+$ and $\alpha_2^{CO}\beta_2^{CO}$ mixed and incubated at $0\,°C$ for 5 min prior to quenching at $-23\,°C$; c, $\alpha_2^{CO}\beta_2^+$ and $\alpha_2^+\beta_2^+$ mixed and incubated at $0\,°C$ for 5 min prior to quenching at $-23\,°C$. (From Perrella et al., 1981.)

dissociation of deoxy Hb tetramer into dimers. This constant was found to be $0.3\ h^{-1}$ at pH 7.2, $24\,°C$, but it differs by one order of magnitude from estimates made by Nagel and Gibson (1971). Similar results have

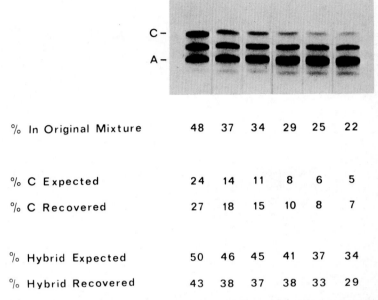

% In Original Mixture	48	37	34	29	25	22
% C Expected	24	14	11	8	6	5
% C Recovered	27	18	15	10	8	7
% Hybrid Expected	50	46	45	41	37	34
% Hybrid Recovered	43	38	37	38	33	29

Fig. 5.19. Demonstration of the hybrid hemoglobin $\alpha_2 \beta^A \beta^C$. Mixtures containing varying amounts of oxyhemoglobin A and oxyhemoglobin C were deoxygenated and then applied to the gels anaerobically. Sixty minutes later they were photographed and scanned at 555 nm. The amount of C and hybrid hemoglobins recovered (determined from the scanning data) is compared with that expected from the binomial distribution $a^2 + 2ab + b^2 = 1$. (From Bunn and McDonough, 1974.)

been obtained by Perrella et al. (1978) with liganded Hbs, by performing IEF at sub-zero temperature ($-20\,^{\circ}$C). The fact that short-lived intermediates of biochemical interest can be stabilized and analyzed by IEF at such low temperatures could be of general interest in the study of enzyme reaction intermediates.

Globin chain IEF. For a correct evaluation of thalassemia syndromes, heme-free globins have to be separated by chromatographic or electrophoretic means and their relative ratios assessed. Righetti et al. (1979) have devised a new method for globin chain separation by IEF in the presence of 8 M urea and detergents (1–3% Nonidet P-40). The detergent, probably by binding to denatured globins, greatly increases the separation between β- and γ-chains, thus allowing for a proper densitometric and fluorographic quantitation of their relative ratios (Fig. 3.38). In addition, the detergent splits the γ-zone into two bands, which have been demonstrated to be the two phenotypes of the human fetal γ-globin chains, differing in the content of a neutral amino acid in position 136 in the polypeptide, this residue being Ala in A_γ and Gly in G_γ (Comi et al., 1979; Saglio et al., 1979). According to Righetti et al. (1980) this 'Nonidet effect' might be due to preferential binding of the detergent micelle to the hydrophobic stretch ^{133}Met to ^{141}Leu in A_γ chains (Fig. 5.20). upon binding the detergent could sorb the ^{132}Lys in its Stern-layer or bury it within the micelle, thus inducing a total loss of one proton from otherwise identical-charge phenotypes. This unique separation between A_γ- and G_γ-chains has made possible studies on the switch from G_γ to A_γ-chain synthesis during development and maturation of erythroblasts (Papayannopoulou, 1981). IEF separation of globin chains has also been successfully applied to the antenatal diagnosis of β^0 thalassemia (Gianni et al., 1981) aimed at genetic counseling of couples at risk in thalassemic areas, such as in Sardinia. IEF of denatured chains in 8 M urea/NP-40 gels, has also been instrumental to the study of human embryonic hemoglobins Gower I and Gower II. In pH 6–9 ranges, optimal resolution of α, ζ, ϵ, β and γ-chains could be achieved in the

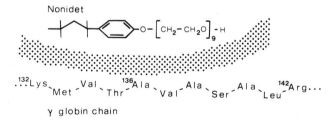

Fig. 5.20. Proposed model for the 'Nonidet effect' in the separation of A_γ and G_γ globin chains by IEF. It is suggested that in A_γ chains the detergent micelle could bind preferentially to the hydrophobic stretch ^{133}Met to ^{141}Leu. Upon binding, the detergent could sorb the ^{132}Lys in its Stern-layer or bury it within the micelle, thus inducing a total loss of one proton from otherwise identical phenotypes. (From Righetti et al., 1980.)

absence of detergent (Presta et al., 1982). The resolution of the human globin chain clusters ($\epsilon - G_\gamma - A_\gamma - \delta - \beta$ and $\zeta - \alpha_2 - \alpha_1$) is extremely important for genetic studies, since it represents one of the most thoroughly studied examples of coordinated developmental regulation of a set of related genes (Weatherall and Clegg, 1981).

Umbilical cord blood screening. For a screening of homozygous β-thalassemia and of β-thalassemia trait, globin chain IEF still represents one of the simplest and most reliable methods, however, the sample preparation for IEF analysis (acid acetone powder) is still a lengthy and cumbersome process. Ideally, the method of choice would combine an IEF analysis performed on intact Hb species. By using umbilical cord blood, where only 3 major hemoglobin components are present (HbF, HbA and HbF acetylated) (HbF_{ac}) it should be possible to perform thalassemia screening provided a good separation is obtained between HbA and HbF_{ac}, which have minute differences in isoelectric points (pI) (Drysdale et al., 1971). Cossu et al. (1982) have reported a modified IEF technique by which it is possible to perform a large screening of a population at risk for β-thalassemia, by substituting the β/γ ratio obtained upon IEF of globin chains with the HbA/HbF_{ac} (or HbF/HbA) ratios obtained by IEF of intact Hbs from cord blood of newborns. In order to improve the separation, the pH 6−8 Ampholine was admixed with an equimolar mixture of 'separators', namely 0.2 M β-alanine and 0.2 M 6-amino caproic acid.

As shown in Fig. 5.21, excellent resolution is obtained among the three bands: the method is simple, can unambiguously detect any thalassemic condition and can be easily performed on a routine basis for each delivery in a neonatal unit. In an extensive screening of newborns in the Sardinia island, it was found that all the samples containing less than 10% HbA at birth were associated with β-thalassemia heterozygosity (Cossu et al., 1982).

IEF of minor hemoglobins. In man it is especially important to measure three of these minor Hb components, since changes in their levels can play a role in the diagnosis of some diseases. For example, the HbA_2 ($\alpha_2 \delta_2$) level is about twice as high in the blood of carriers of the β-thalassemia trait as in the blood of non-carriers (Kunkel et al., 1957).

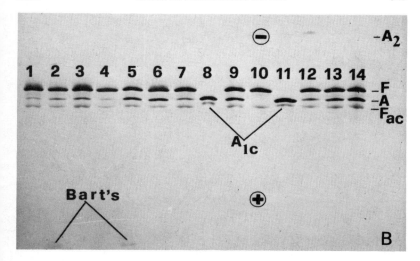

Fig. 5.21. IEF of cord blood hemolysates in a pH 6–8 gradient added with 0.2 M
β-alanine and 0.2 M 6-aminocaproic acid. Experimental: 6% T, 4% C gel; run: 15
min at 400 V, 90 min at 15 W, 1500 V max; protein load: 20 μg. Levels of HbA in
samples: 1: 9.1%; 2: 11.4%; 3: 14.7%; 4: 7.9%; 5: 31.1%; 6: 28.9%; 7: 18.9%;
9: 31.3%; 10: zero; 13: 28%; 14: 34.3%. Samples 8 and 11: normal adult lysates
(notice the bands of HbA$_{1c}$); sample 12: adult β^+ thalassemic homozygous. (From
Cossu et al., 1982.)

HbF ($\alpha_2\gamma_2$), which is the major component in fetal blood and at birth, is
normally present at trace levels in the adult but may be elevated in
genetically determined or acquired disorders, including thalassemia,
hereditary persistence of HbF and leukemias (Weatherall et al., 1974).
The third minor component of importance in pathology is HbA$_{1c}$, which
is a glycosylated HbA whose level is elevated in diabetes mellitus (Rahbar,
1968). Precise quantitation of HbA$_{1c}$ is of great interest since it may
provide the best means of assessing accurately the degree of hypergly-
cemia on a long-term basis (Gabbay, 1976). In 1978, no less than four
articles appeared in the press describing the separation and quantitation
by IEF of HbA and HbA$_{1c}$ (Schoos et al., 1978; Spicer et al., 1978;
Jeppsson et al., 1978; Beccaria et al., 1978). Since the two species had
minute pI differences, it was necessary to resort to the use of separators,

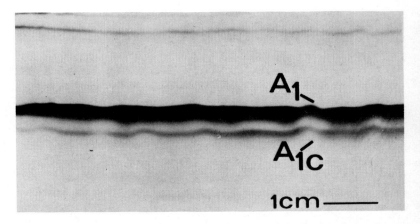

Fig. 5.22. IEF in an Immobiline pH gradient (nominal range pH 6.5–7.5) of adult lysates from diabetic patients. The run lasted overnight at 200 V/cm. About 300 μg Hb were applied per track, resulting in lateral diffusion of Hb in a continuous line. The cathode was uppermost. (From Gianazza and Righetti, unpublished.)

such as β-alanine or the dipeptide His-Gly. Even so, the two bands are often separated by barely 1 mm, which makes quantitation by common densitometers a very difficult task. Much better results are obtained if the separation is performed in immobilized pH gradients: as shown in Fig. 5.22, the distance between the two bands is of the order of several mm, allowing for a proper densitometric evaluation.

5.5. Immobiline® does it

You cannot believe how glad I am that this rather complicated story is drawing to an end. I should like to end it up with a message of bright hope for the future. Tired of drifting pH gradients? Upset with these shifting spots on 2-D maps? Irritated with conductivity and buffering capacity gaps? And what about those skewed and wavy sample zones? Cheer up: Immobiline does it. Do you remember we discussed the possibility you might have to burn your copy of this book if immobilized pH gradients did not work? Well, you have saved your money. There are now seven Immobiline chemicals available in the market: their pK values and

TABLE 5.3
Properties of Immobiline®[c]. Apparent pK-values[a] ($I = 10^{-2}$)

	H$_2$O		Polyacrylamide gel[b] $T = 5\%$ $C = 3\%$		Polyacrylamide gel[b] $T = 5\%$ $C = 3\%$ Glycerol 25% w/v		Physical state at room temperature
	10°C	25°C	10°C	25°C	10°C	25°C	
Acids with carboxyl as buffering group							
Immobiline pK 3.6	3.57	3.58	–	–	3.68 ± 0.02	3.75 ± 0.02	solid
Immobiline pK 4.4	4.39	4.39	4.30 ± 0.02	4.36 ± 0.02	4.40 ± 0.03	4.47 ± 0.03	solid
Immobiline pK 4.6	4.60	4.61	4.51 ± 0.02	4.61 ± 0.02	4.61 ± 0.02	4.71 ± 0.03	solid
Bases with tertiary amine as buffering group							
Immobiline pK 6.2	6.41	6.23	6.21 ± 0.05	6.15 ± 0.03	6.32 ± 0.08	6.24 ± 0.07	solid
Immobiline pK 7.0	7.12	6.97	7.06 ± 0.07	6.96 ± 0.05	7.08 ± 0.07	6.95 ± 0.06	solid
Immobiline pK 8.5	8.96	8.53	8.50 ± 0.06	8.38 ± 0.06	8.66 ± 0.09	8.45 ± 0.07	liquid
Immobiline pK 9.3	9.64	9.28	9.59 ± 0.08	9.31 ± 0.07	9.57 ± 0.06	9.30 ± 0.05	liquid

[a] pK values measured with glass surface electrode given without any corrections.
[b] Mean values of 10 determinations. Due to the slow response of the electrode the pK values for the amines are uncertain.
[c] From Bjellqvist et al. (1982a,b).

physico chemical data are given in Table 5.3. These compounds are acrylamide derivatives with the general structure:

$$CH_2=\underset{\underset{H}{|}}{C}-\underset{\underset{O}{\|}}{C}-\underset{\underset{H}{|}}{N}-R$$

where R contains either a carboxylic group or a tertiary amino group. During gel polymerization, these buffering species are efficiently incorporated in the gel. The distance between the double bond and the group taking part in the protolytic equilibrium has in all cases been chosen long enough so that the influence of the double bond on the dissociation constant can be neglected. As a result, the pK difference between the free and bound Immobiline is mainly due to the presence of the polyacrylamide matrix and to temperature variations during the IEF run. Immobiline-based pH gradients can be cast in the same way as a conventional polyacrylamide gradient gels using a density gradient to stabilize the Immobiline concentration gradient (e.g., see Fig. 3.33 for a typical gradient set up, except that, in most cases, a linear pH gradient will be preferred to the exponential gradient depicted in this figure). Let us take a simple example: generation of a pH-gradient with two Immobilines, one buffering and one non-buffering. Since in this case one of the Immobilines can be regarded as fully ionized, given the pK of the buffering Immobiline and the molar ratio buffering/non buffering Immobiline, the pH of the solution can be calculated with the Henderson–Hasselbalch equation. Thus, if the buffering species is an acid (A) and C_A and C_B are the molar concentrations of the acidic and basic Immobiline, respectively, the pH will be given by:

$$pH = pK_A + \log \frac{C_B}{C_A - C_B}$$

while in the case of a base, it will be:

$$pH = pK_B + \log \frac{C_B - C_A}{C_A}$$

If the buffering Immobiline is kept constant, the resulting pH-gradient generated by increasing or decreasing amounts of added non-buffering Immobiline (varied in a linear fashion) will correspond to a segment of the titration curve of the buffering Immobiline. Thus, the best pH-gradients with respect to linearity and buffering capacity will be those with midpoint corresponding to the inflection point of the titration curve (i.e. to the pK of the buffering species). In this case a width of 0.4 pH units in the immobilized gel gradient will give less than ±5% variation

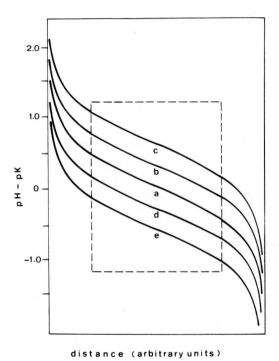

distance (arbitrary units)

Fig. 5.23. pH gradients generated with one buffering and one non buffering Immobiline with linear gradient mixing: (a) titration curve centered at the pK of the buffering ion; (b) curve with inflection point centered at pK + 0.3 pH units; (c) curve with inflection point centered at pK + 0.6 pH units; (d) curve with inflection point centered at pK − 0.3 pH units; (e) curve with inflection point centered at pK − 0.6 pH units. (From Bjellqvist et al., 1982.)

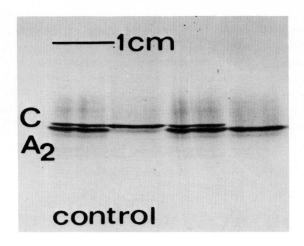

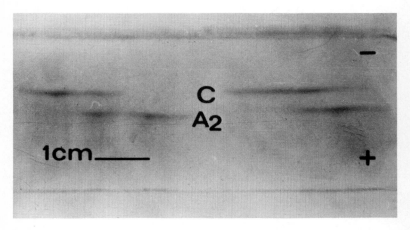

Fig. 5.24. IEF of Hbs C and A_2 in a pH 6–9 gradient as such (upper) as compared with an Immobiline pH gradient (lower panel) spanning a nominal pH 7.7 to 8.2 range. Hb C was from a homozygous patient while HbA_2 was purified chromatographically from adult blood. The cathod was uppermost. (From Gianazza and Righetti, unpublished.)

in the slope $d(pH)/dx$. In practice gradients as wide as 1.2 pH-units can be used if they are centered on the pK of the buffering group, but there will be some deviation from linearity at the two extremes. If the concentration of the buffering Immobiline is also varied linearly, the inflection point can be moved to higher or lower pH-values than what corresponds to the equivalence point of the titration curve. In this case, a family of titration curves, displaced parallel one to the other by as much as 1.2 pH units along the pH axis, can be generated displaying still a useful linear segment (boxed area in Fig. 5.23).

In practice, how should one operate? Suppose you want to focus hemoglobins. It would be best to select Immobiline pK 7.0 and prepare a 10 mM solution of it, both in the dense and light gradient solutions. Then take a non buffering Immobiline (e.g., pK 3.6) and titrate the dense solution down to pH 6.6. The light solution will be titrated to pH 7.6 and then the gel cassette will be filled up with a linear density (and thus pH) gradient. Upon polymerization, the gel will contain grafted on it an almost linear pH gradient spanning the pH interval 6.6 to 7.6 (excellent for resolving most Hb species) (Fig. 5.12). Unlike Svensson–Rilbe type gradients, this pH gradient will pre-exist the electrophoretic process but, unlike Kolin type gradients, it will stand indefinitely the passage of the current (as a precaution though: mark the gel polarity, so as to be sure to apply the anode at the acidic gel end and the cathode at its basic end). Do you realize we are back to square zero? The year 1982 will mark the new foundation of IEF and, by Jupiter, step in and carve out your share.

Post scriptum: For those skeptical fellows out there: have a look at Fig. 5.24, I know you will change your mind.

References

Ackers, G.K. (1964) Biochemistry *3*, 723–730.

Ackers, G.K. (1967) J. Biol. Chem. *242*, 3237–3241.

Addison, G.M., Beamish, M.R., Hales, C.N., Hodgkins, M.R., Jacobs, A. and Lee-wellen, P. (1972) J. Clin. Pathol. *25*, 326–333.

Ahlgren, E., Eriksson, K.E. and Vesterberg, O. (1967) Acta Chem. Scand. *21*, 937–944.

Akedo, H., Mori, Y., Kobayashi, M. and Okada, M. (1972) Biochem. Biophys. Res. Commun. *49*, 107–113.

Albanese, A.A. (1940) J. Biol. Chem. *134*, 467–471.

Albertsson, P.A. (1960) The Partitioning of Cell Particles and Macromolecules, Academic Press, New York.

Alexander, A., Cullen, B., Emigholz, K., Norgard, M.V. and Monahan, J.J. (1980) Anal. Biochem. *103*, 176–183.

Allen, J.C. and Humphries, C. (1975): In: Isoelectric Focusing; Arbuthnott, J.P. and Beeley, J.A., eds. (Butterworths, London) pp. 347–354.

Allen, R.C. and Maurer, H.R. (eds.) (1974) Electrophoresis and Isoelectric Focusing in Polyacrylamide Gel (de Gruyter, Berlin).

Allen, R.C. (1978) J. Chromatogr. *146*, 1–32.

Allen, R.C. (1980) Electrophoresis *1*, 32–37.

Allen, R.C. and Arnaud, P. (eds.) (1981) Electrophoresis '81 (de Gruyter, Berlin).

Almgren, M. (1971) Chem. Scripta *1*, 69–75.

Alper, C.A., Hobart, M.J. and Lachmann, P.J. (1975) In: Isoelectric Focusing; Arbuthnott, J.P. and Beeley, JA., eds. (Butterworths, London) pp. 306–312.

Alpert, E., Drysdale, J.W., Isselbacher, K.J. and Schur, P.H. (1972) J. Biol. Chem. *247*, 3792–3798.

Altland, K. (1977) In: Electrofocusing and Isotachophoresis; Radola, B.J. and Graesslin, D., eds. (de Gruyter, Berlin) pp. 295–301.

Altland, K. and Kaempfer, M. (1980) Electrophoresis *1*, 57–62.

Ambler, J. (1978a) Clin. Chim. Acta *85*, 183–191.

Ambler, J. (1978b) Clin. Chim. Acta *88*, 63–70.

Ambler, J. and Walker, G. (1979) Clin. Chem. *25*, 1320–1322.

Ames, G.F.L. and Nikaido, K. (1976) Biochemistry *15*, 616–623.

An der Laan, B. and Chrambach, A. (1980) Electrophoresis *1*, 23–27.

Andersson, L.O., Borg, H. and Mikaelsson, M. (1972) FEBS Lett. *20*, 199–202.

Anderson, N.G. and Anderson, N.L. (1977) Proc. Natl. Acad. Sci. USA *74*, 5421–5426.

Anderson, N.G. and Anderson, N.L. (1978a) Anal. Biochem. *85*, 331–340.

Anderson, N.G. and Anderson, N.L. (1978b) Anal. Biochem. *85*, 341–354.

Anderson, N.L. and Hickman, B.J. (1979) Anal. Biochem. *93*, 312–320.

Anderson, N.L. (1981) In: Electrophoresis '81; Allen, R.C. and Arnaud, P., eds. (de Gruyter, Berlin) pp. 309–316.

Anker, H.S. (1970) FEBS Lett. *7*, 293–296.

Ansorge, W. and De Mayer, L. (1980) J. Chromatogr. *202*, 45–53.

Araki, C. (1956) Bull. Chem. Soc. Jpn. *29*, 543–544.

Arbuthnott, J.P. and Beeley, J.A. (eds.) (1975) Isoelectric Focusing (Butterworths, London).

Arnaud, P., Wilson, G.B., Koistinen, J. and Fudenberg, H. (1977) J. Immunol. Methods *16*, 221–231.

Arnott, S., Fulmer, A., Scott, W.E., Dea, I.C.M., Moorhouse, R. and Rees, D.A. (1974) J. Mol. Biol. *90*, 269–284.

Arosio, P., Yokota, M. and Drysdale, J.W. (1976) Cancer Res. *36*, 1735–1739.

Arosio, P., Yokota, M. and Drysdale, J.W. (1977) Br. J. Haematol. *36*, 201–209.

Arosio, P., Gianazza, E. and Righetti, P.G. (1978a) J. Chromatogr. *166*, 55–64.

Arosio, P., Adelman, T.G. and Drysdale, J.W. (1978b) J. Biol. Chem. *253*, 4451–4458.

Arvidson, S. and Wadström, T. (1973) Biochim. Biophys. Acta *310*, 418–420.

Asimov, I. (1952) Foundation; Foundation and Empire; Second Foundation (Panther Books, New York).

Awdeh, Z.L., Williamson, A.R. and Askonas, B.A. (1968) Nature *219*, 66–67.

Awdeh, Z.L. (1969) Sci. Tools *16*, 42–43.

Azuma, J.I., Kashimura, N. and Komano, T. (1977) Anal. Biochem. *81*, 454–457.

Bagshaw, J.C., Drysdale, J.W. and Malt, R.A. (1973) Ann. N.Y. Acad. Sci. *209*, 363–371.

Ballou, B., Sundharadas, G. and Bach, M.L. (1974) Science *185*, 531–532.

Banyard, S.H., Stammers, D.K. and Harrison, P.M. (1978) Nature *271*, 282–284.

Barger, B.O., White, F.C., Pace, J.L., Kemper, D.L. and Ragland, W.L. (1976) Anal. Biochem. *70*, 327–335.

Bark, J.E., Harris, M.J. and Firth, M. (1976) J. Forensic Sci. Soc. *16*, 115–120.

Barrett, T. and Gould, H.J. (1973) Biochim. Biophys. Acta *294*, 165–170.

Barrollier, V.J., Watzke, E. and Gibian, H. (1958) Z. Naturforsch. *13b*, 754–755.

Bassett, P., Beuzard, Y., Garel, M.C. and Rosa, J. (1978) Blood *51*, 971–980.

Bassett, P., Braconnier, F. and Rosa, J. (1982) J. Chromatogr. *227*, 267–304.

Baumann, G. and Chrambach, A. (1975) Anal. Biochem. *64*, 530–536.

Beccaria, L., Chiumello, G., Gianazza, E., Luppis, B. and Righetti, P.G. (1978) Am. J. Haematol. *4*, 367–374.

Beeley, J.A., Stevenson, S.M. and Beeley, J.G. (1972) Biochim. Biophys. Acta *285*, 293–300.

Behnke, J.N., Dagher, S.M., Massey, T.H. and Deal, W.C., Jr. (1975) Anal. Biochem. *69*, 1–9.

Bengtsson, G. and Olivecrona, T. (1977) In: Isoelectric Focusing and Isotachophoresis; Radola, B.J. and Graesslin, D., eds. (De Gruyter, Berlin) pp. 189–195.

Beuzard, Y., Courvalin, J.C., Cohen-Solal, M., Garel, M.C. and Rosa, J. (1972) FEBS Lett. *27*, 76–80.

Bhakdi,S., Knüfermann, H. and Wallach, D.F.H. (1975) In: Progress in Isoelectric Focusing and Isotachophoresis; Righetti, P.G., ed. (Elsevier, Amsterdam) pp. 281–291.

Bianchi Bosisio, A., Loehrlein, C., Snyder, R.S. and Righetti, P.G. (1980) J. Chromatogr. *189*, 317–330.

Bianchi Bosisio, A., Snyder, R.S. and Righetti, P.G. (1981) J. Chromatogr. *209*, 265–272.

Bianchi, U. and Stefanelli, A. (1970) Atti Accad. Naz. Lincei, Rend. Ser. VIII, *93*, 539–542.

Bibring, T. and Baxandall, J. (1978) Anal. Biochem. *85*, 1–14.

Bier, M. and Egen, N.B. (1979) In: Electrofocus/78; Haglund, H., Westerfeld, J.C. and Ball Jr., J.T., eds. (Elsevier, Amsterdam) pp. 35–48.

Bier, M., Egen, N.B., Allgyer, T.T., Twitty, G.E. and Mosher, R.A. (1979) In: Peptides: Structure and Biological Function; Gross, E. and Meienhofer, J., eds. (Pierce Chemical Co., Rockford, Il) pp. 79–89.

Bier, M., Mosher, R.A. and Palusinski, O.A. (1981) J. Chromatogr. *211*, 313–335.

Binion, S. and Rodkey, L.S. (1981) Anal. Biochem. *112*, 362–366.

Biscoglio de Jimenez Bonino, M.J., Cascone, O., Arnao de Nué, A.I., Santomé, J.A., Sanchez, D., Oré, R. and Villavicencio, M. (1981) Int. J. Peptide Protein Res. *17*, 374–379.

Bishop, R. (1979) Sci. Tools *26*, 2–8.

Bjellqvist, B., Ek, K., Righetti, P.G., Gianazza, E., Görg, A., and Postel, W. (1982a) In: Electrophoresis '82; Stathakos, D., ed. (de Gruyter, Berlin) in press.

Bjellqvist, B., Ek, K., Righetti, P.G., Gianazza, E., Görg, A., Postel, W. and Westermeier, R. (1982b) J. Biochem. Biophys. Methods. *6*, 317–339.

Blakesley, R.W. and Boezi, J.A. (1977) Anal. Biochem. *82*, 580–582.

Blanicky, P. and Pihar, O. (1972) Coll. Czech. Chem. Commun. *37*, 319–325.

Bloomster, T.G. and Watson, D.W. (1981) Anal. Biochem. *113*, 79–84.

Bobb, D. (1973) Ann. N.Y. Acad. Sci. *209*, 225–236.

Boddin, M., Hilderson, J., Lagrou, A. and Dierick, W. (1975) Anal. Biochem. *64*, 293–296.

Bode, H.J. (1980) In: Electrophoresis '79; Radola, B.J., ed. (de Gruyter, Berlin) pp. 39–52.

Bodwell, C.E. and Creed, G.J. (1973) Abstr. 9th Int. Congress Biochem. Stockholm. p. 25.

Boltz, Jr., R.C., Todd, P., Hammerstedt, R.H., Hymer, W.C., Thomson, S.J. and Docherty, J. (1977) In: Cell Separation Methods; Bloemendal, H., ed. (North Holland, Amsterdam) pp. 145–155.

Boltz, Jr., R.C., Miller, T.Y., Todd, P. and Kukulinsky, N.E. (1978) In: Electrophoresis '78; Catsimpoolas, N., ed. (Elsevier, Amsterdam) pp. 345–355.

Boltz, Jr., R.C. and Todd, P. (1979) In: Electrokinetic Separation Methods; Righetti, P.G., Van Oss, C.J. and Vanderhoff, J.W., eds. (Elsevier, Amsterdam) pp. 229–250.

Bonitati, J. (1980) J. Biochem. Biophys. Methods *2*, 341–356.

Bonitati, J. (1981) J. Biochem. Biophys. Methods *4*, 49–69.

Bosshard, H.F. and Datyner, A. (1977) Anal. Biochem. *82*, 327–333.

Bossinger, J., Miller, M.J., Vo, K.P., Geiduschek, E.P. and Xuong, N.H. (1979) J. Biol. Chem. *254*, 7986–7998.

Boulikas, J. and Hancock, R. (1981) J. Biochem. Biophys. Methods *5*, 219–228.

Bours, J. (1973a) Exp. Eye Res. *15*, 299–319.

Bours, J. (1973b) Exp. Eye Res. *16*, 487–499.

Bours, J. (1973c) Exp. Eye Res. *16*, 501–515.

Bours, J. (1973d) Sci. Tools *20*, 29–34.

Bours, J. (1973e) Sci. Tools *20*, 2–14.

Bours, J. (1976) In: Isoelectric Focusing; Catsimpoolas, N., ed. (Academic Press, New York) pp. 209–228.

Boussios, T. and Bertles, J.F. (1978) In: Electrophoresis '78; Catsimpoolas, N., ed. (Elsevier, Amsterdam) pp. 137–144.

Boyd, J.B. and Mitchell, H.K. (1965) Anal. Biochem. *13*, 28–35.

Brackenridge, C.J. and Bachelard, H.S. (1969) J. Chromatogr. *41*, 242–249.

Bradbury, R. (1953) Fahrenheit 451, Ballantine Books, New York.

Bradley, T.B., Wohl, R.C. and Rieder, R.F. (1967) Science *157*, 1581–1583.

Brakke, M.D., Allington, R.W. and Langille, F.A. (1968) Anal. Biochem. *25*, 30–39.

Brenna, O., Gianazza, E. and Righetti, P.G. (1982) J. Chromatogr. *237*, 293–296.

Brewer, J.M. (1967) Science *156*, 256–257.

Brinkmann, B., Koops, E., Klopp, O., Heindl, K. and Rüdiger, H.W. (1972) Ann. Hum. Genet. *35*, 363–366.

Brogren, C.H. (1977) In: Electrofocusing and Isotachophoresis; Radola, B.J. and Graesslin, D., eds. (de Gruyter, Berlin) pp. 549–558.

Brown, W.D. and Green, S. (1970) Anal. Biochem. *34*, 593–595.

Brown, R.K., Lull, J.M., Lowenkron, S., Bagshaw, J.C. and Vinogradov, S.N. (1976) Anal. Biochem. *71*, 325–332.

Brown, R.K., Caspers, M.L., Lull, J.M., Vinogradov, S.N., Felgenhauer, K. and Nekic, M. (1977) J. Chromatogr. *131*, 223–232.

Bryce, C.F.A. and Crichton, R.R. (1971) J. Biol. Chem. *246*, 4198–4205.

Bull, H.B. (1971) An Introduction to Physical Biochemistry, F.A. Davis Co., Philadelphia, pp. 127–135.

Bunn, H.F. and Drysdale, J.W. (1971) Biochim. Biophys. Acta *229*, 51–57.

Bunn, H.F. and McDonough, M. (1974) Biochemistry *13*, 988–993.

Bunn, H.F., Forget, B.G. and Ranney, H.M. (1977) Hemoglobinopathies, Saunders, Philadelphia.

Bunn, H.F. (1977) In: Biological and Biomedical Applications of Isoelectric Focusing; Catsimpoolas, N. and Drysdale, J.W., eds. (Plenum Press, New York) pp. 29–55.

Burdett, P., Lizana, J., Eneroth, P. and Bremme, K. (1981) In: Electrophoresis '81; Allen, R.C. and Arnaud, P., eds. (de Gruyter, Berlin) pp. 329–342.

Burghes, A.H.M., Dunn, M.J., Statham, H.E. and Dubowitz, V. (1981) In: Electrophoresis '81; Allen, R.C. and Arnaud, P., eds. (de Gruyter, Berlin) pp. 295–308.

Burridge, K. (1978) In: Methods in Enzymology; Ginsburg, V., ed. (Academic Press, New York) pp. 54–60.

Bustos, S.E. and Fung, L. (1981) In: Electrophoresis '81; Allen, R.C. and Arnaud, P., eds. (de Gruyter, Berlin) pp. 317–328.

Butler, J.A.V. and Stephen, J.M.L. (1947) Nature *160*, 469–471.

Callaham, M.F., Poe, W.E. and Heitz, J.R. (1976) Anal. Biochem. *70*, 542–546.

Cann, J.R. (1966) Biochemistry *5*, 1108–1112.

Cann, J.R. and Stimpson, D.I. (1977) Biophys. Chem. *7*, 103–114.

Cann, J.R., Stimpson, D.I. and Cox, D.J. (1978) Anal. Biochem. *85*, 34–49.

Cann, J.R. (1979) In: Electrokinetic Separation Methods; Righetti, P.G., Van Oss,

Cantrell, S.J., Babitch, J.A. and Torres, S. (1981) Anal. Biochem. *116*, 168–173. C.J. and Vanderhoff, J.W., eds. (Elsevier, Amsterdam) pp. 369–387.

Capel, M., Redman, B. and Bourque, D.P. (1979) Anal. Biochem. *97*, 210–228.

Carrell, S., Theilkaes, L., Skvaril, S. and Barandun, S. (1969) J. Chromatogr. *45*, 483–486.

Carter, N.D., Auton, J.A., Welch, S.G., Marshall, W.H. and Fraser, G.R. (1976) Hum. Hered. *26*, 4–7.

Casassa, E.F. (1967) J. Polymer Sci. *B5*, 773–780.

Caspers, M.L., Posey, Y. and Brown, R.K. (1977) Anal. Biochem. *79*, 166–180.

Caspers, M.L. and Chrambach, A. (1977) Anal. Biochem. *81*, 54–62.

Catsimpoolas, N. (1969a) Anal. Biochem. *26*, 480–482.

Catsimpoolas, N. (1969b) Clin. Chim. Acta *23*, 237–238.

Catsimpoolas, N. and Meyer, E.W. (1969) Arch. Biochem. Biophys. *132*, 279–285.

Catsimpoolas, N. (1970) Separ. Sci. *5*, 523–544.

Catsimpoolas, N. (ed.) (1973a) Isoelectric Focusing and Isotachophoresis, Ann. N.Y. Acad. Sci. *209*.

Catsimpoolas, N. (1973b) Separ. Sci. *8*, 71–121.

Catsimpoolas, N. (1973c) Ann. N.Y. Acad. Sci. *209*, 65–79.

Catsimpoolas, N. (1973d) Anal. Biochem. *54*, 66–78.

Catsimpoolas, N. (1973e) Anal. Biochem. *54*, 79–87.

Catsimpoolas, N. (1973f) Anal. Biochem. *54*, 88–94.

Catsimpoolas, N. and Griffith, A. (1973) Anal. Biochem. *56*, 100–120.

Catsimpoolas, N. (1975a) Separ. Sci. *10*, 55–76.

Catsimpoolas, N. (1975b) In: Progress in Isoelectric Focusing and Isotachophoresis; Righetti, P.G., ed. (Elsevier, Amsterdam) pp. 77–92.

Catsimpoolas, N. (ed.) (1976) Isoelectric Focusing, Academic Press, New York.

Catsimpoolas, N. and Drysdale, J.W. (eds.) (1977) Biological and Biomedical Applications of Isoelectric Focusing, Plenum Press, New York.

Catsimpoolas, N. and Griffith, A. (1977) In: Electrofocusing and Isotachophoresis; Radola, B.J. and Graesslin, D., eds. (de Gruyter, Berlin) pp. 469–479.

Catsimpoolas, N. (ed.) (1978) Electrophoresis '78, Elsevier, Amsterdam.

Catsimpoolas, N., Stamatopoulou, A. and Griffith, A.L. (1980) In: Electrophoresis '79; Radola, B.J., ed. (de Gruyter, Berlin) pp. 503–515.

Chamberlain, J.P. (1979) Anal. Biochem. *98*, 132–140.

Chamoles, N. and Karcher, D. (1970a) Clin. Chim. Acta *30*, 337–341.

Chamoles, N. and Karcher, D. (1970b) Clin. Chim. Acta *30*, 359–364.

Chapuis-Cellier, C., Francina, A. and Arnaud, P. (1980) Prot. Biol. Fluids *27*, 743–746.

Chapuis-Cellier, C. and Arnaud, P. (1981) Anal. Biochem. *113*, 325–331.

Charlionet, R., Martin, J.P., Sesboüé, R., Madec, P.J. and Lefebvre, F. (1979) J. Chromatogr. *176*, 89–101.

Charlionet, R., Morcamp, C., Sesboüé, R. and Martin, J.P. (1981): J. Chromatogr. *205*, 355–366.

Chidakel, B.E., Nguyen, N.Y. and Chrambach, A. (1977) Anal. Biochem. *77*, 216–225.

Chilla, R., Doering, K.M., Domagk, G.F. and Rippa, M. (1973) Arch. Biochem. Biophys. *159*, 235–239.

Chlumechka, V., D'Obrenen, P. and Colter, J.S. (1973) Canad. J. Biochem. *51*, 1521–1530.

Choo, K.H., Cotton, R.G.H. and Danks, D.M. (1980) Anal. Biochem. *103*, 33–38.

Chrambach, A. (1966) Anal. Biochem. *15*, 544–552.

Chrambach, A. and Rodbard, D. (1971) Science *172*, 440–451.

Chrambach, A. and Rodbard, D. (1972) Separ. Sci. *7*, 663–703.

Chrambach, A., Doerr, P., Finlayson, G.R., Miles, L.E.M., Sherins, R. and Rodbard, D. (1973) Ann. N.Y. Acad. Sci. *209*, 44–64.

Chrambach, A., Jovin, T.M., Svendsen, P.J. and Rodbard, D. (1976) In: Methods of Protein Separation, vol. 2; Catsimpoolas, N., ed. (Plenum Press, New York) pp. 27–144.

Chrambach, A. and Nguyen, N.Y. (1979) In: Electrokinetic Separation Methods; Righetti, P.G., Van Oss, C.J. and Vanderhoff, J.W., eds. (Elsevier, Amsterdam) pp. 337–368.

Chrambach, A. (1980) Mol. Cell Biochem. *29*, 23–46.

Christensen, J.J., Hill, J.O. and Izatt, M.R. (1971) Science *174*, 459–467.

Chuat, J.C. and Pilon, C. (1977) Anal. Biochem. *82*, 258–261.

Cleland, W.W. (1964) Biochemistry *3*, 480–482.

Comi, P., Giglioni, B., Ottolenghi, S., Gianni, A.M., Ricco, G., Mazza, U., Saglio, G., Camaschella, C., Pich, P.G., Gianazza, E. and Righetti, P.G. (1979) Biochem. Biophys. Res., Commun. *87*, 1–8.

Consden, R., Gordon, A.H. and Martin, A.J.P. (1946) Biochem. J. *40*, 33–40.

Constans, J., Viau, M., Gouaillard, C., Bouisson, C. and Clerc, A. (1980) In: Electrophoresis '79; Radola, B.J., ed. (de Gruyter, Berlin) pp. 701–710.

Conway-Jacobs, A. and Lewin, L.M. (1971) Anal. Biochem. *43*, 394–400.

Cossu, G., Manca, M., Gavina, P.M., Bullitta, R., Bianchi Bosisio, A., Gianazza, E. and Righetti, P.G. (1982) Am. J. Haematol. *13*, 149–157.

Cotton, R.G.H. and Milstein, C. (1973) J. Chromatogr. *86*, 219–221.

Coutelle, R. (1971) Acta Biol. Med. Germ. *27*, 681–691.

Cox, J.L., King, H. and Berg, C.P. (1929) J. Biol. Chem. *81*, 755–762.

Creighton, T.E. (1979) J. Mol. Biol. *129*, 235–264.

Crowle, A.J. and Cline, L.J. (1977) J. Immunol. Methods *17*, 379–385.

Czempiel, W. (1977) In: Electrofocusing and Isotachophoresis; Radola, B.J. and Graesslin, D., eds. (de Gruyter, Berlin) pp. 405–412.

Dahlberg, A.E., Dingman, C.W. and Peacock, A.C. (1969) J. Mol. Biol. *41*, 139–150.

Dale, G. and Latner, A.L. (1968) Lancet (April) *20*, 847–848.

Dale, G. and Latner, A.L. (1969) Clin. Chim. Acta *24*, 61–68.

Danno, G.I. (1977) Anal. Biochem. *83*, 189–193.

Datyner, A. and Finnimore, E. (1973) Anal. Biochem. *52*, 45–55.

Davies, H. (1970) Prot. Biol. Fluids *17*, 389–396.

Davies, H. (1975) In: Isoelectric Focusing; Arbuthnott, J.P. and Beeley, J.A., eds. (Butterworths, London) pp. 97–113.

Davis, B.J. (1964) Ann. N.Y. Acad. Sci. *121*, 404–427.

Dean, B. (1979) Anal. Biochem. *99*, 105–111.

Dean, R.T. and Messer, M. (1975) J. Chromatogr. *105*, 353–358.

Delincée, H. and Radola, B.J. (1970) Biochim. Biophys. Acta *200*, 404–407.

Delincée, H. and Radola, B.J. (1971) Prot. Biol. Fluids *18*, 493–497.

Delincée, H. and Radola, B.J. (1975) Int. J. Radiat. Biol. *28*, 565–579.

Delincée, H. and Radola, B.J. (1978) Anal. Biochem. *90*, 609–623.

Delincée, H. (1980) In: Electrophoresis '79; Radola, B.J., ed. (de Gruyter, Berlin) pp. 165–171.

Denckla, W.D. (1974) J. Clin. Invest. *53*, 572—581.

Denckla, W.D. (1975) U.S. Patent 3,901,780.

Denckla, W.D. (1976) U.S. Patent 3,951,777.

Denckla, W.D. (1977) In: Electrofocusing and Isotachophoresis; Radola, B.J. and Graesslin, D., eds. (de Gruyter, Berlin) pp. 423—432.

De Olmos, J. (1969) Brain Behav. Evol. *2*, 213—237.

Dewar, J.H. and Latner, A.L. (1970) Clin. Chim. Acta *28*, 149—152.

Diezel, W., Kopperschläger, G. and Hofmann, E. (1972) Anal. Biochem. *48*, 617—620.

Dirksen, M.L. and Chrambach, A. (1972) Separ. Sci. *7*, 747—772.

Doi, E. and Ohtsuru, C. (1974) Agr. Biol. Chem. *38*, 1747—1748.

Domagk, G.F., Doering, K.M. and Chilla, R. (1973) Eur. J. Biochem. *38*, 259—264.

Domagk, G.F., Alexander, W.R. and Doering, K.M. (1974) Hoppe-Seyler's Z. Physiol. Chem. *355*, 781—786.

Douzou, P. (1977) Cryobiochemistry, Academic Press, New York.

Drysdale, J.W. (1970) Biochim. Biophys. Acta *207*, 256—258.

Drysdale, J.W., Righetti, P.G. and Bunn, H.F. (1971) Biochim. Biophys. Acta *229*, 42—53.

Drysdale, J.W. and Righetti, P.G. (1972) Biochemistry *11*, 4044—4052.

Drysdale, J.W. (1974) Biochem. J. *141*, 627—632.

Drysdale, J.W. and Shafritz, D. (1975) In: Progress in Isoelectric Focusing and Isotachophoresis; Righetti, P.G., ed. (Elsevier, Amsterdam) pp. 293—306.

Drysdale, J.W. (1977a) In: Electrofocusing and Isotachophoresis; Radola, B.J. and Graesslin, D., eds. (de Gruyter, Berlin) pp. 241—252.

Drysdale, J.W. (1977b) In: Ciba Foundation Symposium 51 (Elsevier, Amsterdam) pp. 241—252.

Drysdale, J.W. (1979) In: Carcino-Embrionic Proteins; Lehman; F.G., ed. Vol. 1 (Elsevier, Amsterdam) pp. 249—259.

Drzeniek, R., Reichel, C., Wiegers, K.J., Hamann, A. and Hilbring, M. (1980) In: Electrophoresis '79; Radola, B.J., ed. (de Gruyter, Berlin) pp. 475—492.

Dubernet, M. and Riberau-Gayon, P. (1974) Phytochemistry *13*, 1085—1087.

Duckworth, M. and Yaphe, W. (1971) Anal. Biochem. *44*, 636—641.

Du Cros, D.L. and Wrigley, C.W. (1979) J. Sci. Food Agric. *30*, 785—794.

Dunford, H. R. and Stillman, J.S. (1976) Coord. Chem. Rev. *19*, 187—193.

Du Vigneaud, V., Irwing, G.W., Dyer, H.M. and Sealock, R.R. (1938) J. Biol. Chem. *123*, 45—55.

Earland, C. and Ramsden, D.B. (1969) J. Chromatogr. *41*, 259—261.

Edwards, Y.H., Potter, J.E. and Hopkinson, D.A. (1979) Ann. Human. Genet. *42*, 293—302.

Edwards, J.J., Hahn, H.J. and Anderson, N.G. (1980) In: Electrophoresis '79; Radola, B.J., ed. (de Gruyter, Berlin) pp. 383—394.

Edwards, J.J. and Anderson, N.G. (1981) Electrophoresis *3*, 161—168.

Egen, N.B., Twitty, G.E. and Bier, M. (1979) 17th Aerospace Sciences Meeting, New Orleans, LA, Jan. 15—17, document 79—0405.

Ek, K. and Righetti, P.G. (1980) Electrophoresis *1*, 137—140.

Ek, K., Gianazza, E. and Righetti, P.G. (1980) Biochim. Biophys. Acta *626*, 356—365.

Ek, K. (1981) LKB Application Note No. 319, June 1981.

Elliott, R.W., Hohman, C., Romejko, C., Louis, P. and Lilly, F. (1978) In: Electrophoresis '78; Catsimpoolas, N., ed. (Elsevier, Amsterdam) pp. 261–274.

Emond, J.P. and Pagé, M. (1980) J. Chromatogr. *200*, 57–63.

Engel, M.R. (1888) Bull. Soc. Chim. Paris *50*, 102–110.

Epstein, C.J. and Schechter, A.N. (1968) Ann. N.Y. Acad. Sci. *151*, 85–101.

Eriksson, K.E. and Pettersson, B. (1973) Anal. Biochem. *56*, 618–620.

Esen, A. (1981) In: Electrophoresis '81; Allen, R.C. and Arnaud, P., eds. (de Gruyter, Berlin) pp. 73–76.

Fahnestock, S., Erdmann, U. and Nomura, M. (1973) Biochemistry *12*, 220–224.

Fairbanks, G., Steck, T.L. and Wallach, D.L.H. (1971) Biochemistry *10*, 2026–2032.

Fantes, K.H. and Furminger, I.G.S. (1967) Nature *215*, 750–751.

Faupel, M. and Von Arx, E. (1978) J. Chromatogr. *157*, 243–251.

Fawcett, J.S. and Morris, C.J.O.R. (1966) Separ. Sci. *1*, 9–20.

Fawcett, J.S. (1968) FEBS Lett. *1*, 81–82.

Fawcett, J.S. (1970) Prot. Biol. Fluids *17*, 409–412.

Fawcett, J.S. (1973) Ann. N.Y. Acad. Sci. *209*, 112–126.

Fawcett, J.S. (1975a) In: Progress in Isoelectric Focusing and Isotachophoresis; Righetti, P.G., ed. (Elsevier, Amsterdam) pp. 25–37.

Fawcett, J.S. (1975b) In: Isoelectric Focusing; Arbuthnott, J.P. and Beeley, J.A., eds. (Butterworths, London) pp. 23–43.

Fawcett, J.S. (1976) In: Isoelectric Focusing; Catsimpoolas, N., ed. (Academic Press, New York) pp. 173–208.

Felgenhauer, K. and Pak, S.J. (1973) Ann. N.Y. Acad. Sci. *209*, 147–153.

Felgenhauer, K. (1974) Hoppe Seyler's Z. Physiol. Chem. *355*, 1281–1290.

Felgenhauer, K. and Pak, S.J. (1975) In: Progress in Isoelectric Focusing and Isotachophoresis; Righetti, P.G., ed. (Elsevier, Amsterdam) pp. 115–120.

Ferris, T. (1979) The Red Limit, Corgi Books, Great Britain, pp. 23–25.

Finlayson, G.R. and Chrambach, A. (1971) Anal. Biochem. *40*, 292–311.

Fischer, L. (1969) An Introduction to Gel Chromatography, North Holland, Amsterdam.

Fisher, R.A., Edwards, Y.H., Putt, W., Potter, J. and Hopkinson, D.A. (1977) Ann. Hum. Genet. *41*, 139–149.

Flatmark, T. and Vesterberg, O. (1966) Acta Chem. Scand. *20*, 1497–1503.

Flodin, P. and Kupke, P.W. (1956) Biochim. Biophys. Acta *21*, 368–375.

Flodin, P. (1961) J. Chromatogr. *5*, 103–111.

Foster, G.L. and Schmidt, C.L.A. (1926) J. Am. Chem. Soc. *48*, 1709–1713.

Francina, A., Dorleac, E. and Cloppet, H. (1981) J. Chromatogr. *222*, 116–119.

Frater, R. (1970) J. Chromatogr. *50*, 469–474.

Fredriksson, S. (1972) Anal. Biochem. *50*, 575–585.

Fredriksson, S. and Pettersson, S. (1974) Acta Chem. Scand. *28*, 370–370.

Fredriksson, S. (1975) J. Chromatogr. *108*, 153–167.

Fredriksson, S. (1977a) J. Chromatogr. *135*, 441–446.

Fredriksson, S. (1977b) In: Electrofocusing and Isotachophoresis; Radola, B.J. and Graesslin, D., eds. (de Gruyter, Berlin) pp. 71–83.

Fredriksson, S. (1978) J. Chromatogr. *151*, 347–355.

Friesen, J.D., Parker, J., Watson, R.J., Bendiak, D., Reeh, S.V. and Pedersen, S. (1976) Mol. Gen. Genet. *148*, 93–98.

Fries, E. (1976) Anal. Biochem. *70*, 124–135.

Funatsu, M., Hara, K., Ishguro, M., Funatsu, G. and Kubo, K. (1973): Proc. Jpn. Acad. *49*, 771–776.

Gabbay, K.H. (1976) N. Engl. J. Med. *295*, 443–447.

Gabriel, O. (1971) In: Methods in Enzymology; Colowick, S.P. and Kaplan, N.O., eds. (Academic Press, New York) *22*, 578–590.

Gacon, G., Krishnamoorthy, R., Wajcman, H., Labie, D., Tapon, J. and Cosson, A. (1977a) Biochim. Biophys. Acta *490*, 156–163.

Gacon, G., Belkhodja, O., Wajcman, H. and Labie, D. (1977b) FEBS Lett. *82*, 243–246.

Gainer, H. (1971) Anal. Biochem. *44*, 589–605.

Gainer, H. (1973) Anal. Biochem. *51*, 646–650.

Galante, E., Caravaggio, T. and Righetti, P.G. (1975) In: Progress in Isoelectric Focusing and Isotachophoresis; Righetti, P.G., ed. (Elsevier, Amsterdam) pp. 3–12.

Galante, E., Caravaggio, T. and Righetti, P.G. (1976) Biochim. Biophys. Acta *442*, 309–315.

Garrels, J.S. (1979) J. Biol. Chem. *254*, 7961–7977.

Gasparic, V. and Rosengren, A. (1974) Sci. Tools *21*, 1–2.

Gasparic, V. and Rosengren, A. (1975) In: Isoelectric Focusing; Arbuthnott, J.P. and Beeley, J.A., eds. (Butterworths, London) pp. 178–181.

Gelfi, C. and Righetti, P.G. (1981a) Electrophoresis *2*, 213–219.

Gelfi, C. and Righetti, P.G. (1981b) Electrophoresis *2*, 220–228.

Gelsema, J.W. and De Ligny, C.L. (1977) J. Chromatogr. *130*, 41–50.

Gelsema, J.W., De Ligny, C.L. and Van Der Veen, N.G. (1977) J. Chromatogr. *140*, 149–155.

Gelsema, J.W., De Ligny, C.L. and Van Der Veen, N.G. (1978) J. Chromatogr. *154*, 161–174.

Gelsema, J.W. and De Ligny, C.L. (1979) J. Chromatogr. *178*, 550–554.

Gelsema, J.W., De Ligny, C.L. and Van Der Veen, N.G. (1979) J. Chromatogr. *173*, 33–41.

Gelsema, J.W., De Ligny, C.L. and Blanken, W.M. (1980a) J. Chromatogr. *198*, 301–316.

Gelsema, J.W., De Ligny, C.L., Blanken, W.M., Hamer, R.J., Roozen, A.M.P. and Bakker, J.A. (1980b) J. Chromatogr. *196*, 51–58.

Gianazza, E., Pagani, M., Luzzana, M. and Righetti, P.G. (1975) J. Chromatogr. *109*, 357–364.

Gianazza, E., Righetti, P.G., Bordi, S. and Papeschi, G. (1977) In: Electrofocusing and Isotachophoresis; Radola, B.J. and Graesslin, D., eds. (de Gruyter, Berlin) pp. 173–179.

Gianazza, E. and Righetti, P.G. (1978) Biochim. Biophys. Acta *540*, 357–364.

Gianazza, E. and Righetti, P.G. (1979a) In: Electrokinetic Separation Methods; Righetti, P.G., Van Oss, J.C. and Vanderhoff, J.W., eds. (Elsevier, Amsterdam) pp. 293–311.

Gianazza, E. and Righetti, P.G. (1979b) Prot. Biol. Fluids *27*, 715–718.

Gianazza, E., Astorri, C. and Righetti, P.G. (1979a) J. Chromatogr. *171*, 161–169.

Gianazza, E., Chillemi, F., Gelfi, C. and Righetti, P.G. (1979b) J. Biochem. Biophys. Methods *1*, 237–251.

Gianazza, E. and Righetti, P.G. (1980) In: Electrophoresis '79; Radola, B.J., ed. (de Gruyter, Berlin) pp. 129–140.

Gianazza, E., Gelfi, C. and Righetti, P.G. (1980a) J. Biochem. Biophys. Methods *3*, 65–75.

Gianazza, E., Chillemi, F. and Righetti, P.G. (1980b) J. Biochem. Biophys. Methods *3*, 135–141.

Gianni, A.M., Giglioni, B., Comi, P., Ottolenghi, S., Ferrari, M., Furbetta, M., Angius, M. and Cao, A. (1981) Hemoglobin *5*, 349–355.

Giddings, J.C., Kucera, E., Russell, C.P. and Myers, M.N. (1968) J. Phys. Chem. *72*, 4397–4408.

Giddings, J.C. (1969) Separ. Sci. *4*, 181–188.

Giddings, J.C. and Dahlgren, H. (1971) Separ. Sci. *6*, 345–356.

Gidez, L.I., Swaney, J.B. and Murnane, S. (1977) J. Lipid Res. *18*, 59–68.

Gierthy, J.F., Ellem, K.A.O. and Kongsvik, J.R. (1979) Anal. Biochem. *98*, 27–35.

Giometti, C.S. and Anderson, N.G. (1980) In: Electrophoresis '79; Radola, B.J., ed. (de Gruyter, Berlin) pp. 395–404.

Giometti, C.S., Anderson, N.G., Tollaksen, S.L., Edwards, J.J. and Anderson, N.L. (1980) Anal. Biochem. *102*, 47–58.

Godolphin, W.J. and Stinson, R.A. (1974) Clin. Chim. Acta *56*, 97–103.

Godson, G.N. (1970) Anal. Biochem. *35*, 66–76.

Goerth, K. and Radola, B.J. (1980) In: Electrophoresis '79; Radola, B.J., ed. (de Gruyter, Berlin) pp. 555–563.

Gonenne, A. and Ernst, R. (1978) Anal. Biochem. *87*, 28–38.

Goode, R.L. and Balwin, J.N. (1973) Prep. Biochem. *3*, 349–361.

Goodman, D. and Matzura, H. (1971) Anal. Biochem. *42*, 481–489.

Gordon, A.H. (1975) Electrophoresis of Proteins in Polyacrylamide and Starch Gels (Elsevier, Amsterdam) pp. 14–15.

Görg, A., Postel, W. and Westermeier, R. (1977) Z. Lebensm. Unter. Forsch. *164*, 160–162.

Görg, A., Postel, W. and Westermeier, R. (1978) Anal. Biochem. *89*, 60–70.

Görg, A., Postel, W. and Westermeier, R. (1979a) Z. Lebensm. Unter. Forsch. *168*, 25–28.

Görg, A., Postel, W. and Westermeier, R. (1979b) GIT Lab. Med. *2*, 32–40.

Görg, A., Postel, W. and Westermeier, R. (1980a) In: Electrophoresis '79; Radola, B.J., ed. (de Gruyter, Berlin) pp. 67–78.

Görg, A., Postel, W., Westermeier, R., Gianazza, E. and Righetti, P.G. (1980b) J. Biochem. Biophys. Methods *3*, 273–284.

Görg, A., Postel, W., Westermeier, R., Gianazza, E. and Righetti, P.G. (1981) In: Electrophoresis '81; Allen, R.C. and Arnaud, P., eds. (de Gruyter, Berlin) pp. 257–270.

Gorovsky, M.A., Carlson, K. and Rosenbaum, T.L. (1970) Anal. Biochem. *35*, 359–370.

Graesslin, D. and Weise, H.C. (1974) In: Electrophoresis and Isoelectric Focusing in Polyacrylamide Gel; Allen, R.C. and Maurer, H.R., eds. (de Gruyter, Berlin) pp. 199–205.

Grant, G.M. and Leaback, D.H. (1970) ISCO Applications Research Bulletin No. 3.

Grassmann, W. (1950) Z. Angew. Chem. *62*, 1170–1176.

Green, M.R., Pastewka, J.V. and Peacock, A.C. (1973) Anal. Biochem. *56*, 43–50.

Groome, N.P. and Belyavin, G. (1975) Anal. Biochem. *63*, 249–254.

Gropp, C., Havemann, K. and Lehmann, F.C. (1978) Cancer Res. *42*, 2802–2808.

Grossbach, U. (1972) Biochem. Biophys. Res. Commun. *49*, 667–672.

Grubb, A. (1973) Anal. Biochem. *55*, 582–592.

Grubhofer, N. and Borja, C. (1977) In: Electrofocusing and Isotachophoresis;

Radola, B.J. and Graesslin, D., eds. (de Gruyter, Berlin) pp. 111–120.

Guerrasio, A., Saglio, G., Mazza, U., Pich, P., Camaschella, C., Ricco, G., Gianazza, E. and Righetti, P.G. (1979) Clin. Chim. Acta 99, 7–11.

Hackenberg, H. and Klingenberg, M. (1980) Biochemistry 19, 548–555.

Haff, L.A., Lasky, M. and Manrique, A. (1979) J. Biochem. Biophys. Methods 1, 275–286.

Haglund, H. and Tiselius, A. (1950) Acta Chem. Scand. 4, 957–965.

Haglund, H. (1971) In: Methods of Biochemical Analysis; Glick, D., ed. (Wiley, Interscience, New York) Vol. 19, pp. 1–104.

Haglund, H. (1975) In: Isoelectric Focusing; Arbuthnott, J.P. and Beeley, J.A., eds. (Butterworths, London) pp. 3–22.

Haller, W. (1965) Nature, 206, 693–696.

Halliday, J.M., McKeering, L.V. and Powell, L.W. (1976) Cancer Res. 36, 4486–4490.

Hames, B.D. (1981) In: Gel Electrophoresis of Proteins; Hames, B.D. and Rickwood, D., eds. (IRL Press Ltd., London) pp. 50–55.

Hammerstedt, R.H., Keith, A.K., Boltz, Jr., R.C. and Todd, P.W. (1979) Arch. Biochem. Biophys. 194, 565–580.

Handschin, U.E. and Ritschard, N.J. (1976) Anal. Biochem. 71, 143–148.

Hannig, K. (1961) Z. Anal. Chem. 181, 244–254.

Hannig, K. (1969) In: Modern Separation Methods of Macromolecules and Particles; Gerritsen, T., ed. (Wiley, Interscience, New York) pp. 45–69.

Hannig, K. (1972) In: Techniques of Biochemical Biophysical Morphology; Glick, D. and Rosenbaum, R., eds. (Wiley, New York) Vol. 1, pp. 191–232.

Hannig, K. (1978) In: Electrophoresis '78; Catsimpoolas, N., ed. (Elsevier, Amsterdam) pp. 69–76.

Hansen, J.N. (1976) Anal. Biochem. 76, 37–44.

Hansen, J.N., Pheiffer, B.H. and Boehnert, J.A. (1980) Anal. Biochem. 105, 192–201.

Harada, S. (1975) Clin. Chim. Acta 85, 275–283.

Harada, S., Agarwal, D.P. and Goedde, H.W. (1978a) Hum. Genet. 44, 181–185.

Harada, S., Agarwal, D.P. and Goedde, H.W. (1978b) Hum. Genet. 40, 215–220.

Harada, S., Agarwal, D.P. and Goedde, H.W. (1980) In: Electrophoresis '79; Radola, B.J., ed. (de Gruyter, Berlin) pp. 687–693.

Harell, D. and Morrison, M. (1979) Arch. Biochem. Biophys. 193, 158–168.

Harpel, B.M. and Kueppers, F. (1980) Anal. Biochem. 104, 173–174.

Harris, H., Hopkinson, D.A. and Robson, E.B. (1974) Hum. Genet. 37, 237–253.

Harris, H. and Hopkinson, D.A. (1976) Handbook of Enzyme Electrophoresis in Human Genetics, North Holland, Amsterdam.

Harrison, P.M., Hoare, R.J., Hay, T.J. and McCara, I.G. (1974) In: Iron in Biochemistry and Medicine; Jacobs, A. and Worwood, M., eds. (Academic Press, London) pp. 73–85.

Hayase, K. and Kritchevsky, D. (1973) Clin. Chim. Acta 46, 455–464.

Hayase, K., Reisher, R. and Miller, B.F. (1973) Prep. Biochem. 3, 221–241.

Hayes, M.B. and Wellner, D. (1969) J. Biol. Chem. 244, 6636–6644.

Hearn, M.T.W. and Lyttle, D.J. (1981) J. Chromatogr. 218, 483–495.

Hebert, J.P. and Strobbel, B. (1974) LKB Application Note No. 151.

Heideman, M.L. (1965) Ann. N.Y. Acad. Sci. 121, 501–524.

Held, W.A. and Nomura, M. (1973) Biochemistry 12, 3273–3281.

Helenius, A. and Simons, K. (1975) Biochim. Biophys. Acta *415*, 29–79.

Hendeskog, G. (1975) J. Chromatogr. *107*, 91–98.

Hill, M.J., James, A.M. and Maxted, W.R. (1963) Biochim. Biophys. Acta *75*, 402–410.

Hirsch, E.M. and Gray, I. (1976) J. Cell Biol. *70*, 358–365.

Hjelmeland, L.M., Nebert, D.W. and Chrambach, A. (1978) In: Electrophoresis '78; Catsimpoolas, N. ed. (Elsevier, Amsterdam) pp. 29–56.

Hjelmeland, L.M., Nebert, D.W. and Chrambach, A. (1979) Anal. Biochem. *95*, 201–208.

Hjelmeland, L.M. and Chrambach, A. (1981) Electrophoresis *2*, 1–11.

Hjelmeland, L.M., Allenmark, S., An Der Laan, B., Jackiw, B.A., Nguyen, N.Y. and Chrambach, A. (1981) Electrophoresis *2*, 82–90.

Hjertén, S. (1961) Biochim. Biophys. Acta *53*, 514–520.

Hjertén, S. (1962a) Arch. Biochem. Biophys. Suppl. *1*, 147–151.

Hjertén, S. (1962b) Biochim. Biophys. Acta *62*, 445–449.

Hjertén, S. (1967) J. Chromatogr. *9*, 122–219.

Hjertén, S., Jersted, S. and Tiselius, A. (1969) Anal. Biochem. *27*, 108–115.

Hjertén, S. (1974) J. Chromatogr. *101*, 281–288.

Hjertén, S. (1976) In: Methods of Protein Separation; Catsimpoolas, N., ed. (Plenum, New York) Vol. 2, pp. 219–231.

Hoare, R.J., Harrison, P.M. and Hoy, T.G. (1975) Nature *255*, 653–654.

Hobart, M.J. (1975) In: Isoelectric Focusing; Arbuthnott, J.P. and Beeley, J.A., eds. (Butterworths, London) pp. 275–280.

Hocevar, B.J. and Northcote, D.H. (1957) Nature *179*, 488–490.

Hoch, H. and Barr, C.H. (1955) Science *122*, 243–244.

Hohn, T. and Pollmann, W. (1963) Z. Naturforsch. *18B*, 919–925.

Holbrook, I.B. and Leaver, A.G. (1976) Anal. Biochem. *75*, 634–636.

Holmes, R. (1967) Biochim. Biophys. Acta *133*, 174–177.

Holmquist, L. and Broström, H. (1979) J. Biochem. Biophys. Methods *1*, 117–127.

Holtlund, J. and Kristensen, T. (1978) Anal. Biochem. *87*, 425–432.

Horejsi, V. (1979) J. Chromatogr. *178*, 1–13.

Hovanessian, A. and Awdeh, Z.L. (1975) In: Progress in Isoelectric Focusing and Isotachophoresis; Righetti, P.G., ed. (Elsevier, Amsterdam) pp. 205–211.

Humphryes, K.C. (1970) J. Chromatogr. *49*, 503–510.

Hurley, P.M., Catsimpoolas, N. and Wogan, G.N. (1978) In: Electrophoresis '78; Catsimpoolas, N., ed. (Elsevier, Amsterdam) pp. 283–296.

Ikeda, K. and Suzuki, S. (1912) U.S. Patent 1,015,891.

Imada, M. (1978) Anal. Biochem. *89*, 292–296.

Inouye, M. (1971) J. Biol. Chem. *246*, 4834–4838.

Ishitani, K., Listowsky, I., Hazard, J. and Drysdale, J.W. (1975) J. Biol. Chem. *250*, 5446–5449.

Just, W.W. and Werner, G. (1977a) In: Cell Separation Methods; Bloemendal, H., Jacobs, S. (1971) Prot. Biol. Fluids *19*, 499–502.

Jacobs, S. (1973) Analyst *98*, 25–33.

Janik, B. and Dane, R.G. (1981) In: Electrophoresis '81; Allen, R.C. and Arnaud, P., eds. (de Gruyter, Berlin) pp. 77–82.

Janson, J.C. (1972) Ph.D. Thesis, University of Uppsala, Sweden.

Jensen, M., Oski, F.A., Nathan, D.G. and Bunn, H.F. (1975) J. Clin. Invest. *55*, 469–476.

Jeppsson, J.O. and Berglund, S. (1972) Clin. Chim. Acta *40*, 153–158.

Jeppsson, J.O., Franzen, B. and Nilsson, V.O. (1978) Sci. Tools *25*, 69–73.

Johansson, B.G. and Hjertén, S. (1974) Anal. Biochem. *59*, 200–213.

Johnson, E.A. and Mulloy, B. (1976) Carbohydr. Res. *51*, 119–127.

Johnson, L.K., Liebke, M.M. and O'Brien, T.S. (1976) Anal. Biochem. *76*, 311–320.

Johnson, M. (1978) Ann. Clin. Lab. Sci. *8*, 195–200.

Jonsson, M. and Pettersson, E. (1968) Sci. Tools *15*, 2–5.

Jonsson, M., Pettersson, E. and Rilbe, H. (1969) Acta Chem. Scand. *23*, 1553–1559.

Jonsson, M., Pettersson, S. and Rilbe, H. (1973) Anal. Biochem. *51*, 557–576.

Jonsson, M., Fredriksson, S., Jontell, M. and Linde, A. (1978) J. Chromatogr. *157*, 235–242.

Jonsson, M. and Rilbe, H. (1980) Electrophoresis *1*, 3–14.

Jonsson, M., Stahlberg, J. and Fredriksson, S. (1980) Electrophoresis *1*, 113–118.

Jonsson, M. and Fredriksson, S. (1981) Electrophoresis *2*, 193–203.

Josephson, R.V., Maheswaran, S.K., Morr, C.V., Jenness, R. and Lindorfer, R.K. (1971) Anal. Biochem. *40*, 476–482.

Jovin, T.M., Chrambach, A. and Naughton, M.A. (1964) Anal. Biochem. *9*, 351–362.

Just, W.W. (1980) Anal. Biochem. *102*, 134–144.

Just, W.W., Leon-V, J.O. and Werner, G. (1975a) In: Progress in Isoelectric Focusing and Isotachophoresis; Righetti, P.G., ed. (North-Holland, Amsterdam) pp. 265–280.

Just, W.W., Leon-V, J.O. and Werner, G. (1975b) Anal. Biochem *67*, 590–601.

Just, W.W. and Werner, G. (1977a) In: Cell Separation Methods, Blocmendal, H., ed. (North-Holland, Amsterdam) pp. 131–142.

Just, W.W. and Werner, G. (1977b) In: Electrofocusing and Isotachophoresis; Radola, B.J. and Graesslin, D., eds. (de Gruyter, Berlin) pp. 481–493.

Just, W.W. and Werner, G. (1979) In: Electrokinetic Separation Methods; Righetti, P.G., Van Oss, C.J. and Vanderhoff, J.W., eds. (Elsevier, Amsterdam) pp. 143–167.

Kaiser, K.P., Bruhn, L.C. and Belitz, H.D. (1974) Z. Lebensm. Unters. Forsch. *154*, 339–347.

Kalous, V. and Vacik, J. (1959) Chem. Listy *53*, 35–37.

Kaltschimdt, E. and Wittmann, H.G. (1970a) Anal. Biochem. *36*, 401–412.

Kaltschimdt, E. and Wittmann, H.G. (1970b) Proc. Natl. Acad. Sci. USA *67*, 1276–1282.

Kaplan, L.J. and Foster, J.F. (1971) Biochemistry *10*, 630–636.

Kaplan, N.O. (1968) Ann. N.Y. Acad. Sci. *151*, 382–399.

Karlsson, C., Davies, H., Ohman, J. and Andersson, U.B. (1973) LKB Application Note No. 75.

Keck, K., Grossberg, A.L. and Pressman, D. (1973a) Eur. J. Immunol. *3*, 99–102.

Keck, K., Grossberg, A.L. and Pressman, D. (1973b) Immunochemistry *10*, 331–335.

Kenrick, K.G. and Margolis, J. (1970) Anal. Biochem. *33*, 204–207.

Kilmartin, J.V., Fogg, J., Luzzana, M. and Rossi Bernardi, L. (1973) J. Biol. Chem. *248*, 7039–7043.

Kim, W.J. and White, T.T. (1971) Biochim. Biophys. Acta *242*, 441–445.

Kint, J.A. (1975) Anal. Biochem. *67*, 679–683.

Kinzkofer, A. and Radola, B.J. (1981) Electrophoresis *2*, 174–183.

Kiryukhin, I.F. (1972) Biul. Eksp. Biol. Med. *74*, 120–122.

Klose, J., Nowak, J. and Kade, W. (1980) In: Electrophoresis '79; Radola, B.J., ed.

(de Gruyter, Berlin) pp. 297–312.

Knüfermann (1977) In: Electrofocusing and Isotachophoresis; Radola, B.J. and Graesslin, D., eds. (de Gruyter, Berlin) pp. 395–404.

Koch, H.J.A. and Backx, J. (1969) Sci. Tools *16*, 44–47.

Kohn, J. (1957) Nature *180*, 986–988.

Kohnert, K.D., Schmid, E., Zuhlke, H. and Fiedler, H. (1973) J. Chromatogr. *76*, 263–267.

Kolin, A. (1954) J. Chem. Phys. *22*, 1628–1629.

Kolin, A. (1955a) J. Chem. Phys. *23*, 407–410.

Kolin, A. (1955b) Proc. Natl. Acad. Sci. USA *41*, 101–110.

Kolin, A. (1958) In: Methods of Biochemical Analysis; Glick, D., ed (Wiley, Interscience, New York) Vol. 6, pp. 259–288.

Kolin, A. (1970) In: Methods in Medical Research; Olson, R.E., ed. (Year Book Med. Publ. Inc., Chicago) Vol. 12, pp. 326–358.

Kolin, A. (1976) In: Isoelectric Focusing; Catsimpoolas, N., ed. (Academic Press, New York) pp. 1–11.

Kolin, A. (1977) In: Electrofocusing and Isotachophoresis; Radola, B.J. and Graesslin, D., eds. (de Gruyter, Berlin) pp. 3–33.

Kömpf, J., Bissbort, S., Gussmann, S. and Ritter, H. (1975) Humangenetik *27*, 141–143.

Kopwillen, A., Chillemi, F., Righetti, A.B.B. and Righetti, P.G. (1973) Prot. Biol. Fluids *21*, 657–665.

Korant, B.D. and Lonberg-Holm, K. (1974) Anal. Biochem. *59*, 75–82.

Kostner, G., Albert, W. and Holasck, A. (1969) Hoppe-Seyler's Z. Physiol. Chem. *350*, 1347–1352.

Krishnamoorthy, R., Bianchi Bosisio, A., Labie, D. and Righetti, P.G. (1978) FEBS Lett. *94*, 319–323.

Kronberg, H., Zimmer, H.G. and Neuhoff, V. (1980) Electrophoresis *1*, 27–32.

Kronberg, H., Zimmer, H.G. and Neuhoff, V. (1981) In: Electrophoresis '81; Allen, R.C. and Arnaud, P., eds. (de Gruyter, Berlin) pp. 413–423.

Kronenberg, L.H. (1979) Anal. Biochem. *93*, 189–195.

Kühnl, P. (1979) Ärztl. Lab. *25*, 34–43.

Kühnl, P. (1981) In: Electrophoresis '81; Allen, R.C. and Arnaud, P., eds. (de Gruyter, Berlin) pp. 563–571.

Kühnl, P., Langanke, U., Spielmann, W. and Neubauer, M. (1977) Hum. Genet. *40*, 79–86.

Kühnl, P., Schmidtmann, U. and Spielmann, W. (1977a) Hum. Genet. *35*, 219–223.

Kühnl, P., Schwabenland, R. and Spielmann, W. (1977b) Hum. Genet. *38*, 99–106.

Kühnl, P., Spielmann, W. and Loa, M. (1978) Vox Sang. *35*, 401–404.

Kühnl, P., Spielmann, W. and Weber, W. (1979) Hum. Genet. *46*, 83–87.

Kumarasamy, R. and Symons, R.M. (1979) Anal. Biochem. *95*, 359–364.

Kunkel, H., Ceppellini, R., Müller-Eberhard, U. and Wolf, J. (1957) J. Clin. Invest. *36*, 1615–1620.

Kunkel, H.G. and Slater, R.J. (1952) Proc. Soc. Exp. Biol. Med. *80*, 42–51.

Kurian, P., Gersten, D.M., Suhocki, P.V. and Ledley, G. (1981) Electrophoresis *2*, 184–186.

Lääs, T. (1972) J. Chromatogr. *66*, 347–355.

Lääs, T. and Olsson, I. (1981) Anal. Biochem. *114*, 167–172.

Lambin, P. and Fine, J.M. (1979) Anal. Biochem. *98*, 160–168.

Langton, R.W., Cole IX, J.S. and Quinn, P.F. (1975) Arch. Oral. Biol. 20, 103–106.

Laskey, R.A. and Mills, A.D. (1975) Eur. J. Biochem. 56, 335–340.

Laskey, R.A. and Mills, A.D. (1977) FEBS Lett. 82, 314–317.

Laskey, R.A. (1980) In: Methods in Enzymology; Grossman, L. and Moldave, K., eds. (Academic Press, New York) Vol. 65, pp. 363–370.

Lasky, M. and Manrique, A. (1980) Electrophoresis 1, 112–119.

Latner, A.L. (1973) Ann. N.Y. Acad. Sci. 209, 281–298.

Latner, A.L. (1975) Advan. Clin. Chem., 17, 193–215.

Latner, A.L. and Emes, V. (1975) In: Progress in Isoelectric Focusing and Isotachophoresis; Righetti, P.G., ed. (Elsevier, Amsterdam) pp. 223–233.

Latner, A.L., Parsons, M.E. and Skillen, A.W. (1970) Biochem. J. 118, 299–302.

Laurell, C.B. (1966) Anal. Biochem. 15, 45–52.

Laurent, T.C. and Killander, J. (1964) J. Chromatogr. 14, 317–325.

Leaback, D.H. and Robinson, H.K. (1974) FEBS Lett. 40, 192, 195.

Leaback, D.M. and Robinson, H.K. (1975) Biochem. Biophys. Res. Commun. 67, 248–254.

Leaback, D.H. and Rutter, A.C. (1968) Biochem. Biophys. Res. Commun. 32, 447–453.

Leaback, D.H. and Wrigley, C.W. (1976) In: Chromatographic and Electrophoretic Techniques; Smith, I., ed., 4th edition (Heinemann Med. Books, London) Vol. II, pp. 272–320.

Leader, D.P. (1980) J. Biochem. Biophys. Methods 3, 247–248.

Legocki, R.P. and Verma, D.P.S. (1981) Anal. Biochem. 111, 385–392.

Leise, E.M. and Le Sane, T. (1974) Prep. Biochem. 4, 395–410.

Lemkin, P.F. and Lipkin, L.E. (1981) In: Electrophoresis '81; Allen, R.C. and Arnaud, P., eds. (de Gruyter, Berlin) pp. 401–411.

Lewis, G.N. and Randall, M. (1921) J. Amer. Chem. Soc. 43, 1112–1116.

Li, Y.T. and Li, S.C. (1973) Ann. N.Y. Acad. Sci. 209, 187–197.

Lim, R., Huang, J.J. and Davis, G.A. (1969) Anal. Biochem. 29, 48–57.

Loening, U.E. (1967) Biochem. J. 102, 251–257.

Lostanlen, D., Gacon, G. and Kaplan, J.C. (1980) Eur. J. Biochem. 112, 179–183.

Ludwig, C.S. (1856) Sitzber. Akad. Wiss. Wien. 20, 539–556.

Lundahl, P. and Hjertén, S. (1973) Ann. N.Y. Acad. Sci. 209, 94–111.

Lundin, H., Hjalmarsson, S.G. and Davies, H. (1975) LKB Application Note No. 194.

Luner, S.J. and Kolin, A. (1970) Proc. Natl. Acad. Sci. USA 66, 898–903.

Luner, S.J. and Kolin, A. (1972) U.S. Patent 3,664,939.

Lutin, W.A., Kyle, C.F. and Freeman, J.A. (1977) In: Electrofocusing and Isotachophoresis; Radola, B.J. and Graesslin, D., eds. (de Gruyter, Berlin) pp. 93–106.

MacInnes, D.A. (1939) The Principles of Electrochemistry, Reinhold, New York.

Macko, V. and Stegemann, H. (1969) Hoppe-Seyler's Z. Physiol. Chem. 350, 917–919.

Macko, V. and Stegemann, H. (1970) Anal. Biochem. 37, 186–190.

Magnusson, R.P. and Jackiw, A. (1979) J. Biochem. Biophys. Methods 1, 65–68.

Maher, J.R., Trendle, W.O. and Schultz, R.L. (1956) Naturwissenschaften 43, 423–427.

Maizel, J.V. (1971) In: Methods in Virology; Maramorosch, K. and Koprowski, H., eds. (Academic Press, New York) Vol. 5, pp. 179–191.

Malik, N. and Berrie, A. (1972) Anal. Biochem. 49, 173–176.

Mandel, B. (1971) Virology 44, 554–560.

Manske, W., Bohn, B. and Brossmer, R. (1977) In: Electrofocusing and Isotacho-phoresis; Radola, B.J. and Graesslin, D., eds. (de Gruyter, Berlin) pp. 497–502.

Margolis, J. and Wrigley, C.W. (1975) J. Chromatogr. *106*, 204–210.

Marshall, R.C. and Blagrove, R.J. (1979) J. Chromatogr. *172*, 351–356.

Marshall, T. and Latner, A.L. (1981) Electrophoresis *2*, 228–235.

Martin, A.J.P. and Hampson, F. (1978) J. Chromatogr. *159*, 101–110.

Massey, T.H. and Deal Jr., W.C. (1973) J. Biol. Chem. *248*, 56–61.

Matsudaira, P.T. and Burgess, D.R. (1978) Anal. Biochem. *87*, 386–396.

Matsumura, T. and Noda, H. (1973) Anal. Biochem. *96*, 39–45.

Mauk, M.R. and Girotti, A.W. (1974) Biochemistry *13*, 1757–1761.

Maxam, A.M. and Gilbert, W. (1977) Proc. Natl. Acad. Sci. USA *74*, 560–565.

Mayer, J.W. (1976) Anal. Biochem. *76*, 369–373.

McConkey, E.H. (1979) Anal. Biochem. *96*, 39–45.

McDonell, M.W., Simon, M.N. and Studier, F.W. (1977) J. Mol. Biol. *110*, 119–145.

McDuffie, N.M., Dietrich, C.P. and Nader, H.B. (1975) Biopolymers *14*, 1473–1486.

McGillivray, A.J. and Rickwood, D. (1974) Eur. J. Biochem. *41*, 181–190.

McGuire, J.K., Miller, T.Y., Tipps, R.W., Snyder, R.S. and Righetti, P.G. (1980) J. Chromatogr. *194*, 323–333.

Mel, H.C. (1960) Science *132*, 1255–1256.

Mel, H.C. (1964) J. Theor. Biol. *6*, 307–324.

Melino, G., Stefanini, S., Chiancone, E. and Antonini, E. (1978) FEBS Lett. *86*, 136–138.

Merrifield, R.B. (1969) Adv. Enzymol. *32*, 221–241.

Merril, C.R., Switzer, R.C. and Van Keuren, M.L. (1979) Proc. Natl. Acad. Sci. USA *76*, 4335–4339.

Merril, C.R., Goldman, D., Sedman, S.A. and Ebert, M.H. (1981) Science *211*, 1437–1438.

Merril, C.R., Goldman, D. and Van Keuren, M.L. (1982) Electrophoresis *3*, 17–23.

Meselson, M., Stahl, F.W. and Vinogradov, J. (1957) Proc. Natl. Acad. Sci. USA *43*, 581–585.

Miller, D.W. and Elgin, S.R.C. (1974) Anal. Biochem. *60*, 142–148.

Mitchell, H.K. and Herrenberg, L.A. (1957) Anal. Chem. *29*, 1229–1233.

Mitchell, W.M. (1967) Biochim. Biophys. Acta *147*, 171–174.

Molnarova, B. and Sova, O. (1974) Abstr. Commun. 9th Fed. Europ. Biochem. Soc. Budapest, p. 432.

Monte, M., Beuzard, Y. and Rosa, J. (1976) Am. J. Pathol. *66*, 753–760.

Moore, H.D.M. and Hibbit, K.G. (1975) J. Reprod. Fert. *44*, 329–332.

Morris, C.J.O.R. and Morris, P. (1976) Separation Methods in Biochemistry, 2nd ed. (Pitman, London) pp. 853–876.

Morris, C.J.O.R. (1978) J. Chromatogr. *159*, 33–46.

Murel, A., Kirjanen, I. and Kirret, O. (1979) J. Chromatogr. *174*, 1–11.

Nader, H.B., McDuffie, N.M. and Dietrich, C.P. (1974) Biochem. Biophys. Res. Commun. *57*, 488–493.

Nagel, R.L. and Gibson, Q.H. (1971) J. Biol. Chem. *246*, 69–75.

Nakhleh, E.T., Samra, S.A. and Awdeh, Z.L. (1972) Anal. Biochem. *49*, 218–224.

Narayanan, K.R. and Ray, A.S. (1977) In: Electrofocusing and Isotachophoresis; Radola, B.J. and Graesslin, D., eds. (de Gruyter, Berlin) pp. 221–231.

Needleman, S.B., Koening, H. and Goldstone, A.D. (1975) Biochim. Biophys. Acta *379*, 57–73.

Neuhoff, V. (1973) Micromethods in Molecular Biology (Springer Verlag, Berlin) pp. 49–56.

Nguyen, N.Y. and Chrambach, A. (1976) Anal. Biochem. 74, 145–153.

Nguyen, N.Y. and Chrambach, A. (1977a) Anal. Biochem. 79, 462–469.

Nguyen, N.Y. and Chrambach, A. (1977b) Anal. Biochem. 81, 54–62.

Nguyen, N.Y. and Chrambach, A. (1977c) Anal. Biochem. 82, 226–235.

Nguyen, N.Y., Salokangas, A. and Chrambach, A. (1977) Anal. Biochem. 78, 287–294.

Nguyen, N.Y., Rodbard, D., Svendsen, P.J. and Chrambach, A. (1977a) Anal. Biochem. 77, 39–55.

Niitsu, Y. and Listowsky, I. (1973) Arch. Biochem. Biophys. 158, 276–281.

Niitsu, Y., Kohgo, Y., Yokota, M. and Urushizaki, I. (1975) Ann. N.Y. Acad. Sci. 259, 450–458.

Nilsson, P., Wadström, T. and Vesterberg, O. (1970) Biochim. Biophys. Acta 221, 146–148.

Noble, R.L., Yamashiro, D. and Li, C.H. (1977) Int. J. Peptide Prot. Res. 10, 385–393.

Nozaki, Y. and Tanford, C. (1971) J. Biol. Chem. 264, 2211–2217.

Oakley, B.R., Kirsch, D.R. and Morris, N.R. (1980) Anal. Biochem. 105, 361–363.

O'Brien, T.J., Liebke, H.H., Cheung, H.S. and Johnson, L.K. (1976) Anal. Biochem. 72, 38–44.

Ochs, D.C., McConkey, E.H. and Sammons, D.W. (1981) Electrophoresis 2, 304–307.

O'Connell, P.B.H. and Brady, C.J. (1976) Anal. Biochem. 76, 63–76.

O'Farrell, P. (1975) J. Biol. Chem. 250, 4007–4021.

Ogita, Z.I. and Market, C.L. (1979) Anal. Biochem. 99, 233–241.

Ogston, A.G. (1958) Trans. Faraday Soc. 54, 1754–1756.

Olden, K. and Yamada, K.M. (1977) Anal. Biochem. 78, 483–490.

Olsson, I. and Låås, T. (1981) J. Chromatogr. 215, 373–378.

Omoto, K. and Miyake, K. (1978) Jpn. J. Hum. Genet. 23, 119–125.

Ornstein, L. (1964) Ann. N.Y. Acad. Sci. 121, 321–349.

Osterman, L. (1970) Sci. Tools 17, 31–33.

Otavsky, W.I., Bell, T., Saravis, C. and Drysdale, J.W. (1977) Anal. Biochem. 78, 302–307.

Pagé, M. and Belles-Isles, M. (1978) Can. J. Biochem. 56, 853–856.

Palusinski, O.A., Allgyer, T.T., Mosher, R.A., Bier, M. and Saville, D.A. (1981) Biophys. Chem. 13, 193–202.

Panyim, S. and Chalkley, R. (1971) J. Biol. Chem. 346, 7557–7560.

Papayannopoulou, Th., Kurachi, S., Brice, M., Nakamoto, B. and Stamatoyannopoulos, G. (1981) Blood 57, 531–536.

Papeschi, G., Bordi, S., Beni, C. and Ventura, C. (1976) Biochim. Biophys. Acta 453, 192–199.

Park, C.M. (1973) Ann. N.Y. Acad. Sci. 209, 237–256.

Paus, P.N. (1971) Anal. Biochem. 42, 327–376.

Peacock, A.C. and Dingman, C.W. (1967) Biochemistry 6, 1818–1824.

Perrella, M., Heyda, A., Mosca, A. and Rossi-Bernardi, L. (1978) Anal. Biochem. 88, 212–224.

Perrella, M., Samaja, M. and Rossi-Bernardi, L. (1979) J. Biol. Chem. 254, 8748–8750.

Perrella, M., Cremonesi, L., Vannini Parenti, I., Benazzi, L. and Rossi-Bernardi, L. (1980) Anal. Biochem. *104*, 126–132.

Perrella, M., Cremonesi, L., Benazzi, L. and Rossi-Bernardi, L. (1981) J. Biol. Chem. *256*, 11098–11103.

Perutz, M.F. (1970) Nature *228*, 726–730.

Perutz, M.F., Kilmartin, J.V., Nishikura, K., Fogg, J.H., Butler, P.J.G. and Rollema, H.S. (1980) J. Mol. Biol. *138*, 649–670.

Peterson, J.I., Tipton, H.W. and Chrambach, A. (1974) Anal. Biochem. *62*, 274–280.

Peterson, R.F. (1971) J. Agr. Food Chem. *19*, 595–599.

Petrilli, P., Sannia, G. and Marino, G. (1977) J. Chromatogr. *135*, 511–513.

Pettersson, E. (1969) Acta Chem. Scand. *23*, 2631–2635.

Philpot, J. St. L. (1940) Trans. Faraday Soc. *36*, 38–46.

Pittz, E.P., Jones, R., Goldberg, L. and Coulston, F. (1977) Biorheology *14*, 33–42.

Poduslo, J.F. and Rodbard, D. (1980) Anal. Biochem. *101*, 394–406.

Poehling, H.M. and Neuhoff, V. (1980) Electrophoresis *1*, 198–200.

Poehling, H.M. and Neuhoff, V. (1981) Electrophoresis *2*, 141–147.

Pogacar, P. and Jarecki, R. (1974) In: Electrophoresis and Isoelectric Focusing in Polyacrylamide Gel; Allen, R.C. and Maurer, H., eds. (de Gruyter, Berlin) pp. 153–158.

Pollack, S. (1979) Biochem. Biophys. Res. Commun. *87*, 1252–1255.

Porath, J. (1963) J. Applied Chem. *6*, 233–236.

Porath, J., Jansson, J.C. and Låås, T. (1971) J. Chromatogr. *60*, 167–173.

Powell, L., Alpert, E., Isselbacher, K.J. and Drysdale, J.W. (1975) Br. J. Haematol. *30*, 47–61.

Poyart, C.F., Guesnon, P. and Bohn, B.M. (1981) Biochem. J. *195*, 493–501.

Presta, M., Giglioni, B., Ottolenghi, S., Gianni, A.M., Capaldi, A., Trento, M. and Saglio, G. (1982) Am. J. Hematol., in press.

Pruzik, Z. (1974) J. Chromatogr. *91*, 867–872.

Pulleyblank, D.E. and Booth, G.M. (1981) J. Biochem. Biophys. Methods *4*, 339–346.

Quarmby, C. (1981) Electrophoresis *2*, 203–212.

Quast, R. and Vesterberg, O. (1968) Acta Chem. Scand. *22*, 1499–1508.

Quast, R. (1979) In: Electrokinetic Separation Methods; Righetti, P.G., Van Oss, C.J. and Vanderhoff, J.W., eds. (Elsevier, Amsterdam) pp. 221–227.

Quast, R. (1977) In: Electrofocusing and Isotachophoresis; Radola, B.J. and Graesslin, D., eds. (de Gruyter, Berlin) pp. 455–462.

Quinn, J.R. (1973) J. Chromatogr. *76*, 520–522.

Qureshi, R.A. and Punnett, H.H. (1981) In: Electrophoresis '81; Allen, R.C. and Arnaud, P., eds. (de Gruyter, Berlin) pp. 83–87.

Radola, B.J. (1969) Biochim. Biophys. Acta *194*, 335–338.

Radola, B.J. (1973a) Biochim. Biophys. Acta *295*, 412–428.

Radola, B.J. (1973b) Ann. N.Y. Acad. Sci. *209*, 127–143.

Radola, B.J. (1975) In: Isoelectric Focusing, Arbuthnott; J.P. and Beeley, J.A., eds. (Butterworths, London) pp. 182–197.

Radola, B.J. (1975) In: Isoelectric Focusing; Catsimpoolas, N. ed. (Academic Press, New York) pp. 119–168.

Radola, B.J. and Graesslin, D., eds. (1977) Electrofocusing and Isotachophoresis, de Gruyter, Berlin.

Radola, B.J., Tschesche, H. and Schuricht, H. (1977) In: Electrofocusing and

Isotachophoresis; Radola, B.J. and Graesslin, D., eds. (de Gruyter, Berlin) pp. 97–110.

Radola, B.J. (1980a) In: Electrophoresis '79; Radola, B.J., ed. (de Gruyter, Berlin) pp. 79–94.

Radola, B.J. (1980b) Electrophoresis *1*, 43–56.

Radola, B.J., ed. (1980c) Electrophoresis '79, de Gruyter, Berlin.

Radola, B.J., Kinzkofer, A. and Frey, M. (1981) In: Electrophoresis '81; Allen, R.C. and Arnaud, P. eds. (de Gruyter, Berlin) pp. 181–189.

Ragetli, H.W. and Weintraub, M. (1966) Biochim. Biophys. Acta, *112*, 160–167.

Ragland, W.L., Pace, J.L. and Kemper, U.L. (1974) Anal. Biochem. *59*, 24–30.

Rahbar, S. (1968) Clin. Chim. Acta *22*, 296–301.

Rapaport, R.N., Jackiw, A. and Brown, R.K. (1980) Electrophoresis *1*, 122–126.

Rathnam, P. and Saxena, B.B. (1970) J. Biol. Chem. *245*, 3725–3731.

Raymond, S. and Weintraub, L. (1959) Science *130*, 711–713.

Raymond, S. and Nakamichi, M. (1962) Anal. Biochem. *3*, 23–32.

Raymond, S. and Nakamichi, M. (1964) Anal. Biochem. *7*, 225–230.

Reich, G. and Sieber, H. (1966) Z. Chem. *6*, 351–356.

Reisner, A.H., Newes, P. and Bucholtz, C. (1975) Anal. Biochem. *64*, 509–516.

Renart, J., Reiser, J. and Stark, G.R. (1979) Proc. Natl. Acad. Sci. USA *76*, 3116–3120.

Rice, R.H. and Horst, J. (1972) Virology *49*, 602–610.

Rickwood, D. and Hames, B. (eds.) (1981) Gel Electrophoresis of Nucleic Acids, IRL Press, Oxford.

Righetti, P.G. and Drysdale, J.W. (1971) Biochim. Biophys. Acta *236*, 17–28.

Righetti, P.G., Little, E.D. and Wolf, G. (1971) J. Biol. Chem., *246*, 5742–5747.

Righetti, P.G. and Drysdale, J.W. (1973) Ann. N.Y. Acad. Sci. *209*, 163–186.

Righetti, P.G. and Drysdale, J.W. (1974) J. Chromatogr. *98*, 271–321.

Righetti, P.G. and Righetti, A.B.B. (1974) Abstr. Commun. 9th. Fed. Eur. Biochem. Soc. Budapest, p. 432.

Righetti, P.G., Pagani, M. and Gianazza, E. (1975a) J. Chromatogr. *109*, 341–356.

Righetti, P.G., Bianchi Bosisio, A.R. and Galante, E. (1975b) Anal. Biochem. *63*, 423–432.

Righetti, P.G. (ed.) (1975a) Progress in Isoelectric Focusing and Isotachophoresis (Elsevier, Amsterdam).

Righetti, P.G. (1975b) Separ. Purif. Methods *4*, 23–72.

Righetti, P.G. and Drysdale, J.W. (1976) Isoelectric Focusing (North-Holland, Amsterdam).

Righetti, P.G., Gianazza, E. and Bianchi Bosisio, A. (1976) Giorn. Ital. Chim. Clin. *1*, 11–46.

Righetti, P.G. and Caravaggio, T. (1976) J. Chromatogr. *127*, 1–28.

Righetti, P.G. (1977) J. Chromatogr. *138*, 213–215.

Righetti, P.G., Balzarini, L., Gianazza, E. and Brenna, O. (1977a) J. Chromatogr. *134*, 279–284.

Righetti, P.G., Gianazza, E., Brenna, O. and Galante, E. (1977b) J. Chromatogr. *137*, 171–181.

Righetti, P.G., Molinari, B.M. and Molinari, G. (1977c) J. Dairy Res. *44*, 69–72.

Righetti, P.G. and Chillemi, F. (1978) J. Chromatogr. *157*, 243–251.

Righetti, P.G., Krishnamoorthy, R., Gianazza, E. and Labie, D. (1978a) J. Chromatogr. *166*, 455–460.

Righetti, P.G., Gacon, G., Gianazza, E., Lostanlen, D. and Kaplan, J.C. (1978b) Biochem. Biophys. Res. Commun. *85*, 1575–1581.

Righetti, P.G. and Chrambach, A. (1978) Anal. Biochem. *90*, 633–643.

Righetti, P.G. and Gianazza, E. (1978) Biochim. Biophys. Acta *532*, 137–146.

Righetti, P.G., Brown, R.P. and Stone, A.L. (1978c) Biochim. Biophys. Acta *542*, 232–244.

Righetti, P.G. (1979) J. Chromatogr. *173*, 1–5.

Righetti, P.G. and Gianazza, E. (1979) Prot. Biol. Fluids *27*, 711–714.

Righetti, P.G., Krishnamoorthy, R., Lapoumeroulie, C. and Labie, D. (1979a) J. Chromatogr. *177*, 219–225.

Righetti, P.G., Menozzi, M., Gianazza, E. and Valentini, L. (1979b) FEBS Lett. *101*, 51–55.

Righetti, P.G., Van Oss, C.J. and Vanderhoff, J. (eds.) (1979c) Electrokinetic Separation methods, Elsevier, Amsterdam.

Righetti, P.G. (1979d) In: Electrokinetic Separation Methods; Righetti, P.G., Van Oss, C.J. and Vanderhoff, J. eds. (Elsevier, Amsterdam) pp. 389–441.

Righetti, P.G., Gianazza, E. and Bianchi Bosisio, A. (1979e) In: Recent Developments in Chromatography and Electrophoresis; Frigerio, A. and Renoz, L., eds. (Elsevier, Amsterdam) Vol. 9, pp. 1–36.

Righetti, P.G., Gianazza, E., Gianni, A.M., Comi, P., Giglioni, B., Ottolenghi, S., Secchi, C. and Rossi Bernardi, L. (1979f) J. Biochem. Biophys. Methods *1*, 47–57.

Righetti, P.G. (1980) J. Chromatogr. *190*, 275–282.

Righetti, P.G. and Gianazza, E. (1980) In: Electrophoresis '79; Radola, B.J., ed. (de Gruyter, Berlin) pp. 23–38.

Righetti, P.G., Gianazza, E. and Bianchi Bosisio, A. (1980a) In: Recent Developments in Chromatography and Electrophoresis; Frigerio, A. and McCamish, M., eds. (Elsevier, Amsterdam) Vol. 10, pp. 89–117.

Righetti, P.G., Gianazza, E. and Ek, K. (1980b) J. Chromatogr. *184*, 415–456.

Righetti, P.G., Muneroni, P., Todesco, R. and Carini, S. (1980c) Electrophoresis *1*, 37–42.

Righetti, P.G. (1981) In: Electrophoresis '81; Allen, R.C. and Arnaud, P., eds. (de Gruyter, Berlin) pp. 3–16.

Righetti, P.G. and Bianchi Bosisio, A. (1981) Electrophoresis, *2*, 65–75.

Righetti, P.G. and Gianazza, E. (1981) In: Electrophoresis '81; Allen, R.C. and Arnaud, P., eds. (de Gruyter, Berlin) pp. 655–665.

Righetti, P.G. and Hjertén, S. (1981) J. Biochem. Biophys. Methods *5*, 259–272.

Righetti, P.G., Tudor, G. and Ek, K. (1981a) J. Chromatogr. *220*, 115–194.

Righetti, P.G., Brost, B.C.W. and Snyder, R.S. (1981b) J. Biochem. Biophys. Methods *4*, 347–363.

Righetti, P.G., Gelfi, C. and Bianchi Bosisio, A. (1981c) Electrophoresis *2*, 291–295.

Righetti, P.G. and Macelloni, C. (1982) J. Biochem. Biophys. Methods *6*, 1–15.

Rilbe, H. and Pettersson, S. (1968) Separ. Sci. *3*, 209–234.

Rilbe, H. (1970) Prot. Biol. Fluids *17*, 369–382.

Rilbe, H. (1971) Acta Chem. Scand. *25*, 2768–2769.

Rilbe, H. (1973a) Ann. N.Y. Acad. Sci. *209*, 11–22.

Rilbe, H. (1973b) Ann. N.Y. Acad. Sci. *209*, 80–93.

Rilbe, H. and Pettersson, S. (1975a) In: Isoelectric Focusing; Arbuthnott, J.P. and Beeley, J.A., eds. (Butterworths, London) pp. 44–57.

Rilbe, H., Forcheimer, A., Pettersson, S. and Jonsson, M. (1975b) In: Progress in Isoelectric Focusing and Isotachophoresis; Righetti, P.G., ed. (Elsevier, Amsterdam) pp. 51–63.

Rilbe, H. (1976) In: Isoelectric Focusing; Catsimpoolas, N., ed. (Academic Press, New York) pp. 14–52.

Rilbe, H. (1977) In: Electrofocusing and Isotachophoresis; Radola, B.J. and Graesslin, D., eds. (de Gruyter, Berlin) pp. 35–50.

Rilbe, H. (1978) J. Chromatogr. *159*, 193–205.

Rilbe, H. (1981a) Electrophoresis *2*, 261–267.

Rilbe, H. (1981b) Electrophoresis *2*, 268–272.

Riley, R.F. and Coleman, M.K. (1968a) J. Lab. Clin. Med. *72*, 714–720.

Rodbard, D. and Chrambach, A. (1970) Proc. Natl. Acad. Sci. USA *65*, 970–977.

Rodbard, D. and Chrambach, A. (1971) Anal. Biochem. *40*, 95–134.

Rodbard, D., Levitov, C. and Chrambach, A. (1972) Separ. Sci. *7*, 705–723.

Rodbard, D., Chrambach, A. and Weiss, G.H. (1974) In: Electrophoresis and Isoelectrofocusing in Polyacrylamide Gel; Allen, R.C. and Maurer, H.R. eds. (de Gruyter, Berlin) pp. 62–105.

Rodbard, D. (1976) In: Methods of Protein Separation; Catsimpoolas, N., ed. (Plenum Press, New York) Vol. 2, pp. 145–179.

Rose, C. and Harboe, N.M.G. (1970) Prot. Biol. Fluids *17*, 397–400.

Rosén, A., Ek, K. and Aman, P. (1979a) J. Immunol. Methods. *28*, 1–11.

Rosén, A., Ek, K. Aman, P. and Vesterberg, O. (1979b) Prot. Biol. Fluids *27*, 707–710.

Rosén, A. (1980) In: Electrophoresis '79; Radola, B.J. ed. (de Gruyter, Berlin) pp. 105–116.

Rosengren, A., Bjellqvist, B. and Gasparic, V. (1977) In: Electrofocusing and Isotachophoresis; Radola, B.J. and Graesslin, D., eds. (de Gruyter, Berlin) pp. 165–171.

Rosengren, A., Bjellqvist, B. and Gasparic, V. (1978) U.S. patent No. 4,130,470, December 19, 1978.

Rosengren, A., Bjellqvist, B. and Gasparic, V. (1981) German patent No. 2656162, March 9 (1981).

Rüchel, R., Mesecke, S., Wolfrum, D.I. and Neuhoff, V. (1973) Hoppe Seyler's Z., Physiol. Chem. *354*, 1351–1368.

Rüchel, R. and Brager, M.D. (1975) Anal. Biochem. *68*, 415–428.

Rüchel, R. (1977) J. Chromatogr. *132*, 451–468.

Rüchel, R., Steere, R.L. and Erbe, E.F. (1978) J. Chromatogr. *166*, 563–575.

Rücher, W. and Radola, B.J. (1971) Planta *99*, 192–198.

Saglio, G., Ricco, G., Mazza, U., Camaschella, C., Pich, P.G., Gianni, A.M., Gianazza, E., Righetti, P.G., Giglioni, B., Comi, P., Gusmeroli, M. and Ottolenghi, S. (1979) Proc. Natl. Sci. USA *76*, 3420–3424.

Salaman, M.R. and Williamson, A.R. (1971) Biochem. J. *122*, 93–99.

Sammons, D.W., Adams, L.D. and Nishizawa, E.E. (1981) Electrophoresis *2*, 135–141.

Samols, E. and Williams, H.S. (1961) Nature *190*, 1211–1212.

Sanders, M.M., Groppi Jr., V.E. and Browning, E.T. (1980) Anal. Biochem. *103*, 157–165.

Saravis, C.A. and Zamcheck, N. (1979) J. Immunol. Methods *29*, 91–96.

Saravis, C.A., O'Brien, M. and Zamcheck, N. (1979) J. Immunol. Methods *29*,

97–100.

Saravis, C.A., Cunningham, C.G., Marasco, P.V., Cook, R.B. and Zamcheck, N. (1980a) In: Electrophoresis '79; Radola, B.J. ed. (de Gruyter, Berlin) pp. 117–122.

Saravis, C.A., Cantaroa, W., Marasco, P.V., Burke, B. and Zamcheck, N. (1980b) Electrophoresis 1, 191–193.

Satterlee, L.D. and Snyder, H.E. (1969) J. Chromatogr. 41, 417–422.

Savitskii, A.P., Ugarova, N.N. and Berezin, I.V. (1978) Dokl. Akad. Nauk SSSR 241, 977–980.

Schaffer, H.E. and Johnson, F.M. (1973) Anal. Biochem. 51, 577–583.

Schapira, F., Gregori, C., Banrogues, J., Vidailhet, M., Despoisses, S. and Vignerou, C. (1979) Hum. Genet. 46, 89–96.

Schmelzer, W. and Behne, D. (1975) In: Progress in Isoelectric Focusing and Isotachophoresis; Righetti, P.G. ed. (Elsevier, Amsterdam) pp. 257–264.

Schmidt-Ullrich, R. and Wallach, D.F.H. (1977) In: Biological and Biomedical Applications of Isoelectric Focusing: Catsimpoolas, N. and Drysdale, J.W., eds. (Plenum Press, New York) pp. 191–210.

Schoos, R., Schoos-Barbette, S. and Lambotte, C. (1978) Clin. Chim. Acta 86, 61–70.

Schwartz, G.P., Burke, G.T. and Katsoyannis, P.G. (1981) Int. J. Peptide Prot. Res. 17, 243–255.

Seaman, G.V.F. (1975) In: The Red Blood Cell; Surgenor, D.M.N., ed. (Academic Press, New York) Vol. II, pp. 1135–1229.

Secchi, C. (1973) Anal. Biochem. 51, 448–455.

Seiler, N., Thobe, J. and Werner, G. (1970a) Hoppe-Seyler's Z. Physiol. Chem. 351, 865–868.

Seiler, N., Thobe, J. and Werner, G. (1970b) Z. Anal. Chem. 252, 179–182.

Serwer, P. (1980) Biochemistry 19, 3001–3005.

Shaaya, E. (1976) Anal. Biochem. 75, 325–328.

Shafritz, D.A. and Drysdale, J.W. (1975) Biochemistry 14, 61–68.

Sharp, P.A., Sugden, B. and Sanbrook, J. (1973) Biochemistry 12, 3055–3060.

Shaw, C.R. and Prasad, R. (1970) Biochem. Genet. 4, 297–301.

Sherbet, G.V., Lakshmi, M.S. and Rao, K.V. (1972) Exp. Cell. Res. 70, 113–123.

Sherbet, G.V. and Lakshmi, M.S. (1973) Biochim. Biophys. Acta 298, 50–58.

Sherbet, G.V. (1978) The Biophysical Characterization of the Cell Surface (Academic Press, London).

Sherbet, G.V. and Lakshmi, M.S. (1974) Oncology 29, 335–345.

Sherbet, G.V. and Lakshmi, M.S. (1975) In: Isoelectric Focusing; Arbuthnott, J.P. and Beeley, J.A., eds. (Butterworths, London) pp. 338–346.

Sherbet, G.V. and Lakshmi, M.S. (1976) In: Molecular Base of Malignancy; Deutsch, E., Moser, H., Rainer, H. and Stacher, A., eds. (Thieme, Stuttgart) pp. 5–20.

Siciliano, M.J. and Shaw, C.R. (1976) In: Chromatographic and Electrophoretic Techniques; Smith, I., ed. (Heinemann, W. Med. Books Ltd., London) Vol. 2, pp. 185–205.

Siemankowski, R.F., Giambalvo, A. and Dreizen, P. (1978) Physiol. Chem. Physics 10, 415–434.

Simplicio, J., Schwanzer, K. and Maenpa, F. (1975) J. Amer. Chem. Soc. 97, 7319–7323.

Singh, S., Klose, J., Willers, I. and Goedde, M.W. (1978) In: Electrophoresis '78;

Catsimpoolas, N., ed. (North-Holland, Amsterdam) pp. 297–304.
Singh, S., Willers, I., Goedde, H.W. and Klose, J. (1981) In: Electrophoresis '81; Allen, R.C. and Arnaud, P., eds. (de Gruyter, Berlin) pp. 289–294.
Sluyterman, L.A.AE. and Wijdenes, J. (1977) In: Electrofocusing and Isotachophoresis; Radola, B.J. and Graesslin, D., eds. (de Gruyter, Berlin) pp. 463–466.
Sluyterman, L.A.AE. and Elgerson, O. (1978) J. Chromatogr. *150*, 17–30.
Sluyterman, L.A.AE. and Wijdenes, J. (1978) J. Chromatogr. *150*, 31–44.
Sluyterman, L.A.AE. and Wijdenes, J. (1981a) J. Chromatogr. *206*, 429–440.
Sluyterman, L.A.AE. and Wijdenes, J. (1981b) J. Chromatogr. *206*, 441–447.
Smith, I., Lightstone, P.J. and Perry, J.D. (1971) Clin. Chim. Acta *35*, 59–66.
Smithies, O. and Poulik, M.D. (1956) Nature *117*, 1033–1034.
Smyth, C.J. and Arbuthnott, J.P. (1974) Med. Microbiol. *4*, 41–66.
Smyth, C.J., Söderholm, J. and Wadström, T. (1977) LKB, Application Note No. 269, March 1977.
Soave, C., Pioli, F., Viotti, A., Salamini, R. and Righetti, P.G. (1975) Maydica *20*, 83–94.
Söderberg, L., Buckley, D., Hagström, G. and Bergström, J. (1980) Prot. Biol. Fluids *27*, 687–691.
Söderholm, J., Allestam, P. and Wadström, T. (1972) FEBS Lett. *24*, 89–92.
Söderholm, J. and Lidström, P.A. (1975) In: Isoelectric Focusing; Arbuthnott, J.P. and Beeley, JA., eds. (Butterworths, London) pp. 143–146.
Söderholm, J. and Smyth, C.J. (1975) In: Progress in Isoelectric Focusing and Isotachophoresis; Righetti, P.G., ed. (Elsevier, Amsterdam) pp. 99–114.
Söderholm, J., Smyth, C.G. and Wadström, T. (1975) Scand. J. Immunol. *4*, Suppl. 2, 107–113.
Sorensen, S.P.L., Linderstrom Lang, K. and Lund, E. (1926) C.R. Trav. Lab. Carlsberg *16*, 5–15.
Soret, C. (1879) Arch. Sci. Phys. Nath. (Genève) *2*, 48–56.
Spencer, E.M. and King, T.P. (1971) J. Biol. Chem. *246*, 201–208.
Sperber, E. (1946) J. Biol. Chem. *166*, 75–85.
Spicer, K.M., Allen, R.C. and Buse, M.G. (1978) Diabetes *27*, 384–390.
Spies, J.R., Bernton, H.S. and Stevens, M. (1941) J. Am. Chem. Soc. *63*, 2163–2166.
Spragg, J., Kaplan, A.P. and Austen, K.F. (1973) Ann. N.Y. Acad. Sci. *209*, 372–386.
Squire, P.G. (1964) Arch. Biochem. Biophys. *107*, 471–480.
Stathakos, D. (1975) In: Progress in Isoelectric Focusing and Isotachophoresis; Righetti, P.G., ed. (Elsevier, Amsterdam) pp. 65–75.
Stathakos, D., Vellios, A. and Koussoulakos, S. (1980) In: Electrophoresis '79; Radola, B.J., ed. (de Gruyter, Berlin) pp. 517–528.
Steck, G., Leuthard, P. and Bürk, R.R. (1980) Anal. Biochem. *107*, 21–24.
Stefanini, S., Chiancone, E., Vecchini, P. and Antonini, E. (1976) Mol. Cell. Biochem. *13*, 55–61.
Stegemann, H., Franksen, H. and Macko, V. (1973) Z. Naturforsch. *28*, 722–732.
Stein, S., Böhlen, P., Stone, J., Daiman, W. and Udenfriend, S. (1973) Arch. Biochem. Biophys. *155*, 203–208.
Steinhardt, J., Leidy, J.G. and Mooney, J.P. (1972) Biochemistry *11*, 1809–1817.
Stenmann, U.K. and Gräsbeck, R. (1971): Biochim. Biophys. Acta *286*, 243–251.
Stibler, H. (1979) J. Neurol. Sci. *42*, 275–281.
Stimpson, D.I. and Cann, J.R. (1977) Biophys. Chem. *7*, 115–119.
Strain, H.H. (1939) J. Amer. Chem. Soc. *61*, 1291–1295.

Strongin, A.J.A., Baldnev, A.P. and Levin, Z.D. (1973) Sci. Tools 20, 34–35.

Suran, A.A. and Tarver, H. (1975) Arch. Biochem. Biophys. 111, 399–406.

Suzuki, T., Benesch, R.E., Yung, S. and Benesch, R. (1973) Anal. Biochem. 55, 249–254.

Svensson, H. (1948) In: Advances in Protein Chemistry; Anson, M.L. and Edsall, J.T. eds. (Acedemic Press, New York) Vol. IV, pp. 251–295.

Svensson, H. and Brattsten, I. (1949) Arkiv. Kemi 1, 401–409.

Svensson, H. (1961) Acta Chem. Scand. 15, 325–341.

Svensson, H. (1962a) Acta Chem. Scand. 16, 456–466.

Svensson, H. (1962b) Arch. Biochem. Biophys. Suppl. 1, 132–140.

Svensson, H. (1966) J. Chromatogr. 25, 266–273.

Svensson, H. (1967) Prot. Biol. Fluids 15, 515–522.

Swanson, M.J. and Sanders, B.E. (1975) Anal. Biochem., 67, 520–524.

Switzer III, R.C., Merril, C.R. and Shifrin, S. (1979) Anal. Biochem. 98, 231–237.

Taketa, F., Huang, Y.P., Libnoch, J.A. and Dessel, B.H. (1975) Biochim. Biophys. Acta 400, 348–353.

Talbot, D.N. and Yaphantis, D.A. (1971) Anal. Biochem. 44, 246–254.

Talbot, P. (1975a) In: Isoelectric Focusing; Arbuthnott, J.P. and Beeley, J.A., eds. (Butterworths, London) pp. 270–274.

Talbot, P. and Caie, I.S. (1975b) In: Isoelectric Focusing; Arbuthnott, J.P. and Beeley, J.A., eds. (Butterworths, London) pp. 74–77.

Tanaka, T. (1981) Sci. American 244, 124–138.

Tanford, C. (1962) Adv. Protein Chem. 17, 69–95.

Tanford, C., Reynolds, J.A. (1976) Biochim. Biophys. Acta 457, 171–212.

Tariverdian, G., Ritter, H. and Wendt, G.G. (1970) Hum. Genet. 11, 75–77.

Taylor, J., Anderson, N.L., Coulter, B.P., Scandora, E.A., Jr. and Anderson, N.G. (1980) In: Electrophoresis '79; Radola, B.J., ed. (de Gruyter, Berlin) pp. 329–339.

Taylor, J., Anderson, N.L. and Anderson, N.G. (1981) In: Electrophoresis '81; Allen, R.C. and Arnaud, P., eds. (de Gruyter, Berlin) pp. 383–400.

Theorell, H. and Akesson, A. (1941) J. Amer. Chem. Soc. 63, 1804–1808.

Thillet, J., Blouquit, Y., Perrone, F. and Rosa, J. (1977) Biochim. Biophys. Acta 491, 16–22.

Thorsrud, A.K., Haugen, H.F. and Jellum, E. (1980) In: Electrophoresis '79; Radola, B.J., ed. (de Gruyter, Berlin) pp. 425–436.

Thorstensson, A., Sjodin, B. and Karlsson, J. (1975) In: Progress in Isoelectric Focusing and Isotachophoresis; Righetti, P.G., ed. (Elsevier, Amsterdam) pp. 213–222.

Thymann, M. (1980) In: Electrophoresis '79; Radola, B.J., ed. (de Gruyter, Berlin) pp. 123–128.

Tiselius, A. (1941) Svensk Kem. Tidskr. 58, 305–310.

Tolkacho, N.V. (1974) Ukr. Biokhim. 46, 441–445.

Tollaksen, S.L. and Anderson, N.G. (1980) In: Electrophoresis '79; Radola, B.J., ed. (de Gruyter, Berlin) pp. 405–414.

Tollaksen, S.L., Edwards, J.J. and Anderson, N.G. (1981), Electrophoresis 2, 155–160.

Tombs, M.P. (1965) Anal. Biochem. 13, 121–132.

Troitzki, G.V., Saviolov, V.P., Kiritukhin, I.F., Abramov, V.M. and Agitsky, G.Y. 214, 955–958.

Troitzki, G.V., Kiryukhin, I.F., Tolkacheva, N.V. and Agitsky, G.Y. (1974b) Vop. Med. Khim. *20*, 24–31.

Troizki, G.V., Savialov, V.P., Kirjukhin, I.F., Abramov, V.M. and Agitsky, G.J. (1975) Biochim. Biophys. Acta *400*, 24–31.

Tuszynski, G.P., Buck, C.A. and Warren, L. (1979) Anal. Biochem. *93*, 329–338.

Tuttle, A.M. (1956) J. Lab. Clin. Med. *47*, 811–816.

Ui, N. (1971) Biochim. Biophys. Acta *229*, 567–581.

Urushizaki, I., Niitsu, Y., Ishitani, K., Matsuda, M. and Fukuda, M. (1971) Biochim. Biophys. Acta *249*, 187–192.

Valentini, L., Gianazza, E. and Righetti, P.G. (1980) J. Biochem. Biophys. Methods *3*, 323–338.

Valkonen, K., Gianazza, E. and Righetti, P.G. (1980) Clin. Chim. Acta *107*, 223–229.

Valmet, E. (1969) Sci. Tools *16*, 8–13.

Valmet, E. (1970) Prot. Biol. Fluids *17*, 401–407.

Van Orden, D.E. (1971) Immunochemistry *8*, 869–875.

Vesterberg, O. and Svensson, H. (1966a) Acta Chem. Scand. *20*, 820–834.

Vesterberg, O. and Berggren, B. (1966b) Arkiv Kemi *27*, 119–125.

Vesterberg, O., Wadström, T., Vesterberg, K., Svensson, H. and Malmgren, B. (1967) Biochem. Biophys. Acta *133*, 435–445.

Vesterberg, O. (1968) Svensk Kem. Tidskr. *80*, 213–225.

Vesterberg, O. (1969a) Acta Chem. Scand. *23*, 2653–2666.

Vesterberg, O. (1969b) Sci. Tools *16*, 24–27.

Vesterberg, O. (1970a) In: Methods in Microbiology; Norris, J.R. and Ribbons, D.W., eds. Vol. 53 (Academic Press, New York) pp. 595–614.

Vesterberg, O. (1970b) Prot. Biol. Fluids *17*, 383–387.

Vesterberg, O. (1971a) In: Methods in Enzymology; Jakoby, W.B., ed. (Academic Press, New York) Vol. 22, pp. 389–412.

Vesterberg, O. (1971b) Biochim. Biophys. Acta *243*, 345–348.

Vesterberg, O. (1972) Biochim. Biophys. Acta *257*, 11–19.

Vesterberg, O. and Eriksson, R. (1972) Biochim. Biophys. Acta *285*, 393–397.

Vesterberg, O. (1973a) In: VII Symposium on Chromatography and Electrophoresis (Publ. Press Acad. Europeennes, Brussels) pp. 81–98.

Vesterberg, O. (1973b) Sci. Tools *20*, 22–29.

Vesterberg, O. (1973c) Ann. N.Y. Acad. Sci. *209*, 23–33.

Vesterberg, O. (1973d) Acta Chem. Scand. *27*, 2415–2420.

Vesterberg, O. (1975) In: Isoelectric Focusing; Arbuthnott, J.P. and Beeley, J.A., eds. (Butterworths, London) pp. 78–98.

Vesterberg, O. (1976) In: Isoelectric Focusing; Catsimpoolas, N., ed. (Academic Press, New York) pp. 53–76.

Vesterberg, O. and Hansén, L. (1977) In: Electrofocusing and Isotachophoresis; Radola, B.J. and Graesslin, D., eds. (de Gruyter, Berlin) pp. 123–133.

Vesterberg, O. (1978) Int. Lab. May–June, 61–68.

Vesterberg, O. and Hansén, L. (1978) Biochim. Biophys. Acta *534*, 369–373.

Vesterberg, O. (1980) In: Electrophoresis '79; Radola, B.J., ed. (de Gruyter, Berlin) pp. 95–104.

Vincent, R.K., Hartman, J., Barret, A.S. and Sammons, D.W. (1981) In: Electrophoresis '81; Allen, R.C. and Arnaud, P., eds. (de Gruyter, Berlin) pp. 371–381.

Vinogradov, S.N., Lowenkron, S., Andonian, H.R., Bagshaw, J., Felgenhauer, K. and Pak, S.J. (1973) Biochem. Biophys. Res. Commun. *54*, 501–506.

Von Klobusitzky, D. and König, P. (1973) Arch. Exp. Pathol. Pharmakol. *192*, 271−275.

Wadström, J. (1967) Biochim. Biophys. Acta *147*, 441−448.

Wadström, J. and Hisatsune, K. (1970a) Biochem. J. *120*, 725−734.

Wadström, J. and Hisatsune, K. (1970b) Biochem. J. *120*, 735−744.

Wadström, T. and Smyth, C. (1973) Sci. Tools *20*, 17−21.

Wadström, T., Mollby, R., Jeansson, S. and Wretlind, B. (1974) Sci. Tools *21*, 2−4.

Wadström, T. and Smyth, C. (1975a) In: Isoelectric Focusing: Arbuthnott, J.P. and Beeley, J.A., eds. (Butterworths, London) pp. 152−177.

Wadström, T. and Smyth, C. (1975b) In: Progress in Isoelectric Focusing and Isotachophoresis; Righetti, P.G., ed. (Elsevier, Amsterdam) pp. 149−163.

Wadström, T. (1975c) In: Progress in Isoelectric Focusing and Isotachophoresis; Righetti, P.G., ed. (Elsevier, Amsterdam) pp. 389−389.

Wagner, H. and Speer, W. (1978) J. Chromatogr. *157*, 259−265.

Wagstaff, M., Worwood, M. and Jacobs, A. (1978) Biochem. J. *173*, 969−977.

Wajcman, H., Kilmartin, J.V., Najman, A. and Labie, D. (1975) Biochim. Biophys. Acta *400*, 354−364.

Wajcman, H., Krishnamoorthy, R., Gacon, G., Elion, J., Allard, C. and Labie, D. (1976) J. Mol. Med. *1*, 187−197.

Wallevik, K. (1973) Biochim. Biophys. Acta *322*, 75−87.

Walton, K.E., Styer, D. and Gruenstein, E. (1979) J. Biol. Chem. *254*, 795−800.

Walther, F. and Schubert, J.C.F. (1974) Blut *28*, 211−215.

Wardi, A.H. and Michos, G.A. (1972) Anal. Biochem. *49*, 607−613.

Weatherall, D.J., Pembrey, ME. and Pritchard, J. (1974) Clin. Haematol. *3*, 467−472.

Weatherall, D.J. and Clegg, J.B. (1981) The Thalassemia Syndromes, 3rd Edn., Blackwell Scientific, Oxford.

Weller, D.L., Heaney, A. and Sjogren, R.E. (1968) Biochim. Biophys. Acta *168*, 386−388.

Whalen, R.G., Butler-Browne, G.S. and Gross, F. (1976) Proc. Natl. Acad. Sci. USA *73*, 2018−2022.

Whitney III, J.B., Copland, G.T., Skovand, L.C. and Russel, E.S. (1979) Proc. Nat. Acad. Sci. USA *76*, 867−871.

Willard, K.E., Giometti, C.S., Anderson, N.L., O'Connor, J.E. and Anderson, N.G. (1979) Anal. Biochem. 100, 289−298.

Willard, K.E. and Anderson, N.L. (1980): In: Electrophoresis '79; Radola, B.J. ed. (de Gruyter, Berlin) pp. 415−424.

Williams, K.W. and Söderberg, L. (1979) Int. Lab. Jan/Febr., 45−53.

Williams, R.J. (1935) J. Biol. Chem. *110*, 589−596.

Williams, R.R. and Waterman, R.E. (1929) Proc. Exp. Biol. Med. *27*, 56−61.

Williams, R.R. and Truesdail, J.M. (1931) J. Am. Chem. Soc. *53*, 4171−4174.

Williamson, A.R. (1971) Eur. J. Immunol. *1*, 390−394.

Williamson, A.R. (1973) In: Handbook of Experimental Immunology; Weir, D.M., ed. (Blackwell, Oxford) Ch. 8, pp. 1−23.

Williamson, A.R., Salaman, M.R. and Kreth, N.H. (1973) Ann. N.Y. Acad. Sci. *209*, 210−224.

Wilson, C.M. (1973) Anal. Biochem. *53*, 538−545.

Winters, A., Perlmutter, M. and Davies, M. (1975) LKB Application Note No. 198.

Winter, A. and Karlsson, C. (1976) LKB Application Note No. 219.

Winters, A. (1977) In: Electrofocusing and Isotachophoresis; Radola B.J. and

Graesslin, D. eds. (de Gruyter, Berlin) pp. 433–442.

Worwood, M., Wagstaff, M., Jones, B.M., Dawkins, S. and Jacobs, A. (1977) In: Proteins of Iron Transport and Storage in Biochemistry and Medicine; Aisen, P. and Brown, E., eds (Green and Stratton, New York) pp. 79–91.

Wrigley, C. (1968a) Sci. Tools *15*, 17–23.

Wrigley, C. (1968b) J. Chromatogr. *36*, 362–365.

Wrigley, C.W. (1970) Biochem. Genet. *4*, 509–515.

Wrigley, C. and Stepherd, K.W. (1973) Ann. N.Y. Acad. Sci. *209*, 154–162.

Wrigley, C.W. (1977) In: Biological and Biomedical Applications of Isoelectric Focusing; Catsimpoolas, N. and Drysdale, J.W., eds. (Plenum Press, New York) pp. 211–264.

Yeoman, L.C., Taylor, C.W. and Busch, H. (1974) Cancer Res. *34*, 424–428.

Young, C.W. and Bittar, E.S. (1973) Cancer Res. *33*, 2692–2700.

Young, J.L. and Webb, B.A. (1978a) Anal. Biochem. *88*, 619–623.

Young, J.L. and Webb, B.A. (1978b) Sci. Tools *25*, 54–56.

Young, J.L., Webb, B.A., Coutie, D.G. and Reid, B. (1978) Biochem. Soc. Trans. *6*, 1051–1054.

Young, J.L. and Webb, B.A. (1980) Prot. Biol. Fluids *27*, 739–742.

Yu, S.M. and Spring, T.G. (1978) Anal. Biochem. *85*, 287–290.

Zacharius, R.M. and Zell, T.E. (1969) Anal. Biochem. *30*, 148–155.

Zannis, V.I. and Breslow, T.L. (1980): In: Electrophoresis '79; Radola, B.J., ed. (de Gruyter, Berlin) pp. 437–443.

Zech, R. and Zürcher, K. (1973) Life Sci. *13*, 383–389.

Zech, R. and Zürcher, K. (1974) Arzneim. Forsch. *24*, 337–340.

Zechel, K. (1977) Anal. Biochem. *83*, 240–251.

Zeineh, R.A., Nijm, W.P. and Al-Azzawi, F.M. (1975) Am. Lab. *I*, 51–58.

Ziabicki, A. (1979) Polymer *20*, 1373–1381.

Ziegler, A. and Köhler, G. (1976) FEBS Lett. *64*, 48–51.

Subject Index

383